Advances in Distan in Times of Pa

The book *Advances in Distance Learning in Times of Pandemic* is devoted to the issues and challenges faced by universities in the field of distance learning in COVID-19 times. It covers both theoretical and practical aspects connected to distance education. It elaborates on issues regarding distance learning, its challenges, assessment by students and their expectations, the use of tools to improve distance learning, and the functioning of e-learning in Industry 4.0 and Society 5.0 eras. The book also devotes a lot of space to the issues of Web 3.0 in University e-learning, quality assurance, and knowledge management. The aim and scope of this book are to draw a holistic picture of ongoing online teaching activities before and during the lockdown period and present the meaning and future of e-learning from students' points of view, taking into consideration their attitudes and expectations as well as Industry 4.0 and Society 5.0 aspects. The book presents the approach to distance learning and how it has changed, especially in the time of the pandemic that revolutionized education.

Highlights

- The function of online education has changed before and during the time of a pandemic
- E-learning is beneficial in promoting digital citizenship
- Distance learning characteristic in the era of Industry 4.0 and Society 5.0
- Era of Industry 4.0 treats distance learning as a desirable form of education

The book covers both scientific and educational aspects and is useful for university-level undergraduates, postgraduates, and research-grade courses and can be referred to by anyone interested in exploring the diverse aspects of distance learning.

Advances in Distance Learning in Times of Pandemic

Edited by
Joanna Rosak Szyrocka
Justyna Żywiołek
Anand Nayyar
Mohd Naved

CRC Press
Taylor & Francis Group
Boca Raton London New York

CRC Press is an imprint of the
Taylor & Francis Group, an **informa** business
A CHAPMAN & HALL BOOK

Designed cover image: by Authors

First edition published 2023
by CRC Press
6000 Broken Sound Parkway NW, Suite 300, Boca Raton, FL 33487-2742

and by CRC Press
4 Park Square, Milton Park, Abingdon, Oxon, OX14 4RN

CRC Press is an imprint of Taylor & Francis Group, LLC

Library of Congress Cataloging-in-Publication Data
Names: Rosak-Szyrocka, Joanna, editor. | Żywiołek, Justyna, editor. |
Nayyar, Anand, editor. | Naved, Mohd, editor.
Title: Advances in distance learning in times of pandemic / edited by
Joanna Rosak-Szyrocka, Justyna Żywiołek, Anand Nayyar, Mohd Naved.
Description: First edition. | Boca Raton, FL : Chapman & Hall/CRC Press, 2023. |
Includes bibliographical references and index. |
Identifiers: LCCN 2022045251 (print) | LCCN 2022045252 (ebook) |
ISBN 9781032334417 (hardback) | ISBN 9781032344591 (paperback) |
ISBN 9781003322252 (ebook)
Subjects: LCSH: Web-based instruction. | Internet in education. |
Internet in higher education. | Social distancing (Public health) and education.
Classification: LCC LB1044.87 .A384 2023 (print) | LCC LB1044.87 (ebook) |
DDC 371.33/44678–dc23/eng/20221107
LC record available at https://lccn.loc.gov/2022045251
LC ebook record available at https://lccn.loc.gov/2022045252

ISBN: 9781032334417 (hbk)
ISBN: 9781032344591 (pbk)
ISBN: 9781003322252 (ebk)

DOI: 10.1201/9781003322252

Typeset in Minion
by codeMantra

Contents

Preface

"Technology will not replace great teachers
but technology in the hands
of great teachers can be transformational"

George Coures

The focus of the 21st century's educational system is students; therefore, the digitalization of education and its continuous improvement is a necessary condition for the functioning of modern universities. The modern period is characterized by technology. In the current state of the higher education system, every area of life is somehow connected to technology. The emphasis on digital competence continues to grow in popularity in higher education, and COVID-19 has raised social awareness of the need for digital skills because college students in the 21st century are the generation who have grown up with the rapid development of computer networks and who have witnessed the unprecedented development of online media represented by the internet, virtual reality, and artificial intelligence. Higher education, universities, and other institutions employ information and communication technology (ICT) more often now for administrative, educational, and pedagogical responsibilities as well as for curriculum development.

The aim of the book is to present the situation of education both before and after the pandemic; opportunities and challenges faced by universities; and aspects of Web 3.0. The authors analyzed factors affecting the adoption of e-learning systems in management students as well as the level of acceptance of technology. The book is a reference point for readers to seriously think and research distance learning with regard to Industry 4.0 in order to improve quality, university development, and knowledge management because the book provides solutions to various problems faced by universities that are of key importance to their functioning in the world of Society 5.0.

The book comprises 11 chapters. Chapter 1 titled "COVID-19 Impact on Educational Institutions: A Detailed Analysis" analyzes the future of e-learning in higher education institutions and provides an overview of the e-learning process, along with associated information, benefits, and downsides; in addition, the chapter discusses the impact of COVID-19 on the world's economy and behavior of governments toward education spending". Chapter 2 titled "E-Learning as a Desirable Form of Education in the Era of Society 5.0" provides a detailed idea of e-learning platform as a desirable form of education in the age of Industry 4.0 Society 4.0 after covid-19 outbreak". Chapter 3 titled "Education Situation in Online Education before the Pandemic and in the Time of Pandemic" presents a critical description of the global education system to highlight the changes made due to the COVID-19 pandemic and also illustrates the changes taking place in teaching and learning methods before and during the pandemic and also comprehend the role of technology in teaching and learning. Chapter 4 titled "Factors Affecting E-Learning System in Management Studies during COVID-19 in India" discusses the numerous factors that influence management students' adoption of an e-learning system during COVID-19. Chapter 5 titled "New Technology Anxiety and Acceptance of Technology: An Appraisal of MS Teams" stresses the utilization of MS teams as a potential tool for distance learning and also illustrates its importance and impact on education transformation". Chapter 6 titled "The Paradigm Shift in Higher Education and Impact of Distance Learning in Era of Industry 4.0 and Society 5.0" focuses on researching how society has changed since the industrial revolution, and how distance learning and its effects may improve higher education quality to meet student expectations. Chapter 7 titled "The Use of Web 3.0 in University E-learning, Quality Assurance, and Knowledge Management" lists the characteristics of the different web generations and how they impact the various e-learning generations, as well as the numerous issues with Web 3.0 and also elaborates the guidelines for applying Web 3.0 Technologies to higher education for distance learning. Chapter 8 titled "The Influence of Industry 4.0 and 5.0 for Distance Learning Education in Times of Pandemic for a Modern Society" highlights Industry 4.0 and distance learning predictions for the future, the applicability of distance learning in the context of Industry 4.0 and 5.0 in time of pandemic and explores the new technologies in distance learning and also comprehends four major strategies for distance learning paradigm in the context of Industry 4.0. Chapter 9 titled "Innovations of Teaching and Testing English to Students of other Majors Online during COVID-19 Pandemic at Mila University Centre, Algeria" elaborates on the distinction between the theoretical adaptations made when teaching ESP and EAP before the COVID-19 pandemic and how teachers design courses for students of other majors, compares testing technique used before COVID-19 pandemic and during COVID-19 pandemic to elaborate on the use of barcode scanning correction system and some specific tasks both in summative and formative assessments in distance education, and also illustrates the issues and changes made to distance learning after the epidemic in Algerian universities. Chapter 10 titled "Distance Learning and Higher Education Hybridization: Opportunities and Challenges for Students with Disabilities" identifies how distance learning affects students with impairments and establishes effective methods for assisting students with impairments in their transition to college and involvement in academic life and describes some pedagogical strategies and the main principles of distance education and provide some recommendation for higher education institutions for a better accommodation of their services for students with disabilities. Chapter 11 titled "E-Learning: A Catalyst to Promote the Digital Citizenship" discusses e-learning in the era of Industry 4.0 and Society 5.0 and its related perspectives.

We hope that the readers will like the book because being well written with passion and love for the education and learning process, and the book lays a strong foundation to understand distance learning advancements and futuristic guidelines for teaching and learning after the pandemic era.

Joanna Rosak Szyrocka
Justyna Żywiołek
Anand Nayyar
Mohd Naved

Editors

Dr. Joanna Rosak Szyrocka is the Erasmus+ coordinator at the Faculty of Management, Czestochowa University of Technology. She specializes in the fields of quality 4.0, education, distance learning, Industry 4.0, quality management of sports, and IoT issues. She is the author/co-author of over 200 publications in journals and conference materials and guest editor of Energies MDPI, International Journal of Environmental Research, and Public Health MDPI Journal. She is also a review editor of Environmental Economics and Management (ISSN 2296-665X) and a reviewer for a number of prestigious Journals: *IEEE Access, TQM Elsevier, Energies MDPI IJERPH MDPI, Healthcare MDPI, Renewable Energy Elsevier, Resources Policy Elsevier, Frontiers, Sage, Sustainability MDPI, Education Sciences MDPI, Social Sciences MDPI, Applied Sciences*, and the Vice President of Qualitas Foundation, dealing with the promotion of good business practices. She completed a research internship at the University of Żilina, Slovakia, as well as at the Silesian University of Technology, Zabrze, Poland. She has been a participant in multiple Erasmus+ teacher mobility programs in Italy, the UK, Slovenia, Hungary, the Czech Republic, Slovakia, and France.

Dr. Justyna Żywiołek was born in Częstochowa. She completed her master's studies in the field of Metallurgy in 2010. In 2014, she defended her doctoral dissertation with honors in the field of management science, specializing in security science, entitled, *The impact of information process management on the security of knowledge resources in the enterprise*, at the Faculty of Organization and Management of the Lodz University of Technology. Since January 2015, she has been an assistant professor in the Faculty of Management at the Technical University of Częstochowa. She is a respected researcher in Poland and abroad for her experience in management issues: knowledge and information management and their security. She has published over 160 articles and is the author/co-author/editor of seven books. She actively participated in presenting research results at various international conferences. She deals with data protection, is an inspector of personal data protection, and is a leading ISO 27001: 2018 auditor. Ph.D. Eng. Justyna Żywiołek is a member of ICAA (Intelectual Capital Association) and EU-OSHA, as well as in organizations operating in Poland promoting information and knowledge security. She is a guest editor of *Energies MDPI, International Journal of Environmental Research and Public Health MDPI Journal, and Computer Modeling in Engineering & Sciences.* Early in her career, she was on the editorial board of *The Journal of Strategic Information Systems IF 14.682* and the Editorial Advisory Board of the *International Journal of Management and Applied Research IF 3.508.* She has been a participant in multiple Erasmus+ teacher mobility in Slovenia, Hungary, the Czech Republic, Slovakia, and France.

Dr. Anand Nayyar earned a Ph.D. in Computer Science from Desh Bhagat University in 2017 in the area of wireless sensor networks, swarm intelligence, and network simulation. He currently works at the School of Computer Science-Duy Tan University, Da Nang, Vietnam, as an assistant professor, scientist, vice-chairman (Research), and director of the IoT and Intelligent Systems Lab. He is a certified professional with 100+ professional certificates from CISCO, Microsoft, Amazon, EC-Council, Oracle, Google, Beingcert, EXIN, GAQM, Cyberoam, and many more. He has published more than 140+ research papers in various high-quality ISI-SCI/SCIE/SSCI Impact Factor Journals cum Scopus/ESCI indexed Journals, 70+ papers in international conferences indexed with Springer, IEEE Xplore, and ACM Digital Library, 40+ book chapters in various SCOPUS, WEB OF SCIENCE-Indexed Books with Springer, CRC Press, Wiley, IET, Elsevier with 8000+ citations, H-Index: 50, and I-Index: 165. He is a member of more than 60+ associations as a senior and life member including IEEE and ACM. He has authored/co-authored cum edited 40+ books of Computer Science. He is associated with more than 500+ international conferences as a Programme Committee/Chair/Advisory Board/Review Board member. He has 18 Australian patents, 4 German patents, 2 Japanese patents, 11 Indian Design cum utility patents, 3 Indian copyrights, and 2 Canadian copyrights to his credit in the area of wireless communications, artificial intelligence, cloud computing, IoT, and image processing. He has been awarded 40+ awards for Teaching and Research—Young Scientist, Best Scientist, Young Researcher Award, Outstanding Researcher Award, Excellence in Teaching, and many more. He is listed in the top 2% of scientists as per Stanford University (2020 and 2021). He is acting as an associate editor for *Wireless Networks* (Springer), *Computer Communications* (Elsevier), *International Journal of Sensor Networks* (IJSNET) (Inderscience), *Frontiers in Computer Science, PeerJ Computer Science, Human Centric Computing and Information Sciences (HCIS), IET-Quantum Communications, IET Wireless Sensor Systems, IET Networks, IJDST, IJISP, IJCINI,*

and *IJGC*. He is acting as editor-in-chief of *IGI-Global, USA Journal* titled "*International Journal of Smart Vehicles and Smart Transportation (IJSVST)*". He has reviewed more than 2500+ articles for diverse Web of Science and Scopus-indexed Journals. He is currently researching in the area of wireless sensor networks, the Internet of Things, swarm intelligence, cloud computing, artificial intelligence, drones, blockchain, cyber security, network simulation, Big Data, and wireless communications.

Dr. Mohd Naved is a passionate researcher and educator with 16 years of experience and a proven track record of quality research publications and leading teams for the research and overall management of the educational institution. He is a senior member of IEEE and is associated with multiple leading research organizations. He is a machine learning consultant and researcher, currently teaching at Amity University (Noida) for various degree programs in analytics and machine learning. He is actively engaged in academic research on various topics in management as well as on 21st-century technologies. He has published 60+ research articles in reputed journals (SCI/Scopus-indexed/peer-reviewed). He has 16 patents in AI/ML and actively engages in the commercialization of innovative products.

Contributors

Sunday Adeola Ajagbe
Department of Computer Engineering
Ladoke Akintola University of Technology
Ogbomoso, Nigeria

Gbemisola Janet Ajamu
Department of Agricultural Economics and Extension
Landmark University
Omu Aran, Nigeria

Joseph Bamidele Awotunde
Department of Computer Science
Faculty of Information and Communication Sciences
University of Ilorin
Ilorin, Nigeria

Emmanuel Femi Ayo
Department of Mathematical Science
Olabisi Onabanjo University
Ago-Iwoye, Nigeria

Abderrahim Bouderbane
Faculty of Letters and Languages
Department of English
University Centre of Mila
Mila, Algeria

Pooja Darda
School of Management (PG)
Dr. Vishwanath Karad MIT World Peace University
Pune, India

Pratiksha Dixit
Springfields
Department of Science
Moradabad, India

Cristina Dumitru
Department of Education
University of Piteşti
Piteşti, Romania

R. Gomathi
Department of Computer Science (PG)
KGiSL Institute of Information Management
Coimbatore, India

Om Jee Gupta
School of Management (PG)
Dr. Vishwanath Karad MIT World Peace University
Pune, India

V. Harish
PSG Institute of Management
PSG College of Technology
Coimbatore, India

Fathyah Hashim
Graduate School of Business
Universiti Sains Malaysia
Penang, Malaysia

Zahid Hussain
Department of Business Administration
Shaheed Benazir Bhutto University
Nawabshah, Pakistan

K. Jayanthi
Department of Computer Science (PG)
KGiSL Institute of Information Management
Coimbatore, India

Taibat Bolanle Jimoh
National Examinations Council
Ilorin Zonal Office
Ilorin, Nigeria

Henry Jonathan
Health and Safety Management
Modern College of Business and Science
Muscat, Oman

Shad Ahmad Khan
College of Business
University of Buraimi
Al Buraimi, Oman

Shujaat Khan
Internal Audit Department
A'Sharqiyah University
Ibra, Oman

Hesham Magd
Quality Assurance and Accreditation
Faculty of Business and Economic
Modern College of Business and Science
Muscat, Oman

Usha Pathak
Department of Teacher Education
DAV(PG) College
Dehradun, India

Meenal Pendse
School of Management (PG)
Dr. Vishwanath Karad MIT World Peace University
Pune, India

M. Sugantha Priya
Department of Computer Science (PG)
KGiSL Institute of Information Management
Coimbatore, India

Ali Rehman
Internal Audit Department
A'Sharqiyah University
Ibra, Oman

Dahlia Sam
Department of Information Technology
Dr. M.G.R. Educational and Research Institute
Chennai, India

S. Saranya
Department of Computer Science (PG)
KGiSL Institute of Information Management
Coimbatore, India

R. Thiyagarajan
PSG Institute of Management
PSG College of Technology
Coimbatore, India

CHAPTER 1

COVID-19 Impact on Educational Institutions

A Detailed Analysis

Ali Rehman and Shujaat Khan
A'Sharqiyah University

Fathyah Hashim
Universiti Sains Malaysia

CONTENTS

DOI: 10.1201/9781003322252-1

1.1 INTRODUCTION

Governments around the world have worked and continue to work tirelessly to tackle the COVID-19 crisis, which is still far from being resolved (WHO, 2022; United Nations Health, 2022). Around the world, governments make full use of technology to combat COVID-19 and address a wide range of problems that are associated with it. Several changes emerged in response to these problems, such as an increase in online shopping and automated delivery systems, the adoption of both contactless payment systems and digital payments, remote working, online entertainment, and the use of technology in distance learning (Renu, 2021). Several countries around the world have used technological advancements during this pandemic owing to their limitations in some underdeveloped and developing countries (Jackson, 2020). These underdeveloped or developing countries were not able to provide online services for the entire country and were able to provide these services only in large cities, depriving the people of small cities or villages of everything including education (Ndambakuwa & Brand, 2020; Mathrania, Sarveshb, & Umer, 2021).

Many universities in developed countries, such as universities in the USA, England, and many European countries, signed contracts with Zoom and other platforms in response to the impact of COVID-19 and rolled out standard remote learning models. Developing countries were not able to implement such models, and the situation was different for them. Global access to knowledge has changed because of this situation, and the gaps between those who have access to the internet and those who struggle to get it have become even more pronounced (Ndambakuwa & Brand, 2020). These internet- or online-deprived areas impacted students' progress and knowledge. Faculty members were also not able to perform any task and were forced to perform some other jobs in order to survive this period as many lost their jobs or their universities were completely closed (Fies & Hill, 2020). It is mentioned by the World Bank that "students now risk losing $17 trillion in lifetime earnings in present value, or about 14% of today's global gross domestic product (GDP) as a result of COVID-19 pandemic-related school closures" (Bank, 2021).

Globally several governments had temporarily shut HEIs to contain the spread of COVID-19 (Zivkovic et al., 2021). Over 60% of students around the world were affected by these closures. The UNESCO Institute of Statistical Data reports that school closures have affected 67.7% of enrolled students in 144 countries (UNESCO, Education: From disruption to recovery, 2022). Students who were in high schools, i.e., becoming legible for entering higher education institutes (HEIs), were not allowed to appear in exams, and their assessments were made either based on their previous performance or the average of the few exams they appeared in (Gamage, Silva, & Gunawardhana, 2020). This created a mind shift among students as they were allowed to enter the HEIs without their studies and annual assessments. This practice was made so that students do not lose their education year, but it also created dissatisfaction among plenty of high achievers. HEIs were forced to reduce their criteria for recruiting students, and as a result, many HEIs' net profit increased with diminishing standards of education. Table 1.1 presents the pre-COVID-19 and COVID-19 net profit situation of HEIs, and Table 1.2 presents the number of graduates within these periods.

Table 1.1 depicts very few HEIs as there are several for-profit HEIs available across the world, and demonstrating financial data is outside the scope of this study. However, it is

TABLE 1.1 Pre-COVID-19 and COVID-19 %age of Net Profit Margins for Higher Education Institutes

University Name	2020–2021	2019–2020	2018–2019	2017–2018	Source
Oxford University	33%	9%	4%	17%	Oxford University (2021)
Cambridge University	42%	−4%	−4%	20%	Cambridge University (2021)
Imperial College London	2%	6%	−4%	8%	Imperial College London (2021)
University of Edinburgh	20%	13%	−15%	17%	University of Edinburgh (2021)
London School of Economics and Political Science	29%	18%	4%	12%	London School of Economics and Political Sciences (2021)
The University of Warwick	12%	16%	−10%	−11%	The University of Warwick (2021)
Harvard University	28%	−2%	11%	14%	Harvard University (2021)
Stanford University	41%	1%	12%	19%	Stanford University (2021)
Massachusetts Institute of Technology	5%	3%	6%	3%	Massachusetts Institute of Technology (2021)
University of California Berkeley	59%	−7%	−15%	15%	University of California Berkeley (2021)
University of California	16%	−14%	−2%	3%	University of California (2021)
Yale University	8%	5%	6%	5%	Yale University (2021)
Columbia University	61%	2%	5%	21%	Columbia University (2021)
New York University	4%	1%	3%	0%	New York University (2021)
University of Pennsylvania	8%	3%	5%	5%	University of Pennsylvania (2021)

TABLE 1.2 Pre-COVID-19 and COVID-19 Number of Graduates

Country	2019/2020	2018/2019	2017/2018	2016/2017	Source
Oman	21,870	25,194	24,904	24,411	Ministry of Higher Education, Research and Innovation (2020)
United Kingdom	492,355	513,670	508,420	506,825	Fennell (2022)

evident that HEIs earn extra profits during this time. There could be plenty of reasons for such an increase, and a few to highlight are the closure of many HEIs, leaving only those who were able to provide online education, a high number of student recruitment as HEIs were forced to downgrade their recruitment criteria, and an increase in tuition fees.

Table 1.2 demonstrates the 13% and 4% decline in graduation numbers only for Oman and the United Kingdom, respectively. Students were not able to concentrate more due to online education, and for this reason, the graduation rate decreased. There could be other reasons; however, separate studies can be conducted to identify the reasons for such a decline in the COVID-19 period.

COVID-19 enforced the online education platform, which also possesses several benefits in comparison with the traditional physical presence in classrooms. Massive open online courses have positively impacted education, reduced costs, improved content delivery, and increased user accessibility (Liyanagunawardena & Aboshady, 2018).

There are several types/categories of online learning, a big part of which is internet-based learning, requiring both students and faculty to have technical knowledge and skills (Paudel, 2021). Online education provides global access to resources and enables students to enroll/admit to any HEI across the world (Pape, 2010). Students' learning outcomes improved and enhanced during this COVID-19 era through online learning, and it also reduced educational costs in HEIs (Krakoff, 2022). Reduced costs can be associated with accommodation costs, traveling costs, and other related costs.

Using online education, practitioners can access a wealth of knowledge and valid/authentic resources from different domains of knowledge and information. It is believed that online learning contributes to the declining cost of education by distributing the costs of classes and anyone with internet access can get access to online education and receive world knowledge (Liyanagunawardena & Aboshady, 2018). Online courses via online classes provide an excellent way to learn unbounded by location or time, enabling students to access teaching anytime and anywhere. Students have an easier time working online when dealing with individual tasks but not when working in teams (Angelova, 2020).

As students utilize online education, they access and study educational materials, share knowledge and experiences with others, and do exercises online. Similarly, students are guided and facilitated online as they learn and practice. E-learning encourages HEIs' faculty members to develop learning modules to facilitate online learning. These modules developed by faculty members have measurable, clear/well-defined, specific, and observable learning objectives that aim at changing students' behavior, and include content relevant to students, society, industry, and education (Laksana, 2020). With these advantages, online learning has become the principal option today (Mustakim, 2020).

This study will highlight the benefits and related detriments and will also deliberate the future of e-learning in HEIs which arise in the COVID-19 era. There are benefits for several HEIs and their students, whereas many students, especially in developing countries, suffer from this facility and will lag behind their peers. To the best of the knowledge of researchers, there have been limited studies conducted on this topic, and this study will contribute to the enhancement of the existing body of knowledge. This study will be beneficial to HEIs, related educational ministries, and other regulatory bodies.

Objectives of the Chapter

The objectives of this chapter are as follows; these objectives are related to the higher education sector in the COVID-19 era:

- To discuss the advantages, disadvantages, and future of e-learning;
- To discuss the gap created between student learning, faculty teaching, and knowledge sharing;
- And, to discuss the impact of COVID-19 on the world's economy and the behavior of governments toward educational spending.

Organization of the Chapter

The chapter is organized as follows: Section 1.2 defines the literature review. Section 1.3 elaborates on case studies concerning e-learning challenges at the time of COVID-19. Section 1.4 concludes the chapter with future scope.

1.2 LITERATURE REVIEW

This section discusses the role of e-learning during the COVID-19 era and its related advantages, disadvantages, and future of e-learning within HEIs. This section also discusses COVID-19 impact on nations' economies and its related effects on nations' past, present, and future spending on education. Detailed discussion related to the impact of e-learning during COVID-19 times on HEIs and challenges faced by the students is presented. Recent case studies are also presented which highlight the challenges faced by students and HEIs and how the quality of education is impacted.

1.2.1 COVID-19 and Its Impact on Nations' Economies

As of the year 2020, there were 2 million cases (Sobti, Nayyar, & Nagrath, 2021), and to the date of this study, corona cases rose to more than 60 million people across the globe. Governments in all countries took desperate measures to cope with the difficulties and challenges in health, education, and economic situations (Kumar et al., 2021). It is worth mentioning that COVID-19 created the worst economic situation for the entire world after Second World War. Many economies are on the path of recovery and now facing the toughest inflation which could lead to the second great depression. To date, the path toward recovery is uncertain; however, the COVID-19 situation extensively augmented some pre-existing trends, in particular digitalization (Allain-Dupré et al., 2020).

With the high death rates and more focus on the healthcare system (Sharma, et al., 2021), there were many sectors that were deprived of governments' attention which also include HEIs and other education at lower levels. In accordance with the survey conducted by Allain-Dupré et al. (2020), it was identified that 82% of respondents in the European Union are facing high pressure to maintain the budget for education as economies of the countries are prioritized in the areas of healthcare and revival/sustenance of large-scale businesses.

COVID-19 brought about many social and psychological impacts, one of which was panic buying (Yasir Arafat et al., 2020), which created an environment of hoarding/stockpiling. This situation was initiated due to the fact of uncertainties at individual levels and also at government levels. Such a situation exposed many unsustainable areas and resulted in the profiteering and fragility of the economic system (Spash, 2020; Ibn-Mohammed, et al., 2021). Many nations including the US and many countries in Europe were unable to provide facilities or relief to their nationals which resulted in increased loan rates, a decrease in returns on investment, and minimized nationals' confidence in their government.

Negative effects of COVID-19 have ranged from a decline in the GDP of all countries to multidimensional social and environmental issues across society. As a result of the outbreak, socio-economic activity was disrupted in many ways, including millions of people being quarantined, schools and universities being closed, borders being sealed, manufacturing and travel industries being crippled, gatherings including sporting and entertainment events being canceled, and nationalist and protectionist views being re-surfaced (Ibn-Mohammed, et al., 2021; Baker, Bloom, Davis, & Terry, 2020; Basilaia & Kvavadze, 2020; Devakumar, Shannon, Bhopal, & Abubaker, 2020).

The world trade network, as we know, does not exist anymore and countries started trading in blocks. Many European countries stopped/curtailed trade with China and India as they were much affected by this pandemic. On a similar front, China and India reduced their imports and focused on those countries that relied on these two nations' exports. It is worth mentioning that the study of Kiyota (2022) provided contradictory conclusions and stated that there was no or marginal impact on the world trade network, but they also noted that their study covered a concise span of time and results could vary if the current data are also included.

COVID-19, in fact, distorted the world's operating assumptions, revealing the utter incapability of the dominant model of production to cope with unplanned shocks and crises (Pinner, Rogers, & Samandari, 2020). Failures of over-centralization are exposed in supply chains and production networks, and the weakness of global economics emerged while illustrating limitations across different industries (Fernandes, 2020). This has directly impacted employment and increased the risk of food insecurity for millions due to the lockdown and border restrictions (Guerrieri, Lorenzoni, Straub, & Werning, 2022).

Global poverty is likely to increase for the first time since 1998 due to massive job losses and excessive income inequality. Nearly 49 million people could be pushed into extreme poverty as a result of COVID-19, with Sub-Saharan Africa being hit the hardest (Ibn-Mohammed, et al., 2021). Using the most effective policy instruments, the United Nations Department of Economic and Social Affairs concluded that the COVID-19 pandemic may also increase inequality, exclusion, discrimination, and unemployment globally in the medium and long term. The probable reason for unemployment can be the non-availability of education or the education gap between those who were able to afford education during the pandemic against those who were deprived.

1.2.2 COVID-19's Past, Current, and Future Impact on Education Funding

Since the onset of COVID-19, more than 70% of low and lower-middle-income countries decreased public education budgets. In the year 2018, high-income countries spent USD 500 more on education per child than in low-income countries. Due to COVID-19, this gap in education spending extended further (Paul, 2021). Figure 1.1 demonstrates the spending of 286 countries on education in contrast to their GDP (UNESCO, Government expenditure on education, total (% of GDP), 2021). For summarizing the data, the average GDP is calculated for every country. Furthermore, Table 1.3 demonstrates the trend of budget spending for the pre- and post-COVID-19 situations.

Figure 1.1 presents an alarming situation as countries show less interest in educational spending and budget declines every year.

It is observed in Table 1.3 that, overall, countries reduced their educational budgets. This budget curtailment resulted in educational gaps among students, school closures, and the non-availability of faculty. The reason for such a reduction could be the fact that countries were not ready, and they had no continuity or sustainability plans. Many countries struggle to provide basic health facilities including the US, Italy, and France.

According to the data released by the World Bank, 65% of countries have curtailed their educational budget after the COVID-19 situation (Bank, 2021). It is mentioned by the

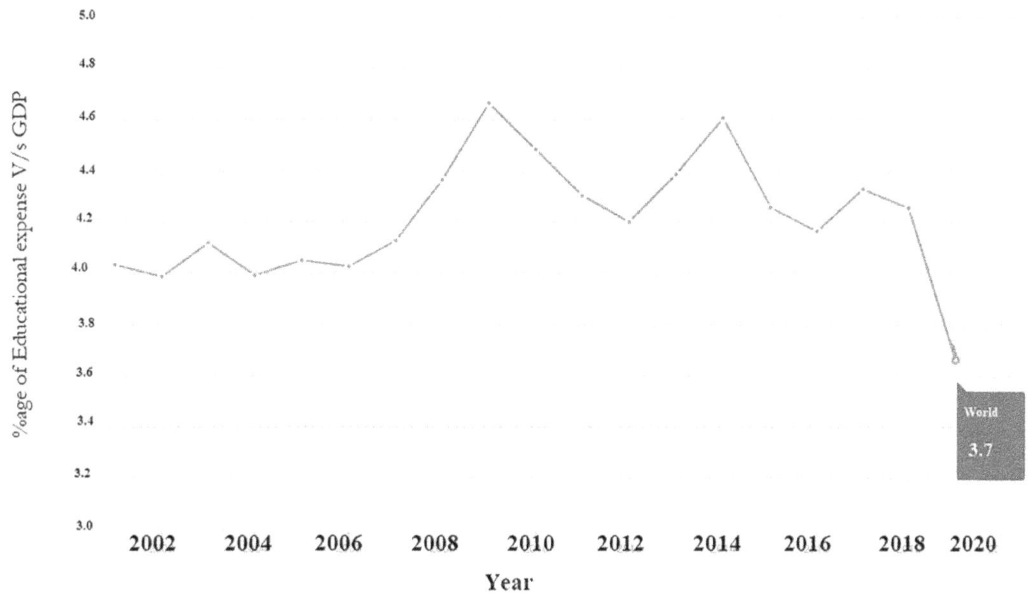

FIGURE 1.1 Region's government spending on education against GDP (%age) (UNESCO, 2021)

TABLE 1.3 Decrease in Education Budget before and after COVID-19 (%Age)

Description	All Countries		Upper and High Middle Income		Lower and Low Middle Income	
	Before COVID-19	After COVID-19	Before COVID-19	After COVID-19	Before COVID-19	After COVID-19
			Total Budget of Education			
Countries' share of the declining education budget	44.80	51.70	41.70	33.30	47.10	64.70
Average decline in countries that decreased their budget	−8.3	−4.6	−6.9	−7.4	−9.1	−3.6
			Share of Education in Total Budget			
Change in percentage points (Average)	−0.50	0.10	−0.30	−0.10	−0.60	0.20
Share decrease (Average)	13.4	13.5	13.7	13.6	13.2	13.4

Source: Al-Samarrai et al. (2021).

United Nations that "the pandemic has caused a global education catastrophe that desperately needs addressing to avoid the COVID-19 generation from becoming a lost generation. Yet, due to COVID-19 and the debt burdens faced by these countries, we are already seeing a contraction in education budgets at a time when countries need to be investing in improving schools and education systems (UN, 2021).

Education Finance Watch identified the challenges related to educational finance and urged to improve the process of funding and its related utilization. In recent times, negligible improvement can be observed in educational reforms. However, prior to COVID-19, 53% of children were unable to read short (age-appropriate) text in low- and middle-income countries, and due to school closures and a decline in educational budgets, it is expected that 63% of these children will not be able to read a short text. This can be termed a learning poverty crisis which is getting worse with every passage of time. It is worth noting that only 9% of children in high-income countries are not able to read age-appropriate text (Schrader-King, Ordon, & Dafalia, 2021). The huge disparity between high-income and low-income countries will grow even further as e-platforms and e-learning become more available in high-income countries.

After the decline in COVID-19 cases, countries with high or middle income were able to increase their educational budget, whereas low- and lower-middle-income countries slashed their educational budgets further (Al-Samarrai et al., 2021). Low- and lower-middle-income countries grappled with piles of debt that they were not able to pay earlier, and due to the closure of all business activities, these loans became extremely high and raised financial vulnerability in these countries (Kirby, Celiks, Essl, & Proite, 2019). There could be several other factors in the non-repayment of loans in which fraud and bribery cannot be ignored. Post-pandemic data demonstrated that one in eight countries pays more on debt servicing than on the combined budget for education, social protection, and health (Alhattab, 2021).

1.2.3 Role of E-Learning in Education

Due to the increasing demand for knowledge, students are expected to identify and retrieve more information in order to support and develop their learning process. HEIs have gradually embraced the concept of e-learning and are now implementing e-learning in their curriculum to accommodate diverse learning needs and provide more interactive materials that will facilitate learning. There are potential benefits to e-learning across the board for all faculty and students. Among several benefits, e-learning improves educational standards and augments students' engagement. It is worth noting that e-learning does not replace the role of the faculty member but contributes toward improving quality and enhancing efficiency (Ayu, 2020).

Today's information age demands that students be able to acquire information and find it on their own. The internet is seen as the key to providing access to more information and giving them the opportunity to find it on their own. E-learning is a method used by many to solve the problem of facilitating lifelong learning through the delivery of resources. E-learning refers to the use of information and communication technologies (ICTs) to enhance and/or support learning in higher education. Additionally, e-learning has been shown to improve computer literacy and productivity skills when students start working (Lin & Lin, 2015).

E-learning can be considered as compulsory and not an extra option (Dhawan, 2020) to be available within HEIs. In today's modern world where all the information is available at one touch, e-learning plays a pivotal role in developing and enhancing students' knowledge. There are two famous e-learning tools utilized by HEIs across the globe. These are MOODLE and Blackboard. There are 213 million users of MOODLE and 150 million users of Blackboard speared around more than 80 countries (Moodle, 2020; Hill, 2021). There are several HEIs that developed their own e-learning tools which are also effective and efficient. There is no doubt that e-learning contributes to the globalization of education and HEIs strive to break geographical and social boundaries to offer distance learning education, which results in the integration of academic standards and views (Babu & Sridevi, 2018).

1.2.4 Role of E-Learning in the COVID-19 Crisis

In the current HEI operational environment, e-learning is becoming a buzzword. With the ever-growing pace of technology, it was necessary to embed e-learning in the HEI mode of study; however, due to COVID-19, HEIs are forced to forego traditional educational methods and switch to e-learning.

Modern technologies are becoming an essential part of knowledge enhancement as we utilize these in our normal daily life (Chhabra, 2021; Tripathy, Mishra, Suman, Nayyar, & Sahoo, 2022; Al-Turjman, Devi, & Nayyar, 2021). These new technologies which enhance knowledge can include YouTube, blogs, Facebook, etc. These technologies are embedded in the e-learning of HEIs which enhances the students' engagement in virtual classrooms (Devi & Nayyar, 2021) and also enables real-time interactions. Students can search and study at the same time, and they can be contacted or they can contact faculty members at any point in time. With the utilization of cameras, video conferencing, and online presentations, learning became more interactive and with recorded lessons, it became easier for students to revise or understand in their own leisure time.

A good nation is built on good education (Baiyere & Li, 2016). Schools, colleges, universities, and other government institutions were closed suddenly due to the COVID-19 outbreak. E-learning platforms have been used by faculty members to deliver/teach education during these challenging times. E-learning refers to an educational process conducted through the use of electronic media. In 1999, e-learning was adopted for the first time during a CBT seminar. It can also be described as online learning (Soni, 2020). This service allows users to share reading materials via email, documents, presentations, or webinars. Education has become more sophisticated with the increasing influence of ICT in today's teaching–learning process. During this lockdown, educators shared study materials and lectures as PowerPoint, PDF, or Word documents on their university web pages, on WhatsApp, or by email. Several platforms were utilized for the delivery of classes online. These platforms include MOODLE, Blackboard, Zoom, Google classroom, etc. (Ayu, 2020; Ile & Ukabam, 2021)

Several HEIs utilized YouTube to upload the lectures and applied WhatsApp for the questions and answers. For the purpose of online education, several singular, multiple, or hybrid services were utilized. Technological developments have created a favorable environment for teaching and learning which changed the teaching approaches for the adoption of this new environment. Through innovative means, faculty members help students improve their learning skills. There are several ICT tools available free of cost, but there are plenty that charge money for their services. It is becoming a widely known fact that technology has influenced people's lives, and e-learning has become something that cannot be ignored. E-learning utilizes technology for accessing educational materials (Salamat, Ahmad, Bakht, & Saifi, 2018; Osman, Ustadi, Kamar, Johari, & Ismail, 2021).

In order to mitigate the impact of COVID-19 on education, specifically for HEIs, e-learning played a vital role in ensuring the continuation of education by facilitating ease in learning, promoting flexibility in the use of pedagogical methods, focusing on the needs of students, encouraging more hands-on learning experiences, allowing students to study independently, and allowing them to learn at their own pace. Education via electronic means is a true tool for achieving academic continuation (Ile & Ukabam, 2021).

1.2.5 Advantages of E-Learning

Due to the COVID-19 pandemic, many sectors have been forced to change the way they do business. Distancing precautions have reduced interpersonal contact to minimize the risk of community transmission in a densely populated community like the university campus. In response, the education sector has implemented "emergency e-learning," which forces rapid changes from traditional learning to online learning (Murphy, 2020). As a result of rapid changes, students who were not introduced to e-learning before the pandemic were forced to adapt to the change in order to receive education (Osman, Ustadi, Kamar, Johari, & Ismail, 2021).

E-learning is an advanced and technological way of getting knowledge by utilizing a computer network-based environment. There are many advantages associated with the development of e-learning which can be considered as crucial for its growth and sustainability. In HEIs, e-learning has been offered to all students regardless of their tenure in HEI; however, it is beneficial to adult students aged 21 years and older (Bourdeaux & Schoenack., 2016). E-learning offers flexibility for these adult students, enabling them to balance their other responsibilities. As mentioned earlier, HEIs have offered e-learning for quite some time, but it was not fully utilized until it was enforced by the COVID-19 situation. According to Al-rahmi, Othman, and Yusuf (2015), e-learning promotes less dependency on time, and students feel more confident in asking questions. Students can access, read, and edit educational materials easily at any time and from anywhere.

The integration of e-learning in the teacher education program mitigates the concerns of insufficient classrooms for faculty members as students can easily take lectures online without any disruption at their convenience (Coman, Laurent, Luiza Meses, Stanciu, & Bularca, 2020). Learning through e-learning is a convenient and cost-effective method of meeting today's learners' needs. E-learning has proven to be fruitful for a number of reasons; a few to mention are that learners may access the courses at any time by subscribing to different platforms or logging in to access the courses at their convenience (Colchester, Hagras, Alghazzawi, & Aldabbagh, 2016); e-learning provides an opportunity to share and offer a variety of teaching and learning materials; in addition to webinars and direct communication with teachers, multiple chat forums and messaging are also an added option; e-learning provides access to certain e-manuals free of charge; learners receive clear and simple instructions through gradual progression, and e-learning provides an actual platform for self-learning (Bajaja & Sharma, 2018).

E-learning provides a wide range of options for sharing information and uploading documents of various formats, which helps to foster the learning–teaching process. Using the web-based system does not require the installation of additional tools, and once the

TABLE 1.4 Types of E-Learning

E-Learning Types	Benefits
Individual courses (one on one classroom session)	No classmates are required, and students take classes on their own
Virtual/simulated classes	Similar to classroom experience but in a virtual/simulated world
Learning games	Process of assimilating and understanding information is conducted through virtual activities that are simulated
Related and blended learning	Combines online and traditional classes
Knowledge management and mobile learning	Online distribution of materials and documents that are destined to educate large numbers of communities, individuals, and organizations

Source: Fischer, Heise, Heinz, Moebius, and Koehler (2014).

content is uploaded, the user can access it anywhere and anytime (Alyoussef, 2021). With the range of technological tools available today, many types of e-learning methods have been developed. Table 1.4 defines the types of e-learning and its related benefits:

E-learning is also distinguished from traditional learning from the point of view of information sources, assessment methods, and quality of learning. In the traditional way of teaching, assessment is conducted on the knowledge provided by the faculty member and is also dependent upon the limited source of information. For e-learning, evaluation can be conducted with the assistance of several online tools (Kumar, Krishnamorti et al., 2021) and students can obtain information from various sources (Coman, Laurent, Luiza Meses, Stanciu, & Bularca, 2020). This practice enhances the quality of education and assists in developing a knowledge-based economy for countries. E-learning develops core online teaching principles such as forcing interactions between faculty and students, efficient and effective feedback, collaborative and active learning, a time-oriented task with proper follow-up, diversified learning, and utilization of technology (Cheung & Cable, 2017). E-learning also eliminates the utilization of paper and physical attendance which contributes to the green community and is also cost-effective.

1.2.6 Disadvantages of E-Learning

E-learning is the mind shift and the culture which is required to be properly implemented by HEIs. People show resistance to change (Rehman et al., 2021), and this resistance can sometimes affect negatively the perceived performance. In a similar way, if e-learning is not implemented properly and students are not aware of its traits, then it will bring disappointment to HEIs' performance which also impacts the future of students (Kumar, Sharma et al., 2021). Faculty members and students both are required to follow ethical norms while adopting e-learning as academic dishonesty cannot be ignored in this environment (McCabe, Butterfield, & Treviñ, 2012). Few scholars argue that summative and formative assessments in the e-learning environment do not reflect actual learning as they are probably tainted by cheating (Dendir & Maxwell, 2020).

E-learning has different understanding and tolerance levels for every student. Students might have accepted the e-learning methods, whereas others might face difficulty to comprehend them. With the variety of online tools available, faculty and students' difficulties could arise with technology itself, such as login, downloading, installation errors, and video and audio problems (Dhawan, 2020). These difficulties arise due to poor internet connection such as uploading and downloading speed and a faulty/old operating system in digital devices such as computers, mobile phones, and laptops. Furthermore, there are many students who are not able to afford these digital devices or pay for a good internet connection.

Another problem associated with e-learning is that it is difficult for students to interact with their lecturers as they feel the need for two-way communication. Due to e-learning platforms, it is very difficult to achieve two-way interaction, and thus, the learning outcome cannot be achieved. E-lectures, for instance, do not allow students to raise their hands and students are uncomfortable asking questions (Dhawan, 2020). Furthermore, and in accordance with recent research, it was identified that the shift from traditional learning to e-learning negatively impacts students' motivation (Sandybayev, 2020).

Education may become more unequal with e-learning (Moralista & Oducado, 2020). Students deprived of devices, connectivity, and technology are less likely to participate in e-learning than privileged students (Biswas & Debnath, 2020). Comparing developing countries to developed ones, dissatisfaction with e-learning is more pronounced in developing countries, which lack economic and technical expertise. As a result, developing countries are less prepared to adopt e-learning as an alternative to traditional teaching and learning, and the COVID-19 pandemic has highlighted in a large way the lack of adequate facilities and technology for implementing e-learning solutions (Adzovie, Jibril, Adzovie, & Nyieku, 2020; Oducado & Soriano, 2021).

There are several obstacles associated with e-learning platforms that create hindrances in the process of learning. These obstacles can be considered as non-motivation among students, non-availability of faculty members as students study on their own time, and there is a possibility that faculty time does not match with the student's availability time and the social dilemma of isolation as the virtual world only connects virtually, whereas many students feel that they are not part of the community or they are disregarded by other fellow students (Yusuf & Al-Banawi, 2013). These obstacles/problems can be mitigated by proper planning, implementation of teaching ethics, and developing HEIs' strategy which supports e-learning.

Further to the above and to implement mitigating actions, knowledge, and experience in e-learning are required to be available in the HEI's faculty. Additionally, these obstacles and hindrances are more obvious and noticeable when the HEI provides only/exclusively online education such as in the case of COVID-19 time. In accordance with the recent study conducted by Gateway (2020), it was identified that 67% of respondents utilized online learning platforms for the first time. From this study, it is obvious that faculty and HEIs were not ready and they were forced to provide online education which created lots of difficulties, complications, and problems (Coman, Laurent, Luiza Meses, Stanciu, & Bularca, 2020).

To implement and utilize e-learning HEIs, their faculty and students need to overcome many challenges. Few of the challenges HEIs have to encounter are keeping balance and stability for online courses keeping in mind that e-learning could affect the health of students as they spend many screen hours and are not involved in any other physical activities, focusing and analyzing on students' and faculty's emotional health by providing them with learning support, considering the fact that many students have limited access to the internet; therefore, HEIs are required to deal with every student and faculty independently (Coman et al., 2020) which is very difficult and in fact, not possible. Additionally, in an e-learning environment, HEIs find it difficult to maintain the integrity of their course content, to communicate clearly with their academic community, and to recruit quality students (Marinoni, Land, & Jensen, 2020).

1.2.7 Future of E-Learning

E-learning has paved its way in all sorts of educational areas. It is here to stay, and with the assistance of artificial intelligence (AI), it will mature and will deliver optimum services for knowledge enhancement and knowledge delivery. AI is a modern technology that is more powerful and innovative than any other (Venaik, Kumari, Venaik, & Nayyar, 2022), with the potential to transform the world. By utilizing AI, organizations including HEIs will be able to perform cognitive functions, including comprehending, recognizing, observing, perceiving, learning, and interacting (Rehman, 2022).

In recent years, e-learning has been recognized as an important teaching and learning mode and is also considered one of the most effective and efficient tools. The number of internet users has increased twofold in the last couple of years (Johnson, 2022), and with the utilization of smartphones and other smart technologies, e-learning is not only utilized in HEIs but also in vocational, primary, and secondary schools (Gross & García-Peñalvo, 2017).

E-learning will highlight and promote the culture and concept of self-motivation, independent learning, and goal-oriented learners. E-learning will pave the way for faculty and students to share their work, interact with small and large groups virtually, and collaborate on their work. This will eliminate the barrier of physical availability and the need to be present at specific times. E-learning will assist the users in developing strong academic self-concept, make everyone competent for the utilization of technology, and promote social interaction and knowledge-based learning. Other characteristics that e-learning will draw attention to are self-directedness, the level of motivation, time, and the level of interaction among faculty and students (Goyal, 2012).

In addition to remote and lifelong learning, e-learning offers a promising alternative to traditional classroom learning. E-learning complements classroom learning in many ways. E-learning will continue to grow as an integral part of academic and professional education. The creation of more appealing and effective online learning environments should continue to be explored. Integrating appropriate pedagogical methods, improving system interactivity and personalization, and engaging learners is one way to achieve this. Lifelong learning and on-the-job training are made easier with e-learning. As a form of technology-based learning, e-learning is the delivery of learning materials electronically via a computer network to remote learners. It is crucial for HEIs to ensure that their faculty and students are equipped with the latest information and advanced skills by implementing effective and efficient training methods. Many HEIs around the world are now offering online courses, including degree and certificate programs, in response to this need (Alsuwaida, 2022).

TABLE 1.5 E-Learning Rules of Engagement

E-Learning Impact	Details
Faculty–content	Faculty will no longer rely on past available content and will be obliged to include new knowledge to be delivered to students
Content–content	Competency-based curricula will be introduced which will not be common to all but will be based on the student's capability and delivered course contents
Student–faculty	Different relationships will be developed where the student and faculty both can be contacted at any point in time and can share ideas, which will be available for everyone to understand
Student–student	Interaction among students will be changed as ideas and content will be shared which will be accessible to all
Faculty–faculty	How faculty interact with each other will be digitalized. Knowledge, content, presentations, etc. will be shared online and can be accessed from anywhere
Student–content	Course contents will be tailor-made for the students as per their capabilities, understanding, and progress

Source: Jameel (2019), Centre for Educational Research and Innovation (2019) and Talebian, Mohammadi and Rezvanfar (2014).

The future of e-learning will be advantageous in all areas and will enhance efficiency and effectiveness. E-learners will be able to cover a similar curriculum in half of the time in comparison to traditional learning. Furthermore, with an e-learning platform knowledge will reach many students in a shorter time and across the world. Every employee in HEIs will have access to training whenever and wherever it is convenient for them, at home or at work.

E-learning will oblige to new rules of engagement for effective learning which will far reach the benefits of faculty–student interaction. E-learning will have an impact and develop six rules of engagement that will transform all HEIs across the world as presented in Table 1.5.

E-learning will enable students to actively participate in pursuing and creating knowledge with meaningful context by encouraging the adoption of new technologies. Students will be able to learn collaboratively and communicate using a variety of collective and common tools. While there are many different tools available for e-learning, the architecture of a learning system implies a difficult task to integrate into a scalable, flexible, and capable lasting system. Faculty and HEI will no longer have control over the learning pace using the conventional learning management system (Rawashdeh, Mohammed, Arab, Alara, & Al-Rawashdeh, 2021)

1.3 CASE STUDIES

This section will discuss the case studies conducted in recent years which are related to e-learning in the era of the COVID-19 pandemic.

1.3.1 Case Study of Three Faculty Members from the Two Countries

In the case study conducted by Islam, Nur, and Talukder (2021), three HEI faculty from two different countries shared their first-hand experiences and utilized a quantitative research methodology.

This case study revealed that the use of e-learning created complex and challenging situations within institutions and individual classrooms. The study also identified acceptances, struggles, and negotiations at both the macro- and micro-levels of policy- and decision-making. Ultimately it was concluded that e-learning provided complex, challenging, and dynamic experiences coupled with struggles and agreements at the HEI level and also at the individual level related to online classroom practices (Islam, Nur, & Talukder, 2021).

The above-mentioned case study highlighted the fact that e-learning during the COVID-19 times created challenges for HEIs and also for students. E-learning at this time provided several challenges as HEIs were not ready and students were not prepared.

1.3.2 E-Learning Challenges Faced by Students during COVID-19

A case study and online survey was conducted by Nikou and Maslov (2021) in Finnish HEIs. The study was conducted when the Finnish Government announced the closure of all HEIs and institutions and students were forced to utilize e-learning. The study utilized exploratory research with the aim to address the issues by assessing students' experience with e-learning.

The results of the study demonstrated that with the mediation of e-learning, its related ease of use and usefulness can mitigate the challenges for students' intention for e-learning. Moreover, results show that HEI preparedness has no direct impact directly on students' intention to use e-learning during specified times, although gender and the length of time students spend using e-learning systems do have an impact.

It is evident from the study that students faced challenges during COVID-19 times related to e-learning; however, these challenges were mitigated based on the HEIs' preparedness toward e-learning and their system availability.

1.3.3 Quality of E-Learning during COVID-19

A case study conducted by Idwana, Fayyoumi, Hijazi, and Matar (2021) was related to the quality of e-learning during COVID-19 times and its impact on students' education. A quantitative and qualitative study was conducted for only one HEI.

The results of the study demonstrated that HEIs, their faculty, and students faced several challenges which were related to the availability of good internet speed, availability of equipment, and faculty interaction with students. E-learning provided ease of use but also created difficulties in the learning progression of students. Students did not comprehend the instructor's new role as a guide or facilitator. Several students said they were not being taught by their instructors as the new format for adapting has been implemented. Furthermore, e-learning did not provide the emotional and intellectual support that students requested from instructors (Idwana, Fayyoumi, Hijazi, & Matar, 2021).

The case study highlighted the challenges faced by the students regardless of the HEI providing the e-learning platform and other related facilities.

1.4 CONCLUSION AND FUTURE SCOPE

E-learning was the emerging concept that was enforced on HEIs to adopt at the time of the COVID-19 pandemic. In e-learning, electronic media is used to carry out the educational process. It is also known as online learning. In e-learning, users can share reading materials via email, documents, presentations, or webinars. Nowadays, education has become more sophisticated as e-learning plays an increasingly important role in the teaching–learning process. In this lockdown, educators shared study materials and lectures as PowerPoint or PDF documents on their HEI's websites or by email or WhatsApp. Classes are delivered online through a variety of platforms. Among them, the famous ones are MOODLE, Blackboard, Zoom, and Google classroom.

Economies around the world decreased/reduced the budget for education during this pandemic time and shifted their budgets to other necessary activities such as health services and restoration of business communities. This non-availability of funds forced faculty members to perform other jobs and duties which are not associated with knowledge sharing. Furthermore, countries that provided a limited budget and resources faced difficulties with the availability of devices and the readiness of faculty and students.

Regardless of the availability of funds and faculty readiness, e-learning has played a significant role in mitigating the impact of COVID-19 on education, especially at HEIs, by facilitating ease in learning, encouraging flexibility in the use of pedagogical methods, focusing on the interests of students and allowing students to study independently and learn at their own pace. Educational resources via electronic means are a valuable tool for enabling academic growth. The benefit of e-learning courses is that a learner can filter through the course and only participate in the sections that relate to his/her current position. The majority of HEIs reap significant returns from e-learning investments. Many e-learning customers have quickly realized first-tier benefits such as reduced costs for travel, customer support, human resources overhead, and regulatory compliance and, eventually, second-tier benefits such as increased employee performance that contribute to profitability.

The future of e-learning will benefit HEIs in every way and will enhance efficiency and effectiveness. With e-learning, students can cover the same curriculum in half the time. Moreover, e-learning platforms will allow knowledge to reach a wider audience of students in a shorter time span and across the globe. At HEIs, all employees will have access to training at home or at work whenever it is convenient for them.

COVID-19-enforced e-learning caused resistance, which negatively affected performance perceptions. Students were not aware of the effects of e-learning, and the lack of implementation affected their future as well. Using e-learning also raises concerns about ethical issues such as cheating and dishonest assessments. There are many technical difficulties associated with e-learning, such as login issues, downloading problems, and video and audio issues. Moreover, the lack of two-way communication did not allow students to interact with their faculty members, which negatively affected their motivation.

E-learning has also made education unequal as multiple students around the world lack devices, connections, and technology compared to students who have access to e-learning. In developing countries, which lack both economic and technical expertise, dissatisfaction with e-learning is more pronounced than in developed ones. As a result, developing countries were less prepared to adopt e-learning as a substitute for traditional education, and the COVID-19 pandemic highlighted in a large way the lack of facilities and technology needed to implement e-learning solutions. Furthermore, it was noted that most faculty and students were using the e-learning platforms for the very first time. The faculty and HEIs were obviously unprepared and forced to provide online education, which led to a variety of complexities, challenges, and difficulties.

It is obvious from the above literature that COVID-19 created a vast gap between students' learning, faculty teaching, and knowledge sharing. Many HEIs who were able to adopt e-learning was able to attract more student recruitment and were able to increase their profits. On the other hand, HEIs and students who were deprived of e-learning suffered not only academically, but also impacted their future employability. E-learning in the era of COVID-19 created more difficulties than ease of operations. There are still many obstacles to overcome and many probable risks to be mitigated before e-learning emerges as a knowledge enhancement platform and HEIs utilize its maximum benefits.

This study offers an overview and related details of the process of e-learning as well as its advantages and disadvantages and also discusses the future of e-learning in HEIs. This paper also highlights the economic impact of COVID-19 and how it changed the behavior/norms of nations which impacted the operations and behavior of HEIs and students. This study can be utilized as a guideline for the complete empirical study and can be utilized by policymakers, regulators, HEIs' board, and governance management of HEIs. This study can assist in mitigating the risk associated with e-learning and can also be utilized as a guidance document for developing e-learning policies.

Future Scope

This chapter can take the form of a complete empirical study where data can be analyzed for several HEIs and can also be compared among the regions. Future studies can be conducted toward identifying the nations' spending patterns and identifying the reason for the decline in funding. An empirical study can also be conducted to identify the gaps created due to the availability of online facilities among the privileged ones from those who are deprived, this study can be made between different but comparable countries and also within the different cities of the same country. This study provides the disadvantages which can be mitigated with proper policies, availability of devices, and effective involvement of the nation's educational ministries. The advantages of e-learning can be considered as opportunities and can assist HEIs in adopting e-learning platforms

REFERENCES

Adzovie, D. E., Jibril, A. B., Adzovie, R. H., & Nyieku, I. E. (2020). E-learning resulting from covid-19 pandemic: A conceptual study from a developing country perspective. *7th European Conference on Social Media ECSM 2020* (p. 424). Manchister: ACPIL.

Alhattab, S. (2021). Pre-pandemic data show 1 in 8 countries spends more on debt than on education, health and social protection combined – UNICEF. Retrieved from www.unicef.org: https://www.unicef.org/eap/press-releases/pre-pandemic-data-show-1-8-countries-spends-more-debt-education-health-and-social.

Al-rahmi, W. M., Othman, M. S., & Yusuf, L. M. (2015). The effectiveness of using e-learning in Malaysian higher education: A case study Universiti Teknologi Malaysia. *Mediterranean Journal of Social Sciences*, 6(52), 625–637.

Al-Samarrai, S., Cerdan-Infantes, P., Bigarinova, A., Bodmer, J., & Vital, M. (2021). *Education Finance Watch*. NY: World Bank Group - Education.

Al-Turjman, F., Devi, A., & Nayyar, A. (2021). Emerging technologies for battling covid-19. *Studies in Systems, Decision and Control*, 324, 7–9.

Allain-Dupré, D., Chatry, I., Kornprobst, A., Michalun, M.-V., Lafitte, C., Moisio, A., ... Zapata, I. (2020). *The Territorial Impact of COVID-19: Managing the Crisis Across Levels of Government*. Paris, EU: OECD.

Alsuwaida, N. (2022). Designing and evaluating the impact of using a blended art course and web 2.0 tools in Saudi Arabia. *Journal of Information Technology: Research*, 22, 26–52.

Alyoussef, I. (2021). E-learning system use during emergency: An empirical study during the covid-19 pandemic. *Frontier in Education*, 6, 1–11.

Angelova, M. (2020). Students' attitudes to the online university course of management in the context of COVID-19. *International Journal of Technology in Education and Science*, 4(4), 283–292.

Ayu, M. (2020). Online learning: Leading e-learning at higher education. *The Journal of English Literacy Education*, 7(1), 47–54.

Babu, G. S., & Sridevi, K. (2018). Importance of e-learning in higher education: A study. *International Journal Of Research Culture Society*, 2(5), 84–88.

Baiyere, A., & Li, H. (2016). Application of a virtual collaborative environment in a teaching case. *Surfing the IT Innovation Wave -22nd Americas Conference on Information Systems*. Chicago: AMICS.

Bajaja, R., & Sharma, V. (2018). Smart education with artificial intelligence based determination of learning style. *Procedia Computer Science*, 132, 834–842.

Baker, S. R., Bloom, N., Davis, S. J., & Terry, S. J. (2020). Covid-induced economic uncertainty. *National Bureau Of Economic Research*, 26983, 1–16.

Bank, W. (2021). Learning losses from covid-19 could cost this generation of students close to $17 trillion in lifetime earnings. Retrieved from worldbank.org: https://www.worldbank.org/en/news/press-release/2021/12/06/learning-losses-from-covid-19-could-cost-this-generation-of-students-close-to-17-trillion-in-lifetime-earnings.

Basilaia, G., & Kvavadze, D. (2020). Transition to online education in schools during a SARS-CoV-2 coronavirus (COVID-19) pandemic in Georgia. *Pedagogical Research*, 5(4), 1–9.

Biswas, P., & Debnath, A. K. (2020). Worldwide scenario of unplanned transition to e learning in the time of covid-19 and students' perception: A review. *Mukt Shabd Journal*, 9(6), 2038–2042.

Bourdeaux, R., & Schoenack, L. (2016). Adult student expectations and experiences in an online learning environment. *The Journal of Continuing Higher Education*, 64(3), 152–161.

Cambridge University. (2021). *Annual Reports And Financial Statements*. Cambridge: University of Cambridge.

Centre for Educational Research and Innovation. (2019). *Assessment For Learning – The Case For Formative Assessment*. Paris: Organisation for Economic Co-Operation and Development.

Cheung, C., & Cable, J. (2017). Eight principles of effective online teaching: A decade-long lessons learned in project management education. *Project Management World Journal*, 6, 1–16.

Chhabra, P. (2021). Use of e-learning tools in teaching english. *International Journal of Computing & Business Research*, 1–7.

Colchester, K., Hagras, H., Alghazzawi, D., & Aldabbagh, G. (2016). A survey of artificial intelligence techniques employed for adaptive educational systems within e-learning platforms. *Journal of Artificial Intelligence and Soft Computing Research*, 7(1), 47–64.

Columbia University. (2021). *Columbia University Financial Reports*. Columbia: Columbia University.

Coman, C., Laurent, I. G., Luiza Meses, A.-S., Stanciu, C., & Bularca, M. C. (2020). Online teaching and learning in higher education during the coronavirus pandemic: Students' perspective. *Sustainability*, 12(24), 10367. doi:10.3390/su122410367.

Dendir, S., & Maxwell, R. S. (2020). Cheating in online courses: Evidence from online proctoring. *Computers in Human Behavior Reports*, 2, 1–10.

Devakumar, D., Shannon, G., Bhopal, S. S., & Abubaker, I. (2020). Racism and discrimination in COVID-19 responses. *Lancet Journal*, 395 (10231), 1194.

Devi, A., & Nayyar, A. (2021). Perspectives on the definition of data visualization: A mapping study and discussion on coronavirus (COVID-19) dataset. In D. Al-Turjuan & A. Nayyar (Ed.), *Emerging Technologies for Battling Covid-19* (pp. 223–240). Springer, Cham.

Dhawan, S. (2020). Online learning: A panacea in the time of covid-19 crisis. *Journal of Educational Technology Systems*, 49(1), 1–25.

Fennell, A. (2022). Graduate statistics for the UK; The latest facts and figures on UK graduates. Retrieved from https://standout-cv.com/uk-graduate-statistics#number-of-graduates.

Fernandes, N. (2020). Economic effects of coronavirus outbreak (COVID-19) on the world economy. *IESE Business School Working Paper No. WP-1240-E*, doi:10.2139/ssrn.3557504.

Fies, A., & Hill, J. (2020). Coronavirus pandemic brings staggering losses to colleges and universities. Retrieved from https://abcnews.go.com/Business/coronavirus-pandemic-brings-staggering-losses-colleges-universities/story?id=70359686.

Fischer, H., Heise, L., Heinz, M., Moebius, K., & Koehler, T. (2014). E-learning trends and hypes in academic teaching. methodology and findings of trend study. International Conference e-Learning 2014 (pp. 63–69). Dresden: Dresden University of Technology, Media Center.

Gamage, K. A., Silva, E. K., & Gunawardhana, N. (2020). Online delivery and assessment during COVID-19 safeguarding academic integrity. *Education Sciences*, 10, 1–24.

Gateway, S. E. (2020). Survey on online and distance learning – Results. Retrieved from https://www.schooleducationgateway.eu/en/pub/viewpoints/surveys/survey-on-online-teaching.htm.

Goyal, S. (2012). E-learning: Future of education. *Journal of Education and Learning*, 6(2), 23–242.

Gross, B., & García-Peñalvo, F. (2017). Future trends in the design strategies and technological affordances of e-learning. In M. Spector, B. B. Lockee, & M. D. Childress (Eds.), *Learning, Design, and Technology* (pp. 1–23). Cham: Springer.

Guerrieri, V., Lorenzoni, G., Straub, L., & Werning, I. (2022). Macroeconomic implications of COVID-19-19: Can negative supply shocks cause demand shortages? *American Economic Review*, 112(5), 1437–1474.

Harvard University. (2021). *2020–2021 Annual Financial Report*. Harvard: Harvard University.

Hill, P. (2021). The end of blackboard as a standalone edtech company. Retrieved from https://philonedtech.com/the-end-of-blackboard-as-a-standalone-edtech-company/.

Ibn-Mohammed, T., Mustaphab, K. B., Godsella, J., Adamuc, Z., Babatundede, K. A., Akintadef, D. D., Acquayeg, A., Fujiih, H., Ndiayei, M. M., Yamoahj, F. A., Kohk, S. C. L. (2021). A critical analysis of the impacts of COVID-19 on the global economy and ecosystems and opportunities for circular economy strategies. *Resources, Conservation & Recycling*, 164, 1–22.

Idwana, S., Fayyoumi, E., Hijazi, H., & Matar, I. (2021). COVID-19 pandemic endorses new era of e-learning case study: Hashemite University. *Journal of E-Learning and Knowledge Society*, 17(2), 56–65.

Ile, C., & Ukabam, J. A. (2021). Role of e-learning in ensuring the continuation of training in business teacher education programme during the covid-19 pandemic and beyond in nigeria. *Nigerian Journal of Business Education*, 8(1), 121–128.

Imperial College London. (2021). *Annual Report and Accounts*. London: Imperial College London.

Islam, M. A., Nur, S., & Talukder, M. S. (2021). E-learning in the time of COVID-19: Livedexperiencesofthreeuniversity teachers from two countries. *E-Learning and Digital Media*, 18(6), 557–580.

Jackson, E. A. (2020). *Emerging innovative thoughts on globalization amidst the contagion of COVID-19*. Cham: Springer Nature Publisher, 1–17.

Jameel, A. L. (2019). Tailoring instruction to students' learning levels to increase learning. *Poverty Action Lab (J-PAL)*, doi:10.31485/pi.2522.2019.

Johnson, J. (2022). Worldwide digital population as of April 2022. Retrieved from https://www.statista.com/statistics/617136/digital-population-worldwide/.

Kirby, P., Celiks, S. K., Essl, E., & Proite, A. (2019). *Debt in Low-Income Countries: A Rising Vulnerability*. NY: World Bank Blogs.

Kiyota, K. (2022). The COVID-19 pandemic and the world trade network. *Journal of Asian Economics*, 78(101419), 1–23.

Krakoff, S. (2022). The cost of online education vs. traditional education. Retrieved from https://online.champlain.edu/blog/cost-of-online-education-vs-traditional-education.

Kumar, A., Krishnamurthi, R., Bhatia, S., Kaushik, K., Ahuja, N. J., Nayyar, A., & Masud, M. (2021). Blended learning tools and practices: A comprehensive analysis. IEEE Access, 85151–85197.

Kumar, A., Sharma, K., Singh, H., Srikanth, P., Krishnamurthi, R., & Nayyar, A. (2021). Drone-based social distancing, sanitization, inspection, monitoring, and control room for covid-19. In D. Al-Turjuman (Ed.), *Artificial Intelligence and Machine Learning for COVID-19* (924: pp. 153–173). Cham: Springer.

Laksana, D. N. (2020). The implementation of online learning during covid-19 pandemic: student perceptions in areas with minimal internet access. *Journal of Education Technology*, 4(4), 502–509.

Lin, E., & Lin, H. C. (2015). The effect of teacher-student interaction on students' learning achievement in online tutoring environment. *International Journal of Technical Research and Applications*, 22(22), 19–22.

Liyanagunawardena, T. R., & Aboshady, O. A. (2018). Massive open online courses: A resource for health education in developing countries. *Global Health Promotion*, 25(3), 4–6.

London School of Economics and Political Sciences. (2021). *2020/21 Annual Accounts*. London: London School of Economics and Political Sciences.

Marinoni, G., Land, H. V., & Jensen, T. (2020). *The Impact of Covid-19 On Higher Education Around The World*. Paris: International Association of Universities.

Massachusetts Institute of Technology. (2021). *Report of the Treasurer for the Perid Ended June 2021*. Massachusetts: Massachusetts Institute of Technology (MIT).

Mathrania, A., Sarveshb, T., & Umer, R. (2021). Digital divide framework: online learning in developing countries during the COVID-19 lockdown. *Globalisation, Societies and Education*, 20(5) 1–16.

McCabe, D. L., Butterfield, K. D., & Treviñ, L. K. (2012). *Cheating in College: Why Students Do It and What Educators Can Do about It*. Baltimore: Johns Hopkins University Press.

Ministry of Higher Education, Research and Innovation. (2020). *The Annual Statitics Report of Heigher Education in the Sultanate of Oman for Academic Year 2019/20*. Muscat: Heigher Education Admission Center.

Moodle. (2020). About moodle. Retrieved from https://docs.moodle.org/400/en/About_Moodle.

Moralista, R., & Oducado, R. M. (2020). Faculty perception toward online education in a state college in the philippines during the coronavirus disease 19 (covid-19) pandemic. *Universal Journal of Educational Research*, 8(10), 4736–4742.

Murphy, M. P. (2020). Covid-19 and emergency e-learning consequences of the securitization of higher education for post-pandemic pedagogy. *Contemporary Security Policy*, 41(3), 1–14.

Mustakim. (2020). Efektivitas pembelajaran daring menggunakan media online selama pandemi covid-19; Pada mata pelajaran matematika the effectiveness of e-learning using online media during the covid-19 pandemic in mathematics. *Al Asma: Journal of Islamic Education*, 2(1), 1–12.

Ndambakuwa, S., & Brand, G. (2020). *Commentary: Many Students in Developing Countries Cannot Access Education Remotely*. Chicago: The University of Chicago.

New York University. (2021). *New York University 2020–2021 Consolidated Financial Statements*. New York: New York University.

Nikou, S., & Maslov, I. (2021). An analysis of students' perspectives on e-learning participation – the case of COVID-19 pandemic. *International Journal of Information and Learning Technology*, 38(3), 299–315.

Oducado, R. M., & Soriano, G. P. (2021). Shifting the education paradigm amid the covid-19 pandemic: Nursing students' attitude to e-learning. *Africa Journal of Nursing and Midwifer*, 23(1), 1–14.

Osman, S., Ustadi, M. N., Kamar, H. K., Johari, N. H., & Ismail, N. (2021). The impact of covid-19 crisis upon the effectiveness of e-learning in higher education institution. *Journal of Information Technology Management*, 13(3), 160–177.

Oxford University. (2021). *Financial Statements*. Oxford: University of Oxford.

Pape, L. (2010). Blended teaching and learning. *School Administrator*, 67(4), 16–21.

Paudel, P. (2021). Online education: Benefits, challenges and strategies during and after covid-19 in higher education. *International Journal on Studies in Education*, 3(2), 70–85.

Paul, M. (2021). Two-thirds of poor countries slashed education budget due to COVID-19: World Bank, UNESCO. Retrieved from https://www.downtoearth.org.in/news/governance/two-thirds-of-poor-countries-slashed-education-budget-due-to-covid-19-world-bank-unesco-75667.

Pinner, D., Rogers, M., & Samandari, H. (2020). Addressing climate change in a post-pandemic world. *McKinsey Quarterly*, 51(3), 1–6.

Rawashdeh, A. Z., Mohammed, E. Y., Arab, A. R., Alara, M., & Al-Rawashdeh, B. (2021). Advantages and disadvantages of using e-learning in university education: Analyzing students' perspectives. *The Electronic Journal of e-Learning*, 19(2), 107–117.

Rehman, A. (2022). With the mediation of internal audit, can artificial intelligence eliminate and mitigate fraud? In S. S. Kamwani, E. S. Vieira, M. Madaleno, & G. Azevedo, *Handbook of Research on the Significance of Forensic Accounting Techniques in Corporate Governance* (pp. 232–257). PA: IGI Global.

Rehman, N., Mahmood, A., Ibtasam, M., Murtaza, S. A., Iqbal, N., & Molnár, E. (2021). The psychology of resistance to change: The antidotal effect of organizational justice, support and leader-member exchange. *The Psychology of Resistance to Change*, 12, 1–15.

Renu, N. (2021). Technological advancement in the era of COVID-19. *SAGE Open Medicine*, 9, 1–4.

Salamat, L., Ahmad, G., Bakht, I., & Saifi, I. L. (2018). Effects of e-learning on students' academic learning at university level. *Asian Innovative Journal of Social Sciences & Humanities*, 2(2), 1–12.

Sandybayev, A. (2020). The impact of e-learning technologies on student's motivation: student centered interaction in business education. *International Journal of Research in Tourism and Hospitality*, 6(1), 16–24.

Schrader-King, K., Ordon, K., & Dafalia, G. (2021). Two-thirds of poorer countries are cutting education budgets due to covid-19. Retrieved from https://www.worldbank.org/en/news/press-release/2021/02/22/two-thirds-of-poorer-countries-are-cutting-education-budgets-due-to-covid-19.

Sharma, K., Singh, H., Sharma, D. K., Kumar, A., Nayyar, A., & Krishnamurthi, R. (2021). Dynamic models and control techniques for drone delivery of medications and other healthcare items in COVID-19-19 hotspots.. In D. Al-Turjuman and A. Nayyar (Eds.), *Emerging Technologies for Battling Covid-19* (pp. 1–34). Cham: Springer.

Sobti, P., Nayyar, A., & Nagrath, P. (2021). Time series forecasting for coronavirus (covid-19). In P. Kumar Singh, S. Sood, Y. Kumar, M. Paprzycki, A. Pljonkin, W-C. Hong (Eds.), *International Conference on Futuristic Trends in Networks and Computing Technologies* (pp. 309–320). Singapore: Springer.

Soni, V. D. (2020). Global impact of e-learning during COVID 19. Retrieved from https://papers.ssrn.com/sol3/papers.cfm?abstract_id=3630073.

Spash, C. (2020). "The economy' as if people mattered: revisiting critiques of economic growth in a time of crisis. *Globalization*, 18(7), 1–18 doi:10.1080/14747731.2020.1761612

Stanford University. (2021). *Annual Financial Report 2020–2021*. Stanford: Stanford University.

Talebian, S., Mohammadi, H. M., & Rezvanfar, A. (2014). Information and communication technology (ICT) in higher education: Advantages, disadvantages, conveniences and limitations of applying e-learning to agricultural students in Iran. *Social and Behavioral Sciences*, 152, 300–305.

Tripathy, H. K., Mishra, S., Suman, S., Nayyar, A., & Sahoo, K. S. (2022). Smart COVID-shield: An IoT driven reliable and automated prototype model for COVID-19 symptoms tracking. *Computing*, 104(6), 1233-1254, doi:10.1007/s00607-021-01039-0.

UNESCO. (2021). Government expenditure on education, total (% of GDP). Retrieved from https://data.worldbank.org/indicator/SE.XPD.TOTL.GD.ZS?end=2020&start=1970&view=chart.

UNESCO. (2022). Education: From disruption to recovery. Retrieved from https://en.unesco.org/covid19/educationresponse.

United Nations. (2021). Many countries spending more on debt than education, health and social protection combined. Retrieved from https://news.un.org/en/story/2021/04/1088852.

United Nations Health. (2022). Two years on, COVID-19 pandemic 'far from over'. Retrieved from https://news.un.org: https://news.un.org/en/story/2022/03/1113632.

University of California. (2021). *UCLA Annual Financial Reports 2020–2021*. Los Angeles: University of California, Los Angeles (UCLA).

University of California Berkeley. (2021). UC Berkeley financial reports FY 2020/21. California Berkeley: University of California Berkeley (UCB).

University of Edinburgh. (2021). *Annual Report and Accounts*. Edinburgh: University of Edinburgh.

University of Pennsylvania. (2021). *Annual Financial Report Fiscal year 2021*. Pennsylvania: University of Pennsylvania.

Venaik, A., Kumari, R., Venaik, U., & Nayyar, A. (2022). The role of machine learning and artificial intelligence in clinical decisions and the herbal formulations against covid-19. *International Journal of Reliable and Quality E-Healthcare (IJRQEH)*, 11(1), 1–17.

Warwick. (2021). *Annaul Financial Statements for the Year Ending 31st July 2021*. Warwick: The University of Warwick.

WHO. (2022). I like to be safe: The COVID-19 Pandemic is not over yet. Retrieved from https://www.who.int/southeastasia/outbreaks-and-emergencies/covid-19/What-can-we-do-to-keep-safe/protective-measures/pandemic-not-over.

Yale University. (2021). *2020–2021 Financial Report*. Yale: Yale University.

Yasir Arafat, S.M., Kumar Kar, S., Marthoenis, M., Sharma, P., Hoque Apu, E., & Kabirf, R. (2020). Psychological underpinning of panic buying during pandemic (COVID-19). *Psychiatry Research*, 289. doi:10.1016/j.psychres.2020.113061

Yusuf, N., & Al-Banawi, N. (2013). The impact of changing technology: The case of e-learning. *Contemporary Issues in Education Research*, 6(2), 173–180.

Zivkovic, M., Bacanin, N., Venkatachalam, K., Nayyar, A., Djordjevic, A., Strumberger, I., & Al-Turjman, F. (2021). COVID-19 cases prediction by using hybrid machine learning and beetle antennae search approach. *Sustainable Cities and Society*, 66, 102669.

CHAPTER 2

E-Learning as a Desirable Form of Education in the Era of Society 5.0

K. Jayanthi, M. Sugantha Priya, S. Saranya, and R. Gomathi
KGiSL Institute of Information Management

Dahlia Sam
Dr. M.G.R. Educational and Research Institute

CONTENTS

DOI: 10.1201/9781003322252-2

2.1 INTRODUCTION

The emergence of Industry 4.0 is a shift in how people think, work, and produce in huge businesses from a traditional to a digitalized model. In our daily lives, we can see dramatic growth in digital transformation and new technology tools. Most things were implanted in our day-to-day activities as a result of the COVID-19 pandemic, such as artificial intelligence (AI) in health care, the Internet of Things (IoT) in agriculture, and data processing in all key fields, all of which are Industry 4.0 technologies. Apart from that, there were no differences in the educational field since the COVID-19 epidemic has evolved along with other fields (Collins & Halverson, 2009).

Education 4.0, in general, is an institution dedicated to improving educational quality by focusing on smart-thinking pupils rather than hard workers. Specifically, Education 4.0 promotes the most effective use of new technologies and its ecosystem. Students are encouraged to participate in numerous online activities hosted by national and international universities, as well as soft skill courses and certificate courses, in order to improve their lives (Gleason, 2018; UNESCO, 2020; Salkin et al., 2018; IEEE *Xplore*, 2018).

Students in the era of Industry 4.0 and Society 5.0 began to learn not only from textbooks and notes provided by their mentors but also from a variety of dynamic or internet sources such as online courses, YouTube videos, and various web applications that help them gain in-depth knowledge in any given stream. Following the COVID-19 pandemic, online education and distance learning became realities, and everyone had to adapt. Students in the traditional instructional system are subjected to a significant amount of stress while listening to traditional classrooms. However, because direct confrontation does not allow for education, online or remote-based education has provided some flexibility and speed in conceptualizing concepts and processes.

In this chapter, an elaborate review has been given on the different forms of Education 4.0 that has emerged in the recent past as part of the Industrial Revolution 4.0. What was once dreamed of and discussed in textbooks is now slowly becoming a reality. With further developments in new technologies like AI, Cloud Computing, Fog Computing, etc., more advanced are sure to come. The overall objective of this chapter is to provide a detailed perspective of an e-learning platform as a desirable form of education in the age of Industry 4.0 and Society 4.0 after the COVID-19 outbreak.

Organization of the Chapter

The chapter is organized as follows: Section 2.2 discusses what is education 4.0 and Society 5.0. Section 2.3 elaborates on lifelong learning. A strong discussion is done regarding traditional to lifelong learning in Section 2.4. Section 2.5 enlightens regarding the adoption of new realities in education with respect to blended learning, technology after COVID-19, and five ways to adapt to new realities. Section 2.6 elaborates on open-source learning collaboratives. Section 2.7 stresses integrating social media in education 4.0. Sections 2.8 and 2.9 enlightens smart learning and project-based learning. Section 2.10 stresses on upbringing new digital unit for digital education. Section 2.11 elaborates New Education Policy (NEP), and section 2.12 and 2.13 stress on AI-based learning and future perspectives of learning. Section 2.14 focusses on case studies, and Section 2.15 concludes the chapter with future scope.

2.2 WHAT ARE EDUCATION 4.0 AND SOCIETY 5.0?

Education 4.0 and Society 5.0 are the adoption of new technologies to deliver more personalized education, with an emphasis on students' psychosocial development as well as solutions that improve society's standard of living. Aside from technology equipment, the COVID-19 pandemic underlined the importance of preparing humans for hardship, particularly for exceptionally intelligent people who understand how to leverage the digital revolution as a vehicle for social transformation (World Economic Forum, n.d.).

The fundamental aim is to incorporate Industry 4.0 technology, such as AI, the IoT, machine learning (ML), and automation, into learning and educational institutions. The process of communication between a student and lecturers or professors in a classroom context, such as a school or university, is defined as learning or teaching (Asawa, 2018).

However, as a result of digital transformation in the education industry, major technological breakthroughs are now influencing the learning process. Learners in the Industry 4.0 and Society 5.0 era are technologically sophisticated, and nearly, all of them have access to a digital environment where they can obtain information. As a result of breakthroughs and changes in the type of work they conduct, their desire to learn fluctuates.

In line with the sustainable development goals, the recent education system takes advantage of today's technical resources and strives to build Society 5.0 by solving challenges through value creation and high-quality education around the world (Ponidi et al., 2020; Saputra et al., 2020; Suhono & Sari, 2020).

Education 5.0 will have an impact on all levels and fields of education. The transition to novel situations in engineering, which support a constantly evolving engineering education capable of adjusting to these nonstop technological revolutions, is especially important. Educational institutions that can incorporate this aspect into their teaching methods and performance management can help students develop talents in their respective fields. To participate in Society 5.0, students of this generation should have all of the following abilities as real-world skills.

- Interactive and collaborative work
- Improved interpersonal relationships
- Real-world problem-solving
- A more fluid exchange of thoughts
- Assortment tolerance
- Creativity and productivity

As an example of education 4.0, students and instructors obtain remote access for instruction using learning management system (LMS), and this is becoming more common around the world. Learning and teaching, as well as course key agreement, online chat rooms, conversations, collaborations, peer instruction, and hybrid learning, all take place throughout flexible work hours. Figure 2.1 displays the evolution of the education industry over the years.

2.3 WHAT IS LIFELONG LEARNING?

Lifelong learning is a sort of self-directed education in which individuals study and pursue knowledge throughout their lives, both internally and externally, willingly, and consciously. There is no universally acknowledged definition of lifelong learning (Parker & Prabawa-Sear, (2020)).

Learning that takes place inside a formal educational institution, such as a school, university, or corporate training is the classic definition of education. On the other hand, lifelong learning is changing not only in terms of what people learn but also in terms of how they educate themselves. People are returning to lifelong learning in larger numbers

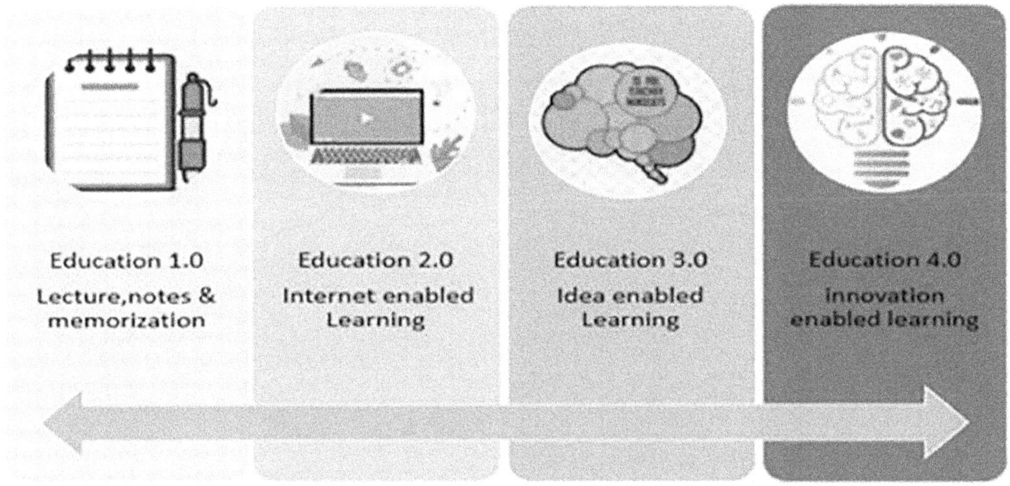

FIGURE 2.1 Evolution of the education industry.

now that it is virtually as common as drinking and eating. The concept of lifelong learning opens up the possibility of a truly revolutionary system of education, particularly in higher education, with established colleges being opened up to those who desire to learn for the rest of their lives.

In our pace with the fast society, we must constantly adjust our learning methods to take advantage of new opportunities, not because we are not good enough but because we have even more opportunities to improve. The students determine what they need to know in order to achieve a specific goal or objective. They then work on building the skills and knowledge they will need to achieve their goals and ambitions. This is easily achieved in lifelong learning (Estriegana et al., 2019; Mukhopadhyay et al., 2020). Figure 2.2 below depicts self-directed education as lifelong learning.

2.4 TRADITIONAL LEARNING TO LIFELONG LEARNING

For more than 200 years, the formative style of teaching has been followed, with teachers lecturing at the front of the classroom while we sat silently and passively in rows to absorb information. This traditional approach to teaching will have to change by adopting a lifelong learning system.

Students and young professionals in their 20s and 30s are flocking to learning centers to improve their skills and knowledge. They also bring new expectations, which will change higher education institutions in a positive way rather than in the old approach.

Youngsters were motivated to learn through lifelong learning to the extent that learning would assist them to accomplish activities or solve real-life challenges. Individuals become increasingly responsible for their own learning behavior as a result of lifelong learning, which is defined as a learning process of ongoing, post-compulsory educational activities.

Teacher-centered learning is a conventional type of learning in which teachers or mentors provide knowledge to students, who really are anticipated to quietly accept the

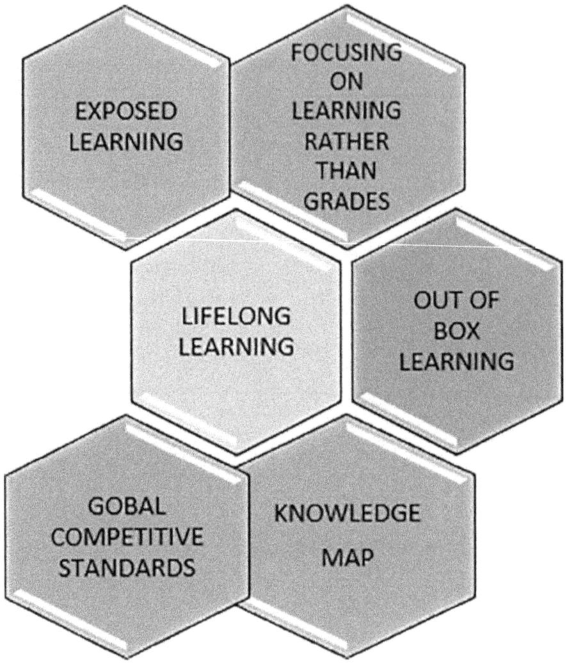

FIGURE 2.2 Self-directed education as a lifelong learning.

knowledge. The student group benefited from a disciplined approach to the study, practice, and research since the instructor undertakes ultimate responsibility for educating a group of students. Some of the perks of one form of education over another may be monetary. When comparing traditional education to online teaching, the most significant consideration is the cost, which is more in traditional education because it establishes engagement with teachers and instructors. On campus, students have the opportunity to meet with their teachers in person to discuss the class, their performance, or a project, and to form significant relationships that will continue for years.

Because of the outbreak of COVID-19, the school environment is being compelled to transition from a blended teaching mode to an online learning teaching model, notwithstanding the benefits of traditional learning. Due to the pandemic crisis, China was the first to announce that online courses would be implemented in their country.

The tale pictured here (Figure 2.3) shows how traditional teaching methods have evolved into lifetime learning.

The Sustainable Advancement of Online Learning in Recent Years has also encouraged teachers in schools and colleges to use a wide range of web-based learning methods, such as learning management systems, World Wide Web technology for studying, information and communication technologies (ICT), and interpersonal P2P learning or m-learning, to improve teaching and learning individually and develop problem-solving skills, thereby increasing the efficiency of the educational process (Ware, 2018).

The following Table 2.1 depicts the basic difference between traditional learning and lifelong learning.

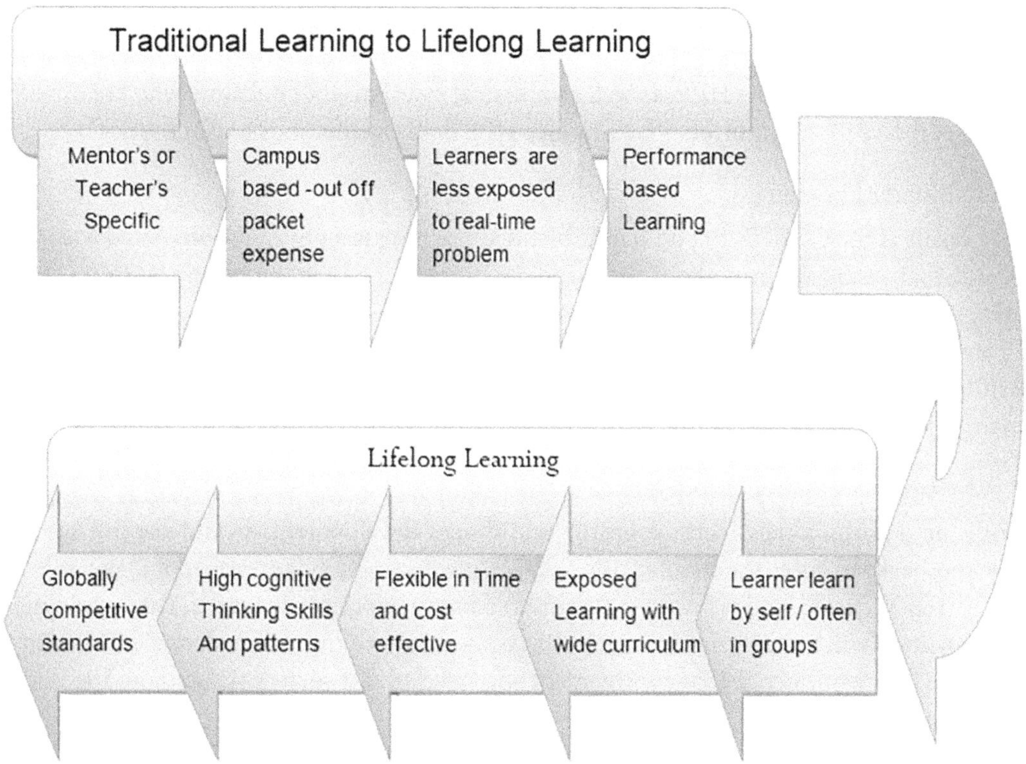

FIGURE 2.3 Traditional learning to lifelong learning.

TABLE 2.1 Traditional vs Lifelong Learning

Traditional Learning	**Life Long Learning**
Teacher-centered learning	Teacher-, instructor-, or guide-centric
Every learner learns the same content	Personalized learning as per students' requirement
Learners might be school-level or college-level	Learners might be young or professional
Curriculum and discipline centered	Problem-solving and competence-based
Learning only is the only profession	Working professionals can be also learners

2.5 ADAPTING TO NEW REALITIES

COVID-19, the most recent new coronavirus disease, has changed the way industry-based and tertiary health professionals' education (HPE) can take place. The recommendation for rigorous and universal social separation has speed up the conversion of course delivery to an online format around the globe. HPE has typically relied on face-to-face learner contact in the form of skills laboratories, simulation training, and industry-based clinical placements, which is an issue. The shift to online-only course delivery has necessitated addressing specific difficulties related to the development and delivery of high-quality curriculum and educational activities (Chen & He, 2013; Luminita, 2011).

Regional, rural, and remote health professionals and academics can offer vital ideas on how to use technology to overcome the tyranny of distance, promote high-quality online HPE, and support the ongoing development of communities of practice in this context. This is the first of a series of articles that will explore the risks and opportunities associated with the current shift to online HPE, as well as practical solutions for educators who are unwilling to embrace more traditional face-to-face HPE delivery methods and activities.

2.5.1 Blended Learning

As a result of the COVID-19 pandemic, blended learning is a novel notion. Amid the pandemic, many schools throughout the world are discovering that combining learning is the greatest way to address educational demands (Ha et al. 2019; Kumar et al., 2021).

Students in blended learning spend half of their time in the classroom and the other half learning at home via virtual learning. It allows schools to keep class sizes small and maintain social distance.

2.5.2 Technology after COVID-19

With the dawn of a new technological era, different developments would see the light of day and begin to enter the market. We can see how technology has played a major role in the current scenario, and we can expect this to continue in the near future. The technological transition will be centered on new business models with minimum contact. This stems from the lasting influence that confinement will have on our society throughout the years.

2.5.3 Contactless Business Models

This trend will be prominent in the growth of businesses, products, and services. The "haphazard" solutions we used before being confined, such as children's homeschooling, will be improved and enhanced. Workspaces will transition from long-term rents to co-working-style environments. Electronically delivered services will progress to a new level. The ability to adapt to pliable and even chameleonic formats will be a feature that will accompany all of these new models.

As automation gains traction in the marketplace, the investment will rise. Near-autonomous warehousing, robots, drone delivery, and 3D manufacturing will all see increased popularity and deployment. This emphasizes the critical role of cyber security, privacy, location, and data analysis in terms of regulation and effectiveness.

2.5.4 Five Ways to Adapt New Reality

Many industries were affected by the global epidemic, yet some businesses were able to adapt and grow better than others despite the adversity. Some of the adaptations they made to adapt to the new reality will continue to benefit them long after the pandemic has passed. These companies will reap the benefits of their efforts for many years to come.

2.5.4.1 Identify New Opportunities

Consider whether there are any new market needs that your company might be able to meet. Consider offering new products or services to assist your clients in adjusting to the new reality. The fitness business – gyms, yoga studios, and other such establishments – is an excellent example of searching and discovering new opportunities. Many fitness studios saw the potential of online exercise programs as their consumers began working from home.

They contacted new corporate clients and offered their services to their staff. Many organizations wanted to offer online fitness and yoga courses to help their employees cope with their sedentary lifestyles and isolation, generating a new potential for the fitness industry (Ekereke & Akpojaro, 2019).

2.5.4.2 Make it Convenient to Pay for Product

During the pandemic, one of the most obvious transformations in the fintech industry occurred when brick-and-mortar businesses recognized that their customers preferred to pay online rather than utilizing a point-of-sale (POS) terminal. Many of our own merchants contacted us to set up a virtual terminal or to integrate a payment solution into their website, allowing their customers to check out from anywhere.

2.5.4.3 Alternative Solutions

Try to make the best of a difficult circumstance. Due to the decreased number of people permitted to operate in the manufacturing area, one of our merchants had to cut its output volume. Rather than becoming dismayed by the new reality, they devised a strategy.

They transformed their lower inventory production volumes into something fantastic by holding a special sale every time new things became available. They introduced a new degree of exclusivity and excitement to their goods, making them even more attractive. Client's "camp" on the brand's website well in advance of each sale to be the first to get their hands on the new inventory (Ben Amor et al., 2020; Brumă 2020).

2.5.4.4 Analyze Your Competitors

Currently, several countries are going through the same thing. Investigate how companies in the same industry are adjusting around the world. Try to figure out what works for them in this new world, and you might get some fresh company ideas.

2.5.4.5 Spot Trends

Companies that allowed their workers to work from home are unlikely to return to their previous state before the pandemic. Many companies have chosen to minimize their office space and allow their workers to work from home, making it a long-term trend worth investigating (Aissaoui et al., 2022; Sharma et al., 2021).

The same may be said for business travel. Companies will most likely look at investing in interactive communications platforms after the epidemic is over now that they have recognized they can save a significant amount of money on business trips while still getting the job done.

2.6 OPEN-SOURCE LEARNING COLLABORATIVES

They observe qualities that are dynamic to the building and maintenance of communities of learners by looking at open-source not just as a software development technique but also as a collaborative construction process. Some of the distinguishing aspects of how open-source communities collaborate to address complicated problems are:

Participants in open source take part in activities that are personally significant to them. Technically motivated individuals construct complex computer systems for their own use in open-source projects. Outsiders may find these systems to be technically obscure, yet it is not unusual for a community to generate something whose usefulness is obvious to participants but opaque to others. What matters instead is the alignment of a community's goals with the software it produces. Projects in the open source must be designed (Siddiqui et al., 2019).

Software design aids in the creation of solutions to problems that are not well-structured. There are rarely any optimal answers or simple concepts of the "correct answer" in complex design tasks. In these situations, learning is difficult. It is not just about gathering information, it is also about framing and addressing problems. We must engage in a dynamic and naturally open-ended process if we want to examine complicated design concerns. The same complicated design challenges that other domains face apply to open-source software.

Problem-solving by framing and presenting solutions. Software is a tangible or almost tangible thing that a company produces. By operationalizing concepts inside the community, the community may assess and modify them, helping to ground abstract concerns. Formal restrictions in the development of computer software also need to be handled.

Open-source software serves as an example of a viable software development technique as well as a model for collaborative object creation. Important principles are exemplified by open-source communities. In the context of collaborative learning, this looks at how open-source activities can be used as a source of inspiration for other projects. In a university course, the establishment of collaborative learning experiences specific public outputs, and the application of students can use collaborative technology to pursue ill-defined projects that are personally relevant to them.

2.6.1 Community Relies on Shared External Factors

Externalizations are useful because they can be easily shared and discussed. They allow abstract concepts to be realized in a concrete way. A community context for problem resolution is provided through a shared external item. Open-source projects encourage developers to submit their contributions since open-source projects can only improve if the external representation software and related artifacts evolve. Sharing something useful and basing your contribution on the shared software helps to reinforce concrete contributions to the achievement of abstract aims.

2.6.2 Collaborative Technology Usage

Computer-based collaborative tools are the major means of communication in most open-source projects. Despite the fact that new collaboration tools are often difficult to embrace, almost every project uses email, newsgroups, and/or web-based discussion groups. Collaborative tools for sharing software and debating the program under development commonly work in tandem. Other virtual communities that rely on computer communication tools can learn from open-source communities' "success tales".

2.6.3 Contributions Are Incremental and Integrated

Open-source communities are good examples of how learning activities can have more fluid borders. Even though we lack the ability to handle ongoing actions, many interesting difficulties persist. The number of participants, available time, geographic location, and other factors all affect learning scenarios. The open-source community is not a static entity. Different people come and go over the years that an open-source project is active, and membership is ill-defined. As a result of their contributions, open-source contributors create something that future generations of developers can build on.

Open-source initiatives, unlike certain design processes where participants are required to start "from scratch" sometimes rediscovering what was already known, are continuing activities where people can utilize not only good ideas but also real elements.

In many collaborative learning contexts, open-source communities exemplify principles that are deemed desirable. Attempting to replicate open-source environments in educational settings, on the other hand, may not be desired. Combining the characteristics that make open-source communities successful with the realities of a face-to-face undergraduate program unique opportunities and problems arise in the learning environment.

This will show you how to do just that. Using tangible public deliverables and a loose organization to promote open-source ideals may be appropriate. In unstructured domains, collaborative design is possible. Future endeavors because of their capacity to draw on previous experience (Ochoa et al., 2017).

2.6.4 Realizing Collaborative Learning with Open-Source Principles

There was no fixed sense of "covering" course material because of the broad, open-ended nature of course topics. Rather, we envisioned the course as a collaborative learning experience in which students would be exposed to a variety of topics. Various conceptual difficulties are addressed through collaboration. One of the objectives of this course's implementation was to consider courses as "seeds" rather than whole things. One might envision the course as a collaborative knowledge-construction experience with a course as the final product that future generations of students might be able to benefit from. Instead of educators completing courses, students do so. Students would take an active role in the development of the course. Instructors utilize a variety of methods to encourage students to enhance their knowledge as shown in Figure 2.4.

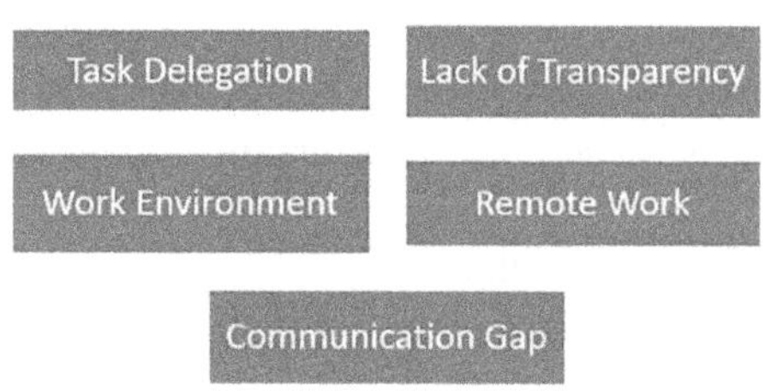

FIGURE 2.4 Team collaboration challenges.

2.6.5 Task Delegation

When people are not given clear objectives and key performance indicators, or when they do not grasp what their teammates contribute to the table, they tend to dislike teamwork.

To foster collaboration, team members must have a compelling motive to support the company's goals. The more compelling the mission, the more easily team members will be inspired to contribute. Team members naturally become as passionate about goals as their leaders when they are given a clear and compelling cause to be involved in. When cooperation pervades an entire organization, huge benefits emerge, ranging from engaged employees to better retention, stronger talent acquisition, faster time to market, and improved profitability (Sung, 2015).

2.6.6 Lack of Transparency

When team members' work is dependent on others, they must communicate their progress, concerns, and roadblocks. Teams cannot build trust until they are transparent.

Create a value-based culture that is motivated by teamwork and professionalism to enhance a firm. When you treat employees like people rather than cogs in a machine and make sure their managers focus on pleasant relationships, you reduce stress, which can contribute to toxic work environments. When people are satisfied and teamwork improves, a company's ability to handle unexpected disruptions improves as well (Trappey et al., 2017). When client preferences change or new technologies enter the picture, teamwork makes it easier to pivot.

2.6.7 Work Environment Changes

Focus on each team member's abilities rather than working around their flaws to empower them. Request that each participant completes a personality test and discuss the results with the group. This is a terrific team-building activity because the findings allow each participant to learn a lot more about their coworkers.

2.6.8 Remote Work

Teams must believe they have the ability to be creative, discuss, and question the status quo in an open, non-judgmental environment in order to thrive. It is also critical to get thoughts and arguments from team members on a frequent basis. They will be more inspired to perform, think outside the box, and exceed expectations if they feel connected and understood.

2.6.9 Communication Gap

Communication is essential for improving collaboration. Use technology to improve internal and external communication by using tools like instant messaging and video conferencing. Collaboration technologies ensure that departments and decision-makers are always on the same page. This is even more important as businesses become more spread.

2.7 INTEGRATING SOCIAL MEDIA

Social media facilitates a variety of interactions with people's imaginations. Through various mediums, such as mobile phones, computers, and email, people can get a sense of other people's lives and accomplishments. These communications are not limited to a single

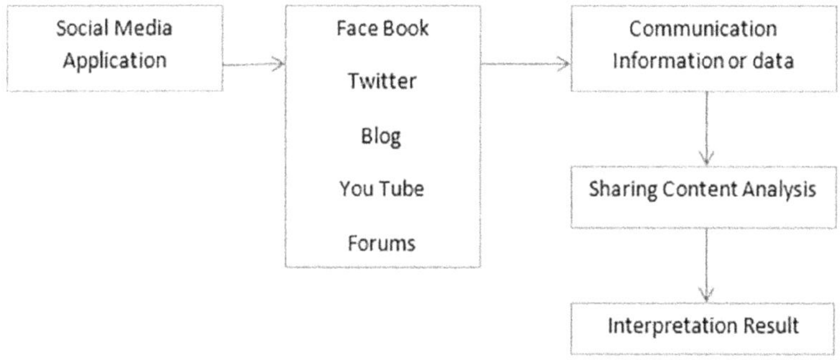

FIGURE 2.5 Social media communication analysis.

person's life. It also applies to professional life (Brunswicker et al., 2015) and offer messages in a variety of formats, including voice, video, animation, and text. Social media is mostly used for social interaction with others, as well as accessing news and information and making decisions (Anderson, 2019). It is one of the most crucial communication platforms for sharing, generating, and disseminating knowledge globally.

Individuals can produce, collaborate, and share material using social media, which is a collection of internet-based applications. Beyond promotion, practitioners can achieve social media's unrealized potential by incorporating it into a bigger social marketing plan. If used effectively, social media can assist firms in increasing their capacity to put the customer at the center of the social marketing process. The goal of is to use a four-step approach to create a framework for strategic thinking in order to successfully use social media as part of a social marketing plan (O'Donovan et al., 2018).

Content sharing, video sharing, photo sharing, social networking, rating, bookmarking, and games are all examples of these applications. Overall, people are spending more time on social media. While social media allows people to connect with one another, it is more than just another form of mass or interpersonal communication, like print, television, or radio. In the case of a public health emergency, social media can provide information quickly and easily.

In a social media communication analysis (Figure 2.5), information or data is sought through various media such as Facebook, Twitter, blogs, YouTube, and forums, and matched content is displayed as the interpretation result based on the analysis. The linked information and dialogue between the data, as well as the relationship between the content and the associated group of information, can be identified and transmitted through the media (Yun et al., 2016).

2.7.1 Diverse Social Media Platforms

Develop your social presence across a range of platforms to appeal to clients that prefer to communicate in different ways. Investigate several social media channels to see which ones would best communicate your message to your target audience. For example, a company that serves a large number of professionals would get the most leads through LinkedIn, while others might get more inquiries from product demonstration videos on YouTube.

Other methods for determining which platforms are appropriate for your brand include assessing the competition and determining which platforms have the most growth potential and audience engagement.

2.7.2 Social Marketing

A four-step methodology, although social media has a lot of appealing features, is not fit for every issue or audience. To guarantee that he or she is ready to fully engage with the audience, the strategic social marketer assesses the possible benefits as well as audience characteristics and the continuing commitment required (Barata et al., 2018).

Step 1: Identify your target audience. Determine which audience segment will be the focus of your social media and marketing efforts. The organization may be reaching out to health partners, or community members may be the focal point. Find out which social media platforms your audience members use.

Step 2: Make a list of why you want to engage with the audience. There are five reasons for this, all of which indicate strategic thinking about how to leverage audience members' enthusiasm and knowledge to enhance products and services. The primary goal is to get feedback and insights from the audience on products, services, habits, and the organization. Social media offers new ways to acquire this data that are not limited by time or geography.

Step 3: Develop a precise plan for engaging the audience and achieving the goals set forth in Step 2. People utilize social media to maintain contact with those with whom they have a personal relationship. The organization must communicate reasons or benefits for the audience to connect with it and form a connection with it.

Step 4: Select a technology. The final stage is to choose a suitable social media application based on audience preferences and practices (e.g., what social media they use and how they use it), the organization's goals, and the recommended approach. Instead of establishing a new social media application, it may be more beneficial for the company to locate existing communities and dialogues and join in the conversation.

When social media and social marketing are combined, they produce a powerful synergy and a potentially more robust model for consumer-oriented health promotion than either strategy can provide on its own. However, firms can only achieve this synergy if they think carefully about how to engage their audiences and include social media in each phase of the social marketing process.

As practitioners travel back and forth between customers and the development of health-related solutions, social media provides a real-time, two-way communication route to guarantee that programs, products, and services are focused on actual concerns and relevant answers. It also ensures a more frequent continuous feedback loop throughout the program implementation phase. Furthermore, combining these two methodologies can aid in ensuring this.

Benefits of Social Media Integration

- Expands your business network
- Easy to manage
- Analytics and reports
- Content aggregation and scheduling
- Improved analysis and forecasts

2.8 SMART LEARNING TECHNOLOGICAL EXEMPLIFICATION

Smart learning, also known as intelligent education, refers to new educational situations in which the focus is on the use of modern and innovative technologies by both students and teachers. This is dependent not just on the software and technology available but also on how they are used in a synergistic manner in the classroom or online. Smart learning is "a self-directed, motivated, adaptable, resource-enriched, and technology-embedded learning concept", according to the Korean Ministry of Education, Science and Technology. These characteristics indicate that smart learning broadens students' and teachers' methodologies, competencies, content, and learning spaces (Kelly, 2002).

Self-directed learning offers the ability to expand learning time while also allowing one to learn at any time and from any location. Intelligent learning's motivational function enhances educational approaches by ensuring empirical and collaborative actions among students and teachers. In turn, the adaptive function expands educational opportunities by allowing for tailored and individualized learning. Furthermore, the variety of forms offered expands the educational content by making numerous educational resources available. Finally, ICT provides almost unlimited access to educational content by providing local and global communication networks (Briones et al., 2011; Thackeray et al., 2012; Thackeray & Neiger, 2009).

In this context, the broadly understood ICT, as well as innovations in the fields of AI, virtual reality (VR), augmented reality (AR), processing of large data sets (Bigdata), 3D visualizations, and ML, are critical for the dynamic development of the concept of intelligent learning. Over the next dozen or so years, educational solutions based on these technologies, incorporated in the cloud computing paradigm with well-defined mobile apps, will govern the way of learning (in a broader sense), and in particular academic education. The operation of regularly used e-learning platforms (LMS class systems) in a cloud model is now almost a standard, which offers up a slew of integration options (Figure 2.6).

Fully virtual teaching support platforms, such as Blackboard Learn Ultra or OpenLMS, are incorporated in the SaaS (Software as a Service) or PaaS (Platform as a Service) cloud model. The possibilities for integration with virtual laboratories, corporate transactional systems, analytical systems, 3D printing systems, and business process simulation platforms are nearly endless (e.g., Marketplace). Virtual modeling and prototyping, which

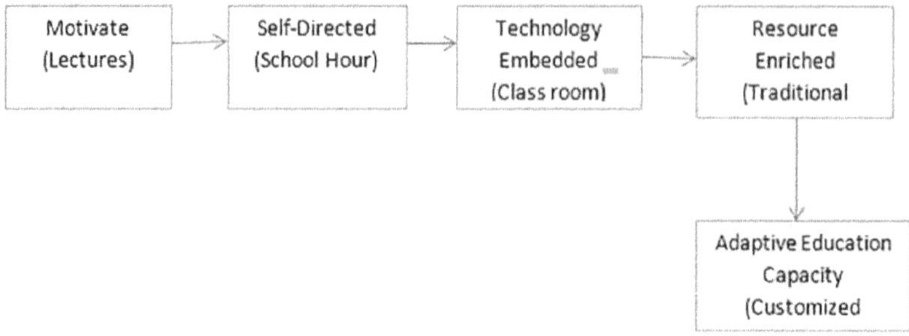

FIGURE 2.6 Traditional school system.

was previously only available to large corporations, is now also available at universities. Furthermore, the virtualization of didactic procedures makes dual education, which combines theoretical (academic) and practical (commercial) components, easier to implement. As a result, the virtual world blurs the line between academics and business (Ellzey et al., 2019).

Respondents say they use their smartphones to connect to the internet frequently, if not exclusively. They understand how to bring cloud computing, hyper-connection, AI, and Big Data as a Service into practice (Knowles, n.d.; Tsybulsky & Muchnik-Rozanov, 2019). Respondents associate smart learning with modern instructional technologies. Smart learning, according to the respondents, will enable the integration of information throughout functional zones such as home–work–virtual classrooms, hence boosting overall learning availability.

The teacher's position will be redefined and the teacher-machine will be introduced as a result of this combination of student functioning areas (home–work–university), communication digitization, and didactics. In the future, the human teacher will play a significant role in the social growth of students at a university, while the artificial teacher will focus on the scientific development of students. AI is a type of software that can serve as a substitute for a human teacher. The introduction of AI will contribute to a number of changes in education, including the creation of an adaptive learning environment tailored to a specific unit's individual learning program, as well as a collaborative environment that will strengthen mutual learning between students through the sharing of knowledge and space (Boaler, 1999).

Future education will combine technology with a new, much more proactive, dynamic, and, in particular, two-way type of knowledge exchange between the teacher and the student. The teacher-duty mentors will be to demonstrate to pupils how to gain knowledge from current sources and how to apply it in practice. Integrated IT solutions (for example, an LMS) will enable continuous communication between students and teachers, as well as the adoption of the SPOC (single point of contact) idea in education. Because the aim model integrating the notion of a smart university and deep integration blurs the line between those who create and transfer knowledge, the mutual interchange of knowledge, and thus learning, between students will become increasingly vital.

2.9 PROJECT-BASED LEARNING

Project-based learning (PBL) is an instructional concept to encourage students based on projects, and it creates opportunities to enhance their skills and experimental knowledge with some challenges and questions in the real world.

High-quality project-based learning aims to teach academic topic knowledge and skills while also fostering deeper understanding. Develop critical thinking, problem-solving, communication, cooperation, and creativity/innovation abilities to help you succeed in the 21st century.

PBL is a question-based active interaction approach that gives a solution to the problem. PBL is a student-centered, non-traditional concept of teacher approach to learning. Learners gain knowledge by interacting with questions that have put out their expected snooping. Mastering the learner's level of education through modernized project-based activities enhances students' skills and increases their competency level (Doppelt, 2003).

To train learners with their expectation level, the trainers (Tutors, Mentors, Experts, etc.,) are in lack of knowledge and there is a need to fulfill the situations in the education environment. To solve this problem, practical lists were prepared and included in the scheme and it has been planned to implement them in the classical education system. Rapid projects have to be assigned to students and need to train them with technical experts. These increase their thinking capability, technical skills, problem-solving ability, leadership skills, team coordination, and outcome-based learning.

Such a representation should be focused on creating an exclusive output product and deriving capacities in skill and production, together with the achievement of some extra learning outcome (Gültekin, 2005). Students increase their inquiry and are guided from end-to-end research under the teacher's direction. Productions are illustrated by making a project to share with select viewers. Trainers support the systematization of the functionalities that will be implemented throughout the study and project phases of PBL (Thomas, 2000). Student selection is an input factor of this approach (Figure 2.7).

The result of PBL is a greater perception of a topic, deeper knowledge, higher-level understanding, and bigger motivation to find out. PBL is a key approach for creating self-regulating thinkers and learners (Tiainen et al., 2007). Students solve real-world problems by scheming their own studies, scheduling their learning, organizing their studies, and implementing a huge amount of learning strategies.

2.9.1 Self-Motivated Learning

First, students need to analyze and understand the problem by making inquiries with questions. For inquiry, students can use their trainers, mentors, technical experts, etc., and also enhance their ideas. Based on their inquiry, a module of the project can be designed. Tasks are assigned to students by modularity. They then come up with what their method will be for making inquiries and spot the resources that they will need to do their research. Subsequently, students opt for a way to present what they have educated in the shape of a project (Annika Andersson, 2008; Bruckman 2002).

To get more opportunities, students have to present their projects to audiences like supervisors, mentors, experts, heads of departments, etc. The audience must be genuine

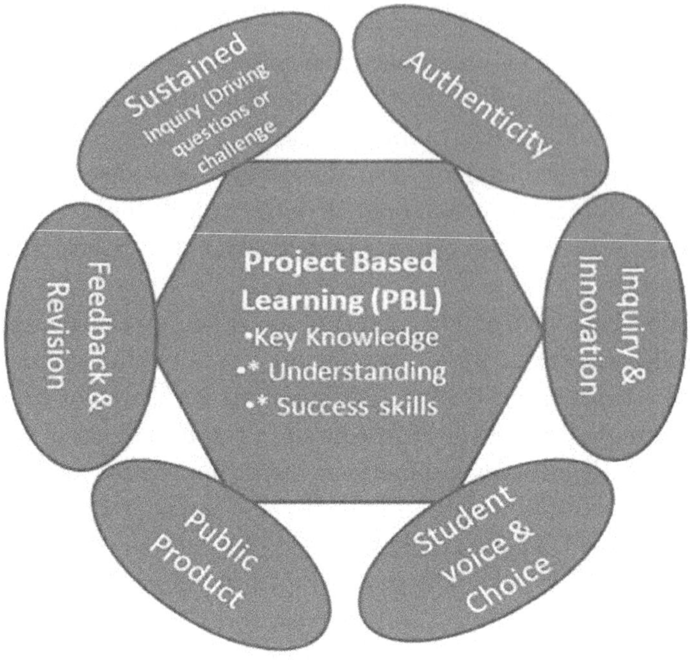

FIGURE 2.7 Project-based learning.

and suitable. For example, if students are doing a project on a computer science inquiry on the topic of the IoT, they may desire to explore how IoT devices work and are automated. A group may choose to build an effective project. The conclusion of the project strength is a challenge to perceive how the devices are implemented with automation, with the class as the spectators.

Responsibility to spectators fixed with a due date keeps students on the path. Learning liability, self-determination, and regulation are three outcomes of PBL. The managerial plan that learners have planned for themselves allows them to stay attentive and on-task.

As learners become more skillful in the PBL approach, they are trained to self-monitor their development during day-by-day agenda setting. At the conclusion of each work part, students need to give daily reports on reaching the day's goal. Students must use their effort time efficiently and stay attentive and on-task to be successful. It is vital for the mentor to present with students frequently to ensure that students are on a path and developing their thoughts and skills entirely. These skills are grave for prospect achievement in both academics and real life.

2.9.2 Curriculum PBL

Most of the university structure gives practice for a student in project activities during the sixth semester for undergraduates and the fourth semester for postgraduate students. They need to (i) understand the project and to gather the requirements, (ii) design the project based on analysis with modules and process flow, (iii) take part in the development of the project based on designing, (iv) perform verification and validation for the project, and (v) get feedback from the teacher and to maintain project activities through the documentation for future enhancement (*Does Digitalization in Higher Education Help to Bridge the Gap between Academia and Industry? An Application to COVID-19- Laura Márquez-Ramos, 2021*, n.d.; Ugur, 2020).

Some Universities/Colleges provide mini projects to enhance their students' skills and abilities in problem-solving and also to face challenges in the real world. They can inquire about questions and can get answers in review meetings with their supervisor/teacher.

2.10 DEDICATED UNIT FOR DIGITAL EDUCATION

Learners need to discover, act together, and try out to have deeper knowledge and experiences. Educators make the students/learners get involved in learning activities. Teachers can produce digital enlightening content that is self-motivated and connects to different learning styles (Abdillah et al., 2021).

2.10.1 How to Create Digital Content?

The following are the six steps for enhancing digital learning content that catches students' interests and produces online education, which are elaborated in Figure 2.8.

Educators can produce the video set for a unit, save the links, and then add them to their lessons. Nowadays, YouTube plays a vital role in getting educational videos through digital learning content providers. K-12 learning was dedicated to various channels. Additionally, the Hapara filter is a tool that makes it secure for learners to utilize YouTube.

These channels, for example, provide films on a variety of academic levels and subjects:

- **StorylineOnline**: This channel features videos of celebrities and actresses reading children's books. It is designed for students in the fourth grade and below.

FIGURE 2.8 Online education with digital content.

- **SciShow or SciShow Kids**: These two channels for older and younger students to explore fascinating scientific themes and concerns.
- **TED-Ed**: This popular channel features short videos on a variety of topics.
- **Mike Likes Science**: Students will enjoy these videos that teach science through music and hip-hop lyrics.
- **Khan Academy or Khan Academy Kids**: Khan Academy focuses on math, science, and history requirements, while Khan Academy Kids caters to younger students.
- **PBS Kids**: This PBS channel features animated videos for children in kindergarten and the first grade.

Educators can also use short movies to asynchronously explain ideas. A screencasting program, such as Screencastify or Loom for Education, is one option. They enable teachers to capture what is on the screen as well as their own faces so that students may see them speak. Using a computer camera or a phone to capture a demonstration is another technique to integrate videos.

2.10.2 Inclusion of Digital Learning with Its Context

How do you create a digital learning material that is audio-based? One way to include audio is to narrate over a Google Slides presentation. Teachers can contact students with a range of learning needs, and students can read or listen to the information.

Teachers can record audio files using programs like Audacity or GarageBand, which can be useful in a variety of situations. Teachers, for example, can leave pupils directions for a project or task by recording a voice message. For students who desire more help, they can also record assessment questions.

In both core and language classes, an audio recording can help students comprehend how to pronounce new vocabulary words. Teachers can record lessons and upload the files for students to listen to the pronunciation as they want.

Teachers can go even further with their creativity by creating a podcast to teach a topic, complete with music and sound effects. They can also make an audiobook out of a story or a passage of text. When adding effects or music, make sure you look for royalty-free audio resources.

2.10.3 Self-Developed Educational Digital Images

Digital visuals are another technique to engage students. What kinds of digital learning resources are there? Illustrations, maps, graphics, pictures, and other instructional images are available. A teacher, for example, can use Google Drawings to construct a brainstorming map, timeline, or plot diagram. It can also be used to demonstrate how to solve a math issue or solve a scientific equation (Baryshnikova et al., 2021).

Another option is to make a text-based visual. Canva is a free tool for making graphs, posters, flyers, and infographics. The image files can be saved and shared, or they can be included in a Google Docs or slideshow. Teachers can also take a photograph of an object, a scene, or a science experiment to show or analyses to students.

2.10.4 Exploring Digital Content in Games

Gamification of lessons and activities is another technique to improve digital teaching strategies. Gamification is used in a variety of digital educational tools. You may develop online learning games or trivia with a tool like Kahoot. To review ideas, there are other online exam builders such as Quizlet. Edpuzzle allows you to add comprehension questions to videos and track learners' understanding if you want to share them.

Making a digital escape room with Google Forms is also a great idea. For example, librarian Sydney Krawiec created a Google Forms escape room for students inspired by Hogwarts. She also offers a video lesson that shows how to put the project together.

2.10.5 Giving Choice for the User

By providing their students with an option, teachers can tailor digital content and offer groups or individuals the option of engaging with a video, audio file, or image, for example. Students are more invested in the topic when they have control over their learning. It is a fantastic approach to increasing active learning in the classroom.

2.10.6 Team-Coordinated Digital Content Creation

Teachers are not required to generate their own digital educational content. Content can be created collaboratively by departments or subject-grade-level teams. Members of the team can focus on one sort of material, such as developing digital graphics or collaborating to make videos, podcasts, or trivia games.

The advantage of digital content is that it can be reused and shared quickly across teams by your district or school. It is also simple to keep the information up to date. Your district or school will provide learners with a more meaningful online experience by allocating time for educators to create interesting content.

2.11 THE NEW EDUCATION POLICY (NEP)

The New Education Policy (NEP) 2020 was announced and released by the Ministry of Education in July 2020. The vision of NEP 2020 is to reshape and transform the education system in India. It mainly aims at the universalization of education from the preschool level to the secondary level.

2.11.1 NEP 2020 and Online Learning

The NEP 2020 gives special emphasis to online education. Pilot studies are being conducted by universities including NITs and IGNOU to maximize the benefits of e-learning in India. Popular online learning platforms like DIKSHA and SWAYAM are planned to be updated with new training materials, in-class resources, assessment aids, profiles, and other features that will allow for seamless engagement with enthusiastic students. It also focuses on the development of public, online, and interoperable infrastructure that could be used by a variety of platforms (Laufer et al., 2021).

The Ministry of Education intends to establish a dedicated unit in order to promote digital learning. This dedicated unit will be made up of professionals in the fields of education, technology, industry, administration, e-governance, etc. who will focus on the school and higher education's online learning needs. Online evaluation and assessments will further be developed. NEP 2020 stresses the development of virtual labs using which students can put their theoretical knowledge into practical experiments. This is also planned to be made available in multiple languages.

The current era is dominated by digital technology, with the internet and the World Wide Web influencing people all over the world. Both the education seekers and the educationalists are equipped by the internet, and they are brought together under one virtual roof. As a result, the concept of a virtual classroom is already gaining importance around the world. The role of digital technology in providing education is becoming critical. Online educational methods are gaining popularity due to their ease of adaptability. In comparison to the centralized classroom education system, internet education is now becoming more accessible to underprivileged people.

2.11.2 COVID-19 Pandemic and Growth of Online Education

During the global pandemic that led to the imposition of lockdowns across the country since March 2020, online classes have become a new normal for both educators and students. This has become a new way of thinking about education. Students have the luxury of attending lessons from any corner of the country. Institutions can reach out to a larger network of students instead of being limited by physical or geographical limits. Students can get their doubts clarified and participate in live conversations from the comfort and safety of their homes.

On the plus side, online learning has improved the technological literacy of both teachers and students. During the lockdown, a variety of mobile phone applications, e-learning tools, and LMSs have arisen as an alternative to traditional classroom instruction. Amid the crisis, several institutions seized every opportunity and started offering free online courses to students thereby advertising themselves. Computers and smartphones were quickly adopted as a necessity by both teachers and students. In the middle of the pandemic, online education emerged as a viable option. In fact, it evolved as a result of a need rather than a choice. As a result, improving the quality and ease of online teaching–learning becomes critical.

On the negative side, although India is a country with physical and cultural diversity, it is also no secret that our country suffers from a significant economic disparity. There are several constraints such as power supply, internet connectivity, non-affordability of necessary gadgets, etc., and access to online education remains a struggle for many. In order to address these issues, government-level planning is required.

2.12 TECHNOLOGY SUPPORT AND AI-BASED LEARNING

Online education's significance is something that cannot be ignored in the present. E-Learning is not an idea that just appeared overnight. People from all walks of life need to continually learn in order to achieve their goals in life. It should not be restricted to only academic settings.

AI is the upcoming thing in the field of education and learning, even though it has not taken root yet. From kindergarten to higher education through corporate training, AI's function and impact will be felt in the upcoming years. It will enhance the learning experience of students by allowing them to construct adaptive learning elements using customized tools. Here are a few examples of how AI is transforming the education industry (Marks et al., 2021).

- **Personalized Content**:
  Smart content can be developed using multiple mediums such as audio, video, and online assistants and can range from digitized guidelines and recommendations to personalized and customized learning interfaces.

- **Intelligent Tutoring Systems**:
  AI has the capability of reducing lectures into flashcards, titbits, and smart guides, but it can also be used to tutor students depending on their specific weak areas or issue areas.

- **Collaboration between Teachers and AI**:
  The objective of AI in education is for AI and teachers to collaborate to get the best results for pupils. For example, they work together to improve efficiency, personalization, and administrative responsibilities.

- **Accessibility for All**:
  AI has the potential to make classrooms accessible to everyone, including people who speak several languages or have visual or hearing impairments.

- **Educators Work Remotely**:
  An actual instructor can be replaced with a robot employing cutting-edge technologies such as gesture recognition. To develop realistic virtual characters and interactions, AI, 3-D gaming, and computer animation are used. This system includes AR as well.

- **Analytical Content**:
  Smart analytics give educators and content producers valuable information about their students' progress. This optimizes the content taught to the students for optimum impact.

- **Admin Tasks can be Automated**:
  Educators devote a substantial amount of time to grading homework and assessments. AI systems can already grade multiple-choice questions, and they are getting close to being able to read written responses.

- **Outside-of-the-Classroom Tutoring**:
  Tutoring and assistance programs are becoming more advanced as a result of AI, and they can now efficiently assist students with homework and test preparation.

2.13 FUTURE OF E-LEARNING

During the COVID-19 period, the major change to an e-learning paradigm gave a realistic and convenient means of learning. Online education is no longer a choice; it has become a necessity. The coronavirus has forced educational institutions to shift from an offline to an online form of instruction. The institutions that were previously resistant to change will be forced to accept modern technologies as a result of this crisis. This disaster has very well demonstrated the financial benefits of online teaching and learning. It has become possible to reach out to a huge number of pupils at any time and from anywhere in the world using online teaching options. All institutions across the world have totally digitalized their operations, recognizing the critical need in this situation. In the middle of the chaos, online learning is emerging as a victor. This is going to get better in the future with the growth of AI and ML.

AI investment is estimated to reach $200+ billion by 2025, a humongous 50% growth in the training and development business in the next 4 years. Many processes that currently fall on people will be handled by AI in the future, and educators will be better trained and capable of handling more complex duties. Companies can use AI to provide tailored learning at scale, saving time for both teachers and employees and allowing them to focus on more important duties. Employees will be more involved and interested in learning, such as adaptive learning, enhanced analytics, and time optimization in the creation process. This will result in better results and higher productivity.

Similarly, cloud technology is going to be used more in the field of education. Users benefit from cloud technology in a variety of ways, including rapid and easy deployment methods, lower costs, consistent resource availability, scalability and elasticity, and the capacity to quickly adapt to changing business needs. In the e-learning area, the usage of cloud computing technology is a constant process by which teaching and learning methodologies adapt to the contemporary technology-based culture. By exploiting the electronic environment in a dynamic and interactive process of learning, teaching, and evaluation, the capabilities given by this form of architecture can help both students and teachers.

The combination of cloud computing and e-learning has resulted in cloud-based e-learning systems, a new form of system. These systems combine the advantages and benefits of cloud computing with e-learning platforms. This combination offers a number of possibilities for increasing the efficiency and simplicity of the use of e-learning systems, as well as contributing to the best possible conditions for using distance learning systems.

Another new e-learning strategy being proposed these days is based on fog computing. The learning materials will be kept on the edge server. The content does not need to be encrypted on the user's device. Instead, it will be handled by the fog server, which increases efficiency. Encryption algorithms can be used to encrypt the course and exam materials, which will also provide improved security. A safe profile matching system can also be provided to help people find colleagues in one's area in a quick and secure manner. As and when more new technologies are being developed, the education system is also going to take new turns.

These are some of the technologies that are sure to penetrate the education system. Apart from these, as humans, we never know what is yet to come. There could be totally a new system with new industrial and societal revolutions. It is the responsibility of educational institutions and educators to keep track of these changes and adapt without any reluctance.

2.14 CASE STUDY

The MicroCollege approach for e-learning is a desirable form of education in the era of Industry 4.0 and Society 5.0 that is adopted in some institutions. With e-learning, students have more control over their learning schedule and pace. Institutions face significant challenges while implementing e-learning or switching from current educational systems. Given that the teacher and the students can only interact visually, providing the students with an online course that is properly created and developed to fit different learning levels and cater to different individual traits is a very challenging task.

For college students and IT professionals, MicroCollege is a concept that offers courses that are ready for the workplace. In general, the idea of MicroCollege is the perfect platform for individuals who want to have an interesting profession following their degree rather than just a job. These programs lay the groundwork for turning students become fully qualified professionals in a matter of months.

At KGiSL MicroCollege, IIM Building, 4th Floor, KGiSL Campus, Saravanampatti, Coimbatore, Tamil Nadu, India 641035, an experiment was conducted to provide a course in software programming – full stack development – as a means of preparing students for the workforce.

After enrolling in the course, students and IT professionals who wish to upgrade their expertise will receive training and pedagogy based on life skills. Courses are provided in a hybrid format at KGiSL MicroCollege, which combines offline and online learning. Basic and theoretical components will be studied via e-learning tools, and practical components will be encouraged to be completed on a college campus if students like to do so. The majority of practical components will be trained hands-on for a full-stack development course. Industrial specialists will assist the participants in this course in acquiring the skills that are most in-demand on the job market.

As a full-stack developer, a student works with back-end technologies like Java, PHP, Ruby on Rails, or Microsoft.NET, as well as database solutions using MySQL, MS-SQL, Oracle, CouchBase, or Mongo. They also work practically with HTML, CSS, and JavaScript, and the numerous tools and frameworks that support them. They will be well prepared for the placement exams when the full stack development courses are finished. They will be hired by the top IT companies after passing the exams.

2.15 CONCLUSION AND FUTURE SCOPE

Until recently, educational institutions relied primarily on traditional learning methods, such as face-to-face lectures in a traditional classroom setting. Even though the government was insisting on the use of blended learning, many institutions were using outdated teaching methodologies. The rapid emergence of the COVID-19 pandemic and the subsequent closing of physical schools and colleges posed a challenge to the global education system, thereby forcing them to switch to the online form of education overnight.

All academic institutions that had previously been hesitant to adopt these pedagogical methods now had no choice but to fully embrace blended teaching–learning. Subsequently, the new fourth industrial revolution Industry 4.0 is also evolving. This mainly focuses on incorporating cyber-physical systems in all fields to reduce the boundary between the real world and the virtual world. Together, Education 4.0 has also been evolving, which is a totally new experience-based education system that responds to the needs of the modern world of today. The usage of e-learning has reduced the number of learning-related issues, including time and money issues that previously limited learners' learning options.

In this chapter, we have seen a detailed review of how Education 4.0, or e-learning, is being adopted as the most desirable form of education. This approach combines technology, individualism, social media, customizations, digital platforms, PBL, self-motivated learning, discovery-based learning, etc. to give the best form of education for the future generation and prepare them to meet the demands of Industry 4.0. It is with no doubt that when forms of education change with the technological advancements of Industry 4.0 and Society 5.0, there will be a great revolution around the globe in all possible sectors.

A clear picture of the future of e-learning and how learning tools will be developed in the ensuing years is evident from a number of current e-learning trends, such as microlearning, gamification, and personalized learning. It may come as a surprise to most people but automatic learning is the next interesting future of e-learning. A student or any individual can choose a task, like playing the keyboard, which requires a lot of practice. Then, they can identify a skilled keyboard player and see what happens in their brain through some machine. The next stage is to begin learning by neuro-feedbacking the captured brain activity and recording it in your brain. While all this is happening, the individual does not have to concentrate on learning too. Theoretically, automated learning is definitely a possible development of how e-learning may evolve in the future.

ACKNOWLEDGMENT

This is a model project fore-learning as a desirable form of education in the era of Industry 4.0 Society 5.0 at KGiSL MicroCollege, IIM Building 4th Floor KGiSL Campus, Saravanampatti, Coimbatore, Tamil Nadu, India 641035. Their support is hereby acknowledged.

REFERENCES

Abdillah, L. A., Handayani, T., Rosalyn, E. R., & Mukti, Y. I. (2021). Collaborating digital social media for teaching science and arabic in higher education during covid-19 pandemic. *Ijaz Arabi Journal of Arabic Learning*, *4*(1), 12–25.

Aissaoui, K., Amane, M., Berrada, M., & Madani, M. A. (2022). A new framework to secure cloud based e-learning systems. In S. Bennani, Y. Lakhrissi, G. Khaissidi, A. Mansouri, & Y. Khamlichi (Eds.), *WITS 2020* (pp. 65–75). Springer. https://doi.org/10.1007/978-981-33-6893-4_7.

Anderson, T. (2019). Challenges and opportunities for use of social media in higher education, *Journal of Learning for Development*, *6*(1), 6–19.

Annika Andersson, Ö. (2008). Seven major challenges for e-learning in developing countries: Case study eBIT, Sri Lanka (Sri Lanka; Tertiary education; support; flexibility; access; academic confidence; localization; interactivity; attitudes) [Peer-Reviewed Article]. *International Journal of Education and Development Using ICT*, *4*(3). http://ijedict.dec.uwi.edu/viewarticle.php?id=472&layout=html.

Asawa, A. (2018). *Artificial Intelligence: The Star of the Digital Galaxy: A Study of Digital Disruption, Innovation, and Economic Transformation*. Independently Published.

Barata, J., Cunha, P. R. D., & Stal, J. (2018). Mobile supply chain management in the industry 4.0 era. *Journal of Enterprise Information Management*, *31*(1), 173–192. https://doi.org/10.1108/jeim-09-2016-0156

Baryshnikova, N., Kiriliuk, O., & Klimecka-Tatar, D. (2021). Enterprises' strategies transformation in the real sector of the economy in the context of the COVID-19 pandemic. *Production Engineering Archives*, *27*(1), 8–15. https://doi.org/10.30657/pea.2021.27.2.

Ben Amor, A., Abid, M., & Meddeb, A. (2020). Secure fog-based e-learning scheme. *IEEE Access*, *8*, 31920–31933. https://doi.org/10.1109/ACCESS.2020.2973325.

Boaler, J. (1999). Mathematics for the moment, or the millennium? (opinion). Retrieved July 1, 2022, from https://www.edweek.org/education/opinion-mathematics-for-the-moment-or-the-millennium/1999/03.

Briones, R. L., Kuch, B., Liu, B. F., & Jin, Y. (2011). Keeping up with the digital age: How the American Red Cross uses social media to build relationships. *Public Relations Review*, *37*(1), 37–43. https://doi.org/10.1016/j.pubrev.2010.12.006.

Bruckman, A. (2002). The future of e-learning communities. *Communications of the ACM 45*(4), 60–63. Retrieved July 1, 2022, from https://dl.acm.org/doi/10.1145/505248.505274.

Brumă, L. M. (2020). Security vulnerabilities in cloud based e-learning. *The International Scientific Conference eLearning and Software for Education*, Bucharest. Retrieved July 1, 2022, from https://www.proquest.com/openview/12e0f3420e778f5bc96227385c0cd884/1?pq-origsite=gscholar&cbl=1876338.

Brunswicker, S., Bertino, E., & Matei, S. (2015). Big data for open digital innovation – A research roadmap. *Big Data Research*, *2*(2), 53–58. https://doi.org/10.1016/j.bdr.2015.01.008.

Chen, Y., & He, W. (2013). Security risks and protection in online learning: A survey. *The International Review of Research in Open and Distributed Learning*, *14*(5). https://doi.org/10.19173/irrodl.v14i5.1632.

Collins, A., & Halverson, R. (2009). *Rethinking Education in the Age of Technology: The Digital Revolution and Schooling in America*. Teachers College. The TEC Series.

Doppelt, Y. (2003). Implementation and assessment of project-based learning in a flexible environment. *International Journal of Technology and Design Education*, *13*(3), 255–272. https://doi.org/10.1023/A:1026125427344.

Ekereke, L., & Akpojaro, J. (2019). Security challenges in accessing e-learning systems a case study of Sagbama Bayelsa state. *International Journal of Innovative Science and Research Technology*, *4*(5). Retrieved July 1, 2022, from https://www.ijisrt.com/security-challenges-in-accessing-e-learning-systems-a-case-study-of-sagbama-bayelsa-state.

Ellzey, J. L., O'Connor, J. T., & Westerman, J. (2019). Projects with underserved communities: Case study of an international project-based service-learning program. *Journal of Professional Issues in Engineering Education and Practice*, *145*(2), 05018018. https://doi.org/10.1061/(ASCE)EI.1943-5541.0000400.

Estriegana, R., Medina-Merodio, J.-A., & Barchino, R. (2019). Student acceptance of virtual laboratory and practical work: An extension of the technology acceptance model. *Computers & Education*, *135*, 1–14. https://doi.org/10.1016/j.compedu.2019.02.010.

Gleason, N. W. (2018). Singapore's higher education systems in the era of the fourth industrial revolution: Preparing lifelong learners. In N. W. Gleason (Ed.), *Higher Education in the Era of the Fourth Industrial Revolution* (pp. 145–169). Springer. https://doi.org/10.1007/978-981-13-0194-0_7.

Gültekin, M. (2005). *The Effect of Project Based Learning on Learning Outcomes in the 5th Grade Social Studies Course in Primary Education*. https://www.scinapse.io.

Ha, N., Nayyar, A., Nguyen, D., & Liu, C. (2019). Enhancing students' soft skills by implementing cdiobased integration teaching mode. In *2019 15th International CDIO Conference*, Aarhus University, Denmark. Retrieved July 9, 2022, from http://www.cdio.org/knowledge-library/documents/enhancing-students%E2%80%99-soft-skills-implementing-cdiobased-integration.

IEEE Xplore. (2018). *Society 5.0: Industry of the Future, Technologies, Methods and Tools*. Wiley eBooks. Retrieved June 23, 2022, from https://ieeexplore.ieee.org/book/8607846.

Kelly, J. (2002). Collaborative learning: Higher education, interdependence, and the authority of knowledge by Kenneth Bruffee: A critical study. *Collaborative Learning*, pp. 91–100.

Knowles, M. S. (n.d.). *From Pedagogy to Andragogy*, Cambridge Adult Education, 59.

Kumar, A., Krishnamurthi, R., Bhatia, S., Kaushik, K., Ahuja, N. J., Nayyar, A., & Masud, M. (2021). Blended learning tools and practices: A comprehensive analysis. *IEEE Access*, *9*, 85151–85197. https://doi.org/10.1109/ACCESS.2021.3085844.

Laufer, M., Leiser, A., Deacon, B., Perrin de Brichambaut, P., Fecher, B., Kobsda, C., & Hesse, F. (2021). Digital higher education: A divider or bridge builder? Leadership perspectives on edtech in a COVID-19 reality. *International Journal of Educational Technology in Higher Education*, *18*(1), 51. https://doi.org/10.1186/s41239-021-00287-6.

Luminita, D. C. (2011). Information security in e-learning platforms. *Procedia - Social and Behavioral Sciences*, *15*, 2689–2693. Retrieved July 1, 2022, from https://www.sciencedirect.com/science/article/pii/S1877042811007178.

Marks, A., Al-Ali, M., Attasi, R., Elkishk, A., & Rezgui, Y. (2021). *Digital Transformation in Higher Education: Maturity and Challenges Post COVID-19* (pp. 53–70). https://doi.org/10.1007/978-3-030-68285-9_6.

Márquez-Ramos, L. (2021). Does digitalization in higher education help to bridge the gap between academia and industry? An application to COVID-19. *Industry and Higher Education*, *35*(6), 630–637. https://doi/org/10.1177/0950422221989190.

Mukhopadhyay, M., Pal, S., Nayyar, A., Pramanik, P. K. D., Dasgupta, N., & Choudhury, P. (2020). Facial Emotion Detection to Assess Learner's State of Mind in an Online Learning System. In *Proceedings of the 2020 5th International Conference on Intelligent Information Technology*. Retrieved July 9, 2022, from https://dl.acm.org/doi/10.1145/3385209.3385231.

Ochoa, S. F., Fortino, G., & Di Fatta, G. (2017). Cyber-physical systems, internet of things and big data. *Future Generation Computer Systems*, *75*, 82–84. https://doi.org/10.1016/j.future.2017.05.040.

O'Donovan, P., Gallagher, C., Bruton, K., & O'Sullivan, D. T. J. (2018). A fog computing industrial cyber-physical system for embedded low-latency machine learning Industry 4.0 applications. *Manufacturing Letters*, *15*, 139–142. https://doi.org/10.1016/j.mfglet.2018.01.005.

Parker, L., & Prabawa-Sear, K. (2020). *Environmental Education in Indonesia: Creating Responsible Citizens in the Global South?* Routledge & CRC Press. Retrieved June 24, 2022, from https://www.routledge.com/Environmental-Education-in-Indonesia-Creating-Responsible-Citizens-in-the/Parker-Prabawa-Sear/p/book/9780367784249.

Ponidi, P., Waziana, W., Kristina, M., & Gumanti, M. (2020). Model of utilizing discovery learning to improve mathematical learning achievements. *Attractive: Innovative Education Journal*, *2*(1), 41. https://doi.org/10.51278/aj.v2i1.27.

Salkin, C., Oner, M., Ustundag, A., & Cevikcan, E. (2018). A conceptual framework for industry 4.0. In A. Ustundag & E. Cevikcan (Eds.), *Industry 4.0: Managing the Digital Transformation* (pp. 3–23). Springer International Publishing. https://doi.org/10.1007/978-3-319-57870-5_1.

Saputra, M. R. Y., Winarno, W. W., Henderi, H., & Shaddiq, S. (2020). Evaluation of maturity level of the electronic based government system in the department of industry and commerce of Banjar regency. *Journal of Robotics and Control (JRC)*, *1*(5), 156–161. https://doi.org/10.18196/jrc.1532.

Sharma, P., Agarwal, K., & Chaudhary, P. (2021). E-learning platform security issues and their prevention techniques: A review. *International Journal of Advance Scientific Research and Engineering Trends*, *6*(8), 51–59.

Siddiqui, S. T., Alam, S., Khan, Z. A., & Gupta, A. (2019). Cloud-based e-learning: Using cloud computing platform for an effective e-learning. In S. Tiwari, M. C. Trivedi, K. K. Mishra, A. K. Misra, & K. K. Kumar (Eds.), *Smart Innovations in Communication and Computational Sciences* (pp. 335–346). Springer. https://doi.org/10.1007/978-981-13-2414-7_31.

Suhono, S., & Sari, D. A. (2020). Developing students' worksheet based educational comic for eleventh grade of vocational high school agriculture. *Anglophile Journal*, *1*(1), 29. https://doi.org/10.51278/anglophile.v1i1.78.

Sung, M. (2015). A study of adults' perception and needs for smart learning. *Procedia - Social and Behavioral Sciences*, *191*, 115–120. https://doi.org/10.1016/j.sbspro.2015.04.480.

Thackeray, R., & Neiger, B. L. (2009). A multidirectional communication model: Implications for social marketing practice. *Health Promotion Practice*, *10*(2), 171–175. https://doi.org/10.1177/1524839908330729.

Thackeray, R., Neiger, B. L., & Keller, H. (2012). Integrating social media and social marketing: A four-step process. *Health Promotion Practice*, *13*(2), 165–168. https://doi.org/10.1177/1524839911432009.

Thomas, J. (2000). *A Review of Research on Project-Based Learning*. Autodesk Foundation, 111 Mclnnis Parkway San Rafael, California 94903.

Tiainen, T., Isomäki, H., Korpela, M., Mursu, A., Paakki, M.-K., & Pekkola, S. (2007). *Proceedings of 30th Information Systems Research Seminar in Scandinavia*. Murikka, Tampere, Finland.

Trappey, A. J. C., Trappey, C. V., Hareesh Govindarajan, U., Chuang, A. C., & Sun, J. J. (2017). A review of essential standards and patent landscapes for the Internet of Things: A key enabler for Industry 4.0. *Advanced Engineering Informatics*, *33*, 208–229. https://doi.org/10.1016/j.aei.2016.11.007.

Tsybulsky, D., & Muchnik-Rozanov, Y. (2019). The development of student-teachers' professional identity while team-teaching science classes using a project-based learning approach: A multi-level analysis. *Teaching and Teacher Education*, *79*, 48–59. https://doi.org/10.1016/j.tate.2018.12.006.

Ugur, N. G. (2020). Digitalization in higher education: A qualitative approach. *International Journal of Technology in Education and Science*, *4*(1), 18–25. https://doi.org/10.46328/ijtes.v4i1.24.

UNESCO. (2020). *Education: From Disruption to Recovery*. UNESCO. https://en.unesco.org/covid19/educationresponse.

Ware, J. L. (2018). Evaluating online and traditional learning environments using cartographic generalization techniques. *Cartographica: The International Journal for Geographic Information and Geovisualization*, *53*(2), 107–114. https://doi.org/10.3138/cart.53.2.2017-0015.

World Economic Forum. (n.d.). Fourth Industrial Revolution. Retrieved June 24, 2022, from https://www.weforum.org/focus/fourth-industrial-revolution/.

Yun, J. J., Lee, D., Ahn, H., Park, K., & Yigitcanlar, T. (2016). Not deep learning but autonomous learning of open innovation for sustainable artificial intelligence. *Sustainability*, *8*(8), 797. https://doi.org/10.3390/su8080797.

CHAPTER 3

Education Situation in Online Education before the Pandemic and in the Time of Pandemic

Hesham Magd and Henry Jonathan
Modern College of Business and Science

Shad Ahmad Khan
University of Buraimi

CONTENTS

DOI: 10.1201/9781003322252-3

3.1 INTRODUCTION

Education is an important facet of life that provides learning through a formalized and systematic manner in giving knowledge, skills, and expertise for employment and sustenance. However, lifelong learning is also one form of education that involves learning through experience, observation, and putting into practice those actions for improving skills. In this sense, education can be formal and informal depending on the approach to learning where the former is more customized, and the latter does not have a stipulated system. Formal education is by large the standard form of imparting education at all levels and is defined by the UNESCO as "education that is institutionalized, intentional, planned through public organizations and recognized private bodies, and, in totality, make up the formal education system of a country" [1]. On the contrary, informal learning is a form of non-formal learning gaining momentum very recently and involves learning through a self-paced approach, self-directed learning, and learning through experience.

The current educational system is transforming, moving toward a more affordable, flexible, and convenient form of teaching and learning that is only brought by online education. Unbelievably, online education has literally expanded the horizons of teaching and learning more recently from the coronavirus disease which has affected the global educational sector. Ironically, online learning has gained importance since the beginning of the 20th millennium, and the online education market is projected to reach $325 billion by the year 2025 according to a study by Forbes. The increasing globalization over the past two decades has caused the education system to rapidly change along with the industrial revolution, which has highlighted the use of digital technologies in teaching and learning. Contrarily, worldwide, the education system in online education has proven to be more beneficial than face-to-face teaching as many countries have reported increased perception among learners with 45% of students in the US, claiming that digital technology in learning is more beneficial [2]. The global e-learning market is projected to raise to $388.23 billion by the year 2026 which was $185.26 billion in the year 2020 (Figure 3.1).

On the other hand, online education has also proven to be environmentally friendly by reducing CO_2 emissions by using less energy on all fronts. Not only that, in the corporate world, online learning has caused 90% of them comfortably use them and there are reports showing that the largest online education movement has taken place in China, with a quarter of a billion full-time students having moved to online learning according to the World Economic Forum. Despite the tremendous developments that are taking place toward online education, World Bank studies indicate that the world is facing a learning crisis that

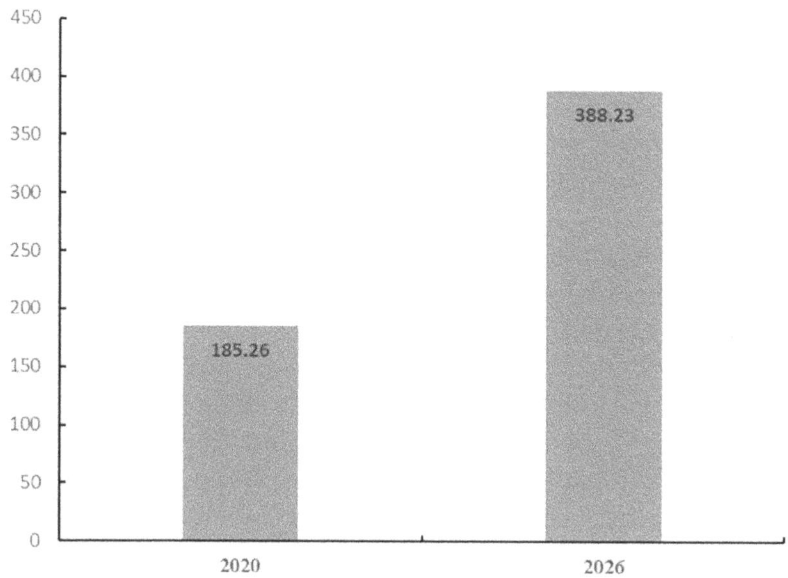

FIGURE 3.1 The future of the global e-learning market in the educational sector.

is prevailing in many educational systems in developing and underdeveloped nations [3]. Moreover, there are also other potential problems associated with online education, especially in the lower economies regarding the use and advancement of digital technologies, which pose a critical challenge to the educational system. Most of the countries in Asia were the first to be affected due to the COVID-19 pandemic [4]; similarly, global surveys on education disruption due to COVID caused vulnerable students to trail behind in gaining knowledge and requisite skills [5].

Historically, pandemics have been known to affect society; however, their impact on education has not been much recorded in the past and present. It is surprising to know that major pandemics and epidemics in the past were known to have brought substantial change in the education system which has led to the establishment of medical schools, public health education, and the most recent COVID-19 introducing digital education [6]. Notwithstanding to say from the analysis that the education system has gone through remarkable changes owing to the crisis of COVID pandemic, and educational providers worldwide have moved to online education, it is very imperative to have a glimpse of the transforming educational system to capture the trend reflecting the online education before and during the pandemic time.

Objectives of the chapter: Therefore, this chapter is written to critically present the educational system prevailing globally and to highlight the changes that were brought into the current educational system due to the COVID-19 pandemic, to describe the distinction or transformation that has taken place in the teaching and learning methods before and during the pandemic in the educational sector globally, and to comprehend the role of technology in the area of teaching and learning tools and the new teaching and learning approaches that have come into existence during the pandemic are elaborated.

3.1.1 Technology Breakthrough in Education Amid the COVID-19 Pandemic

Educational institutions globally rely largely on formal learning methods which is a more reliable form of teaching to impart knowledge and skills and develop an understanding of study content. There are different ways, methods, and techniques in practice in the formal system of learning that are adopted by colleges and universities for a long time and some of them are very significant and efficient methods of teaching. Ever since information technology had an influence on the education system, it brought massive changes in teaching and learning with the growing awareness among educators, decision-makers, and students on the use of technology, which has brought diversity in teaching and learning methods.

Technology intervention has brought the concept of e-learning into the fore of formal education widely in most colleges and universities in developed nations and developing nations. E-learning is not a new concept that involves the use of electronic sources to impart teaching which has expanded the scope of education by teaching and learning outside the classrooms from any remote location with access to an internet facility [7]. There are various ways of teaching used through e-learning such as online including distance education, learning management system, webinars, virtual learning, live recordings, etc. and each of these ways has different platforms for teaching and learning such as Blackboard, Moodle, open edx, Coursera, Skillshare, etc. [8]. Eventually, all these innovative and flexible methods have invariably brought an increase in demand for college education, especially due to the provision and availability of online teaching and learning facility for almost every course in institutions. However, the technology interface and e-learning are not without problems, as with many limitations that hinder the efficiency and effectiveness of the e-learning mode in defiance of information and data on the effectiveness and suitability of utilizing e-learning in the formal education system across many higher education institutions (HEIs) globally and within the region.

With the transformation of the education system for many reasons stated above, more and more institutions are turning toward e-learning facilities; howbeit, it is very important to assess the value and significance of imparting education through innovative mechanisms and evaluate whether teaching and learning critically deliver the required outputs to the student communities.

3.1.2 Technological Boost in the Education System: Pre- and Post-COVID Statistics

The greatest boost the current educational system has received is in technology enhancement to the teaching and learning methodology. HEIs have taken great advantage of embedding technology and tools in teaching, as well as students, have acquired knowledge in learning through digital media. This shift has uninterrupted student communities from learning, which benefited the educational institutions to continue to adapt easily to the e-learning mode of the environment. In other words, technology, especially online education either by use of digital learning or technologically enabled teaching, has metamorphosized teaching and learning as much as online education has become the most sort after and preferably by millions of learners globally. Not only this but there are also many advantages that the online education system can provide over traditional education methods in terms of learning analytics, providing learner performance and progression which can be assessed among international students, etc.

Currently, e-learning has now been used for more than two decades but was not sufficiently explored earlier to the COVID-19 pandemic; but after the pandemic, there was a greater necessity for the educational community to experiment with e-learning approaches in education, and though Europe and North America are leading in online learning since the 1990s, other nations have slowly started to gain the pace only after the spread of COVID-19. Regardless of these, e-learning is not a very old concept in education, which was heard first in the year 1988; then, the concept started to get more concentrated in developed countries.

In 2017, online learning was part of 77% of corporations in the US, which was only 4% in 1995; moreover, corporate e-learning is expected to grow by more than 250% by 2026. E-learning is also leading in the academic sector with the US taking the concept to the forefront in education as more than 80% of college students feel that digital learning technology is aiding them in improving their performance. Globally, studies show that close to 50% of students have taken online education in the past 1 year, and in 2020, there were 120 million learners using massive open online courses, which could lead to boosting online education worldwide. On the other hand, there are countries that are leading online education currently, with 65% of HEIs in the US running under online education, followed by India, China, South Korea, Malaysia, the UK, Australia, and South Africa [9] (Figure 3.2).

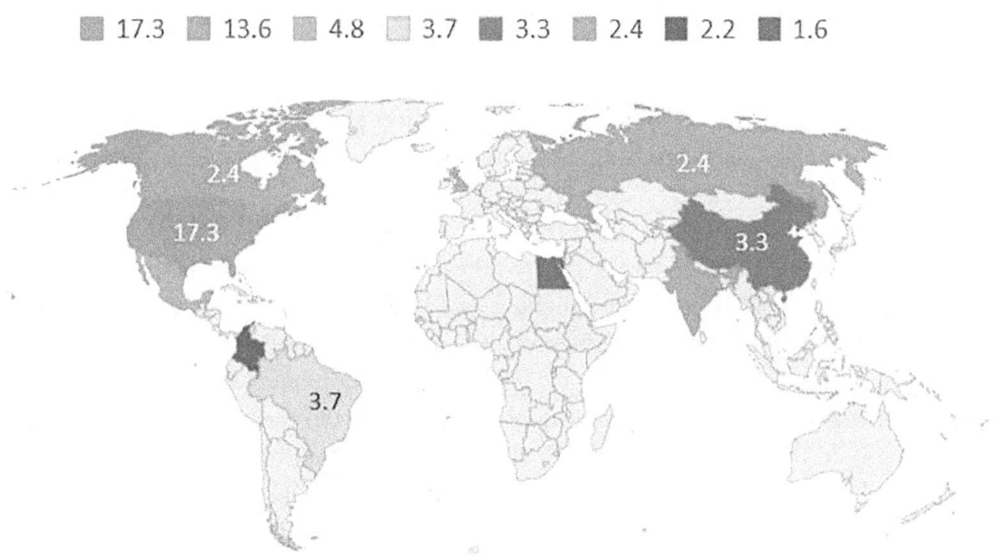

FIGURE 3.2 Top ten countries with the highest registered learners of online courses during the pandemic. (Source: Map reproduced from data obtained from Wood 2022- weforum.)

Organization of the Chapter

The chapter is organized as follows: Section 3.2 discusses the historical perspective and evolution of education. Section 3.3 elaborates on the pandemic crisis's impact on the education system, before and after COVID-19. Section 3.4 highlights online education technological breakthroughs before and during the COVID-19 period. Section 3.5 enlightens the discussion. Section 3.6 concludes the chapter with future scope.

3.2 HISTORICAL PERSPECTIVE AND EVOLUTION OF EDUCATION

The educational system started way back many centuries ago, and the first known school in human history was somewhere around 2000 BC. Some mathematics textbooks were discovered during that time, revealing the existence of schools. Later, between 455 BC to 431 BC, formal education was started in Greek cities to train children for the military, after which Romans were the next to follow the Greek education system from 50 BC to 200 AD. It was during this era, children were taught to read, write, and learn literature, music, math, and history. Further during 400 to 1000 AD referred to as the dark ages, only wealthy families were able to afford education. Then, between 1630 and 1640, the earliest American colleges were established and the first colonial college Harvard was constituted in 1636 to educate the ministers; later, many schools such as Yale, Princeton, and Dartmouth were established. It was late in 1751 when the American Academy was established as an institution to offer formal and higher-order education in history, geography, navigation, surveying, and languages were taught. Nearly after two centuries, between 1946 and 1950, the first computer (ENIAC) vacuum tube computer was built which started the beginning of the computer age. Much later, in 1989, the University of Phoenix became the first university to adopt online education offering fully online undergraduate and graduate degrees [10]. In 1991, the first smartboard was introduced using touchscreen mode to facilitate teaching more easily. By the mid-2000, the online education platform Khan Academy started providing online classes and teaching material to instructors to teach people remotely [11].

3.2.1 Pandemic Implications on the Educational System Pre- and Post-COVID-19

There is no doubt that COVID-19 has affected the education system across the world, more significantly bringing changes to the teaching and learning mechanism in comparison to the pre- and post-COVID times. The biggest change to note was the adoption of e-learning methods such as online teaching and learning mechanism from the traditional teaching practices during the pre-COVID-19 time, which has impacted the learning perceptions of students between the two phases [12]. However, it is important to know that transformation to the education system started quite before the pre-COVID-19 time when there was some emphasis on the use of digital teaching and learning, but there was a lack of seriousness and delays in adoption. Despite these, every economy regardless of the development status has experienced challenges in the education system, deemed to bring educational reforms to develop the right approach to student learning and inculcate the right skills and expertise through education. In addition, the COVID-19 pandemic has propelled educational reforms in many countries augmenting to determine the different approaches to teaching and learning methods.

On the other hand, post-COVID19 implications on the educational system are widely studied in the context of different economies and are well documented by organizations like the World Bank, World Economic Forum, UNESCO, UNICEF, etc. Some serious influence was felt in low-income countries, though high-income countries observed learning losses among students. Countries in Europe and Central Asia have suffered from the closure of educational institutions which has affected more than 185 million students due to the closure or suspension of teaching [13]. On the global front, COVID-19 has disrupted education for more than 1.6 billion students, though most countries have adopted alternative teaching methods such as remote and online modes according to the World Bank; those methods have not fully supplemented the regular learning process. In more than 191 countries, institutions have switched to distance learning either in the form of online or virtual learning or remote learning; however, only a few countries possess the resources to offer education in such a format [14].

One of the serious repercussions arising post-COVID-19 is the learning loss among students due to the disruption of learning, which is not measured or evaluated by most countries. Currently, there are anticipated gaps in skills, knowledge, and expertise that students should possess in the absence of the pandemic. To address the crisis, educational planners, decision-makers, and academic communities must decide on learning recovery to recuperate the losses during the pandemic time. Educational institutions should customize their learning recovery mechanisms to accelerate the learning curve by adopting different approaches. Such measures need changes in the curriculum design, teaching methods, reviewing the performance evaluations, and adopting a blended learning approach to education [15]. In response, the global educational community is finding out ways to bridge the gap and prioritize how the future educational system should transform with digital technology.

3.2.2 Pandemic's Influence on Higher Education

The incidence of coronavirus disease-19 (COVID-19) has affected the entire world and has impacted the educational sector worldwide, causing many HEIs, namely, universities and colleges, to close face-to-face teaching and shift to alternative safe modes of delivering education, which has not only caused a huge economic impact on education but also the global economy [16]. Particularly, this crisis has caused substantial disengagement in the performance evaluation of students through formative and summative assessments. Also, globally, large universities having international students are facing a high risk of continuing their courses in this situation, but administrators must make strategic plans to reduce the impact of COVID-19 on students. Moreover, institutions including the students and instructors experienced difficulties in the form of barriers in moving to online education as almost all institutions have significantly invested in capital to effectively remain on board [17]. To cope with the sudden shutdown of educational institutions and make sustainable alternatives, the most effective and feasible option is to adopt online learning to maximize education in the prevailing conditions; hence, teaching and learning through the online mode using various technology platforms have gained momentum in almost all the educational institutions across the globe.

3.3 THE PANDEMIC CRISIS IMPACT ON THE EDUCATION SYSTEM BEFORE AND DURING COVID-19

Certainly, the COVID-19 pandemic that is still prevailing globally has impacted many sectors but very severely has impacted the educational system, both secondary and higher education. The extent to which the educational system in every economy gets affected during a crisis depends on the allocation of educational funds. In many economies during the COVID-19 crisis, most of the funds were diverted toward more serious causes such as health care and maintenance of the economy. Because of this, many economically weaker countries were deprived of getting educational grants, and scholarships and educational institutions trailed in building infrastructure facilities. To summarize, the pandemic has affected the educational system, directly affecting the learners and indirectly affecting services, and delaying the ability to develop a skilled workforce for the market (Figure 3.3).

3.3.1 Financial Deprivation of Education

Education is basically funded by national governments to support educational institutions in building learners for the world of work. Because of the pandemic, the world economy and the financial situation of every economy weakened, affecting the educational sector among all other sectors. Further studies from the Organization for Economic Cooperation and Development (OECD) report that around 11% of the public expenditure in the OECD countries are devoted to the educational sector before the pandemic time, and as the global economy fell drastically in the year 2020, the share of the education sector substantially dropped. However, the percent share of government expenditure on education substantially varies across different OECD nations, with the highest in Chile at 17% and the least in Greece at 7%. The overall expenditure on the education sector is a result of direct and indirect public transfers that are equated to obtain the total government share in the

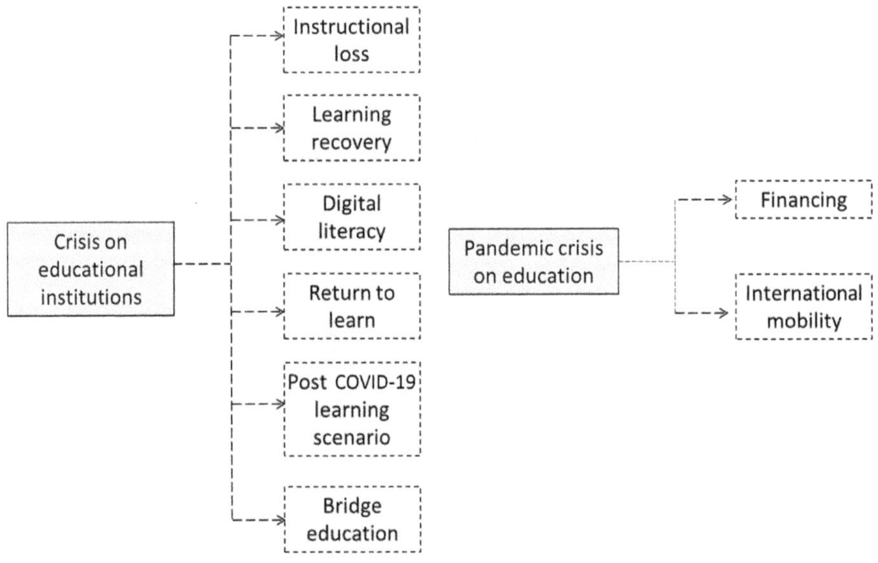

FIGURE 3.3 Challenges posed by the COVID-19 pandemic on the educational sector.

educational sector. Due to the pandemic that has impacted every economy, the sources of public funds have also been disturbed severely which has impacted the educational budgets in many nations [18]. To overcome the impending crisis which has adversely affected many students, some countries like Australia have announced relief packages for HEIs to provide financial support to students, and New Zealand has also made similar arrangements for students to cope with the loss due to the crisis [19]. In response, other countries like the UK, Canada, and the US have also announced similar packages in the way of emergency financial aid, emergency relief funds, and emergency student benefit funds in collaboration with global humanitarian agencies [20]. On the other hand, the lack of financial support to the educational sector has led to many repercussions to both institutions and student communities, by not providing proper infrastructure facilities and mechanisms to augment effective teaching and learning environments, which has resulted in learning loss, decelerated intellectual capacity and ability to think critically, and slowed-down learning capacity, all these being due to the closure of regular teaching and learning in educational institutions. Notwithstanding to say, the financial crisis has especially led students from lower economies to mental imbalance, stress, and anxiety issues according to the studies by various researchers [21].

3.3.2 International Student Mobility

Another major outcome of the pandemic on the educational sector was international student mobility, disrupting international learners from visiting institutions because of lockdowns, closures, restrictions, social distancing, isolation, etc. Statistics show that almost 22% of students from OECD countries are in doctoral programs in leading HEIs across the world, where the pandemic inevitably has affected their learning and international exposure. However, to ensure the continuity of learning and education, higher education has adapted to create virtual learning environments, conduct online teaching, and support learning through technological applications; remote learning is a better alternative and substitute for the learning experience of studying abroad. Many countries such as Australia, Canada, the UK, and the USA which rely heavily on international students were badly hit due to decreasing international student enrollment and a lack of local funding and scholarship support for international students.

3.3.3 Loss of Instructional Hours and Learning Loss

Without any exceptions, educational institutions worldwide have taken steps to fully or partially close face-to-face teaching and learning. Study reports from the OECD show that by the mid of the year 2020, as many as 46 countries in the OECD had closed their institutions and opted for the online education system.

3.4 ONLINE EDUCATION: TECHNOLOGICAL BREAKTHROUGH BEFORE AND DURING THE PANDEMIC

Education from the past to the current century has taken a dramatic transformation in all areas such as teaching, learning, knowledge gain, skill development, etc. The recent COVID-19 pandemic that has engulfed the entire world has certainly caused a breakthrough in teaching and learning methods. The global educational community as a result has moved to online education by successfully adopting digital technologies, creating a virtual teaching and learning environment. Though online teaching and learning was not totally a new methodology in education before the pandemic, online education gained more prominence from the consequences that resulted during the pandemic [22], because of which face-to-face teaching and learning must be replaced with technology-enabled education more precisely for online education.

Digital technology is bound to play a pivotal role in driving online education in the current educational system and would prove very prospective for future learners, while currently, more than 6 million Americans are pursuing education online, and almost 68% of the learners are experienced professionals, with the average of them enrolled are beyond the age of 32 years. Nevertheless, such advancement in online education is not tremendous in the least, and developing nations lacking the necessary infrastructure and financial resources totally depend on and move toward digitally enabled teaching and learning. Within the sphere of online education, there are many digital technology tools that are required to be employed for successfully delivering knowledge, understanding, and skill development, which educational institutions having sound financial resources can afford. Simultaneously, online education has also introduced the concept of digitalization in education, symbolically making way for the use of digital technologies through software applications embedded into teaching and learning content. These technological inclusions are part of online education that presumably aligns with the industrial revolution aiming to develop learners for imbibing 21st-century skills (Figure 3.4).

Impact aspect	Pre-COVID-19	During COVID-19
Financing of education	↑	
International student mobility	↑	
Instructional time	↑	
Remedial learning		↑
Digital literacy		↑
Learner ratio		↑
Vocational education		↑
Learning analytics		↑

FIGURE 3.4 Pros and cons of the COVID-19 pandemic on the educational system.

3.4.1 Evolution of Online Education from the 20th Century to the Present Pandemic

Online education is the more advanced form of education that differs from traditional face-to-face classroom education. Online education is also considered one of the forms of distance education which can take place away from the classroom environment. This mode of education may require internet access unlike other forms of distance education which may not require it. During the 20th century, the education system was formal and non-formal with distance learning, correspondence, and telecourses. However, online education, which was also part of non-formal education in the past, has become the most suitable and appropriate form of education, redefining today's distance education [23]. The present online education has improvised distance education, becoming more internet-based and including digital technologies, tools, and applications in learning [24]. According to educational analysts, the demand for education online is going to increase in the 21st century due to the advancement in technology and innovation. Past figures show that in 2003, there were over 1.9 million students studying online, though most of them were from developed nations. In the US, in 2010, 6.1 million students out of 21 million students enrolled in an online course, which is 20.9% of the total students in higher education.

On the higher education side, the e-learning market has crossed the $250 billion mark and is expected to still grow later in this decade, given the number of higher education students who show a willingness to pursue their education online. Notwithstanding to say, during the pandemic, 98% of HEIs moved to online teaching and learning, while nearly 50% of the institutions invested in technology related to online educational resources. Eventually, these developments saw a huge demand for massive open online courses (MOOCs), attracting students from all over the world to learn online.

3.4.2 Current Trends in the Online Education System

The present education system is symbolically referred to as online education which surfaced predominantly due to the COVID-19 pandemic [25]. Obviously, online education has taken a significant lead more rapidly, which has grabbed the attention of global educational communities in every economy; as a result, the traditional educational system has taken a preceding position to the modern form of teaching and learning. Furthermore, the online education path has gradually become more intensive due to digital technology and tools that got integrated into teaching and learning (Figure 3.5).

In this backdrop, online education is progressively making routes toward intensive technology-enabled systems, providing more avenues for learners to use digital tools in learning such as facilities for open education resources (OERs), virtual reality (VR), artificial intelligence (AI), blended learning, video learning, mobile learning, etc., customizing teaching and learning.

Presently, there is no doubt that the world of education has moved to e-learning with online education as one form to gain interaction with students; in this facet, learning has become technology-enabled, such as learning through virtual games, VR, and augmented reality which are some of the futuristic trends that are going take a lead by the year 2030. Another potential trend that would likely boom in the future of online education is the

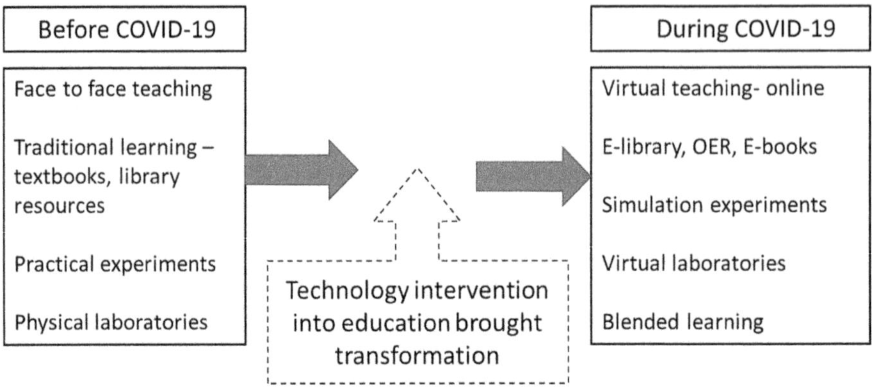

FIGURE 3.5 Education system transformation in pre-COVID-19 and post-COVID-19 phases.

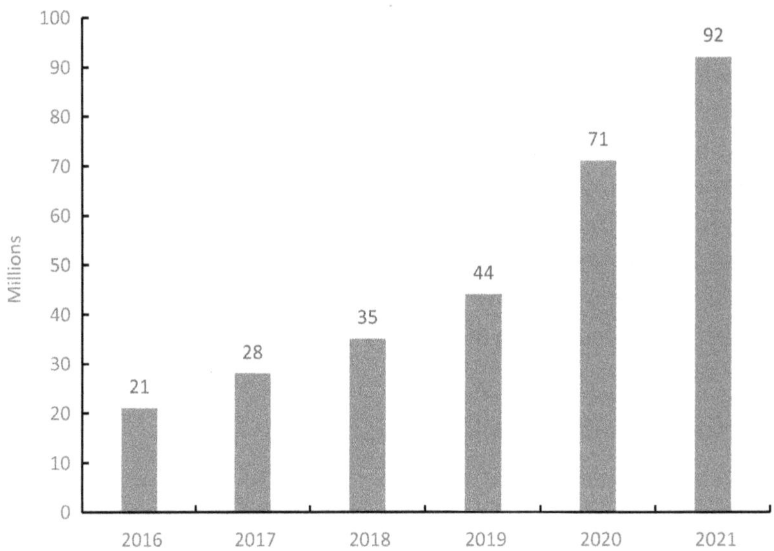

FIGURE 3.6 Number of learners in millions registered for online learning in Coursera. Source: weforum.org 2022 – https://about.coursera.org/press/wp-content/uploads/2021/11/2021-Coursera-Impact-Report.pdf

introduction of neuro-education involving brain-based learning with the use of AI [26]. In addition, to exemplify the growth, there are many online learning platforms that have come to the fore of students worldwide amid the COVID-19 pandemic, and it is interesting to note that the number of students availing online education far exceeds the pre-pandemic levels with almost 20 million registrations in the year 2021 and also not surprising to report that online learning is a now a worldwide new normal with 92 million learners and 189 million enrollments in the year 2021 alone [27] (Figure 3.6).

The highest potential for online education was drawn from the Asia-Pacific region with 28 million learners followed by North America with 20 million and Europe with 16 million

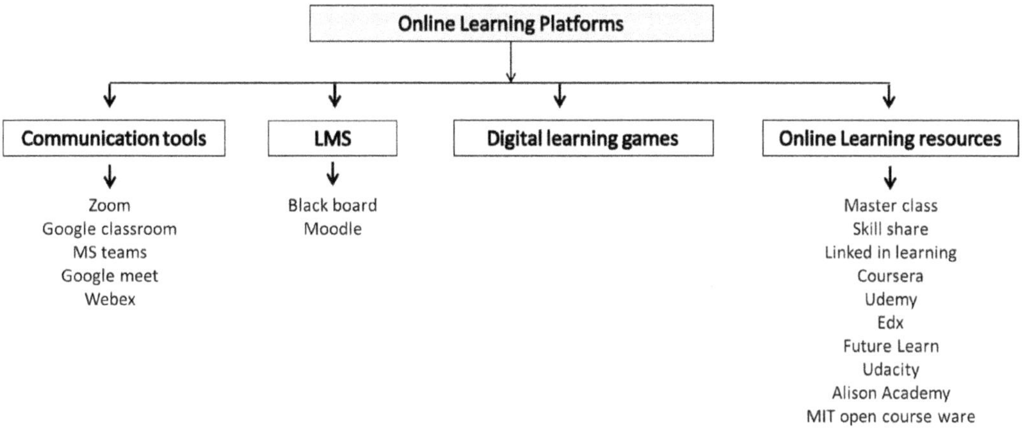

FIGURE 3.7 Online learning platforms.

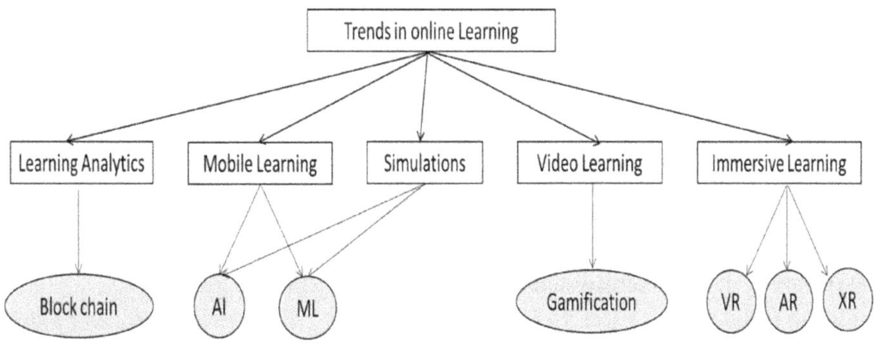

FIGURE 3.8 Digital technologies interfaced with the current trends in online learning.

learners. It is also noteworthy to disclose that owing to the COVID-19 pandemic, online education has become more affordable and accessible to the masses which became more reachable due to many online education platforms today (Figure 3.7). With many online learning platforms that have recently reached the forefront of education, the number of enrollments by learners to various courses has also exponentially increased since the pre-COVID-19 times; though precise figures on global online learners cannot be ascertained, these records are not only limited to undergraduates or graduates but even employed and working professionals have shown interest to online education to upskill and improve their capabilities.

Along with these developments, online education globally became more technology augmented with the use of extended reality (XR), VR, augmented reality, AI, machine learning (ML), and OERs which have diversified the e-learning capabilities in the current scenario. These applications eventually paved the way for learners to have different learning experiences through self-directed learning (SDL), blended learning, adaptive learning, mobile learning, video learning, simulations, etc. which are very potential e-learning trends gaining significance in the global education community (Figure 3.8).

Though the education system now is coupled with technology, which is prominently seen due to the COVID pandemic, the future of education is advancing toward setting new learning trends which will be more promising to meet the demands of future learners by facilitating micro-learning, social learning, virtual classrooms, gamification, immersive learning, video-assisted learning, etc. With this education outlook, brought by COVID-19, the destiny of education especially online education is anticipated to evolve into a more sophisticated, flexible, customized form of education for learners and keep pace with the upcoming industrial revolutions.

3.4.2.1 Blended Learning

One of the noted developments in the concept of online education that has brought to everyone's understanding is blended learning. As the name suggests, it is a method of imparting teaching by blending both technology and traditional classroom-based instruction which will provide better and deeper learning to learners [28]. This teaching concept is also considered hybrid teaching more popular among HEIs and is known to have come into existence with the introduction of internet-based teaching during the 1990s and was adopted in distance education. As learning shifted from teacher-centric to student-centric, the blended learning approach gained more acceptance by student communities due to a variety of reasons. The method of learning mechanism gives more preference to student control over learning and provides sufficient scope for learners to think, analyze, and have their own command of learning. In this self-learning process, learners are more responsible and held accountable for their learning; moreover, blended learning will facilitate adopting different digital technologies such as OERs in the process.

Though the type of education looks more innovative, the approach has very significant drawbacks with respect to the capabilities of learners as this method of learning is not suitable for all types of learners. Sufficient background preparation and modulations in the course of learning should be made by both the instructor and students for effective learning through this method. On the other hand, the future of blended learning as the most prominent e-learning approach is expected to surpass other forms as more than 80% of students are keen to adopt hybrid learning over traditional classroom-based education. Moreover, more than 70% of the students felt their performance was improved much better due to better student engagement and interaction. In the higher education sector, blended learning is gaining more roots as a new form of education that is more improvised to the present century education. In addition, there are many areas in the blended learning approach that are leading ahead in terms of student engagement, student satisfaction, reduction in the dropout rate, increased student retention rates, enhanced community learning, etc. [29]. At this point, it is worthwhile to debate that the blended learning approach as a successful alternative teaching and learning technique over other forms of teaching cannot be ascertained truly as there are many challenges to be overcome by the educational sector in terms of information and communications technology (ICT) literacy and awareness by student communities. Some of the pressing concerns to recommend a blended learning approach in the present education system mainly are the technological infrastructure requirement, design of the course, the expertise of the instructor to streamline the process of learning, learners' acumen in utilizing the dual mode of learning, etc., which definitely confront all HEIs and also the current education system.

Currently, the perspective of blended learning in the present educational system seems more promising to hold simply because it combines online with face-to-face learning, thus addressing the advantages of both technology and interaction. Ideally, in the blended learning approach, online and face-to-face teaching follows indefinite proportions depending on the nature of the course and the methodology of the instructor. Within this process, blended learning most probably can be adopted using four different models based on the way online and face-to-face learning takes place. Studies show that blended learning in four Asian countries, China, Korea, Japan, and Singapore, is ranked by the Economist Intelligence Unit based on their readiness for e-learning.

3.4.2.2 Mobile Learning

Mobile learning, also referred to as M-learning, is another form of learning using electronic devices such as mobile phones, tablets, smartphones, or any other device that is portable and handy, where learning can take place through different learning applications in those devices. M-learning is not much different from e-learning, where the learning process can take place either sometimes with the help of the internet or in offline modes. However, M-learning acts as a substitute for other forms of e-learning due to their limitations, lack of affordability, flexibility, usage preferences among learners, etc., which are some of the capabilities of this mode of learning. On the contrary, mobile learning as one form of e-learning has quickly become the easiest and most convenient method of learning even in low-income countries. Of course, with the increase in mobile users globally and mobile phones becoming more affordable to the common man, learning through mobile became more realistic to the masses, and many corporate organizations have encouraged their employees to undertake professional development through M-learning. Comparatively, M-learning has multiple benefits to learners by providing uninterrupted learning, being self-paced, having normally high retention rates, being affordable and cost-effective, having the ability to learn through multiple devices, etc. Though in comparison, M-learning is not considered much valuable in terms of formal or e-learning among other forms of online education, there are certain components that are best suitable to be learned by this mode such as gamification, simulations, video learning, etc. that are all other technological applications that can sync with this type of learning.

3.4.2.3 Immersive Learning

Another important development in online learning that has recently gained popularity is the immersive learning method. Immersive learning is similar to the blended learning approach where the teaching is integrated with technology but differs in how the digital technology applications are imbibed into the learning process [30]. Immersive learning mainly features using digital media to create a more realistic simulative kind of environment for the learner to get completely involved and immersed in the learning. This method of teaching and learning can be employed both through online and offline modes depending on the application used. While immersive learning is a new and innovative approach to learning, there are different strategies that can be employed under this approach for teaching such as using simulation methods to explain the content, gamification, storytelling, VR, augmented reality, mixed reality, extended reality, etc., which are ways in which immersive learning is being adopted in teaching and learning. These strategies of learning are being used in different branches of science in higher education institutions currently, and the prospects for immersive learning in the future are going to transform the education system profoundly. It is not only going to innovate the educational sector but also has attracted corporates, manufacturing, and service sectors globally in adopting immersive learning in their business operations. There are many studies published on the use of immersive learning methods in the non-educational sector, such as Walmart's leading retailer stores in the United States have used VR tools in the immersive learning method to provide strategic solutions to meet the customer rush in its 20 stores, especially during Black Friday. United Parcel Service (UPS) has also adopted immersive learning methods for parcel delivery drivers to simulate real-life road traffic conditions using VR technology to safely deliver packages. Besides, the immersive learning method is also widely being used in health care and diagnostics, telecommunications, the food industry, IT, automotive, oil and gas sector, etc. in different applications and processes.

3.5 DISCUSSION

The education system obviously has transformed, prominently in the area of teaching and learning methodologies, which has become more technology-driven with the incorporation of the digitalization concept. This is more felt predominantly during the COVID pandemic that has forced educational institutions to adopt online education. Likewise, many HEIs across the world have been driven toward online teaching and learning as a result of COVID, such as educational institutions in Australia have adopted emergency online teaching (EOT), though there are some challenges institutions have to overcome in switching over to online education suddenly on a massive scale [31]. A case study done on electrical engineering students in online learning shows increased participation among students during the online approach; however, the study also indicates that some tasks have to be seriously given consideration to ensure online education becomes more affordable and effective in employing it [32]. On the other front, there are also other advancements in the field of technology application in the education system, predominantly the use of digital technologies in knowledge and skill dissemination. This is not only confined to educational institutions but has also expanded to industries and organizational and corporate entities where different innovative learning and teaching mechanisms are explored in skill development. For instance, there is an increased use of AI in almost every field of science, in academics, and in industries through ML using natural language processing (NLP) to communicate and converse with AI. The Georgia Institute of Technology and Georgia State University have successfully used conversational AI to respond to the more than 10K messages posted by students on academic doubts. Universities have now started to adopt AI applications to the educational sector through conversational AI to interact and respond to student queries related to subject questions, evaluate assessments and feedback, proposing study material through artificial teaching assistants (ATAs). Micro-learning, gamification, extended reality, and blockchain are some of the future trends that are going to take a significant lead in online education, which are likely to be explored extensively by the educational sector as well as the industrial sector.

3.6 CONCLUSION AND FUTURE SCOPE

The education system is one of the sectors that has been greatly affected due to the COVID-19 pandemic, impacting the teaching and learning of millions of learners globally. The sudden disruption and closure of face-to-face teaching and learning support have affected the students' knowledge and skills, eventually leading to learning loss in all age-group learners. The crisis created a significant gap due to a lack of acquiring the requisite skills for the job market. To regain knowledge and skills, educational communities must continue teaching and learning undisruptive to recover from the learning loss. In this journey to impart education without effective teaching and learning processes, online education using digitally enabled technology was the best alternative explored to serve millions of learners. This mode of online teaching and learning in education has gradually momentum to take forward the face-to-face classroom teaching that got disrupted due to the COVID-19 pandemic. This sudden transition in the way teaching and learning are imparted on a wide scale across academic institutions around the world has subsequently imposed challenges and exposed the limitations of teaching and learning taking place in an online environment. As a result, the research study specifically focused on knowing the different limitations and challenges of online teaching and learning experienced in HEIs in the region. While the study provides a wealth of knowledge and understanding and adds substantial information to the literature on e-learning education, the outcomes can help researchers and academicians to further analyze the effectiveness and efficiency of online education in a diversified context.

In response to the changes brought in teaching and learning, most of the HEIs were seen to have taken cautious steps and consistent efforts in understanding the need to shift the teaching and learning mechanism to the online environment; as a result, the majority of the institutions have adopted to put into practice various tools, technologies, and techniques available for teaching in an online environment. However, it should also be noted that shifting to online teaching and learning of all courses and programs in a short period has been a significant challenge faced by academic institutions, especially when such unforeseen situations are not anticipated. Even so, a majority of the educational institutions have adapted to the crisis by providing training/workshops to faculty on how to teach effectively in the online environment, facilitating resources by the management, developing mechanisms and procedures for evaluating students' performance, reviewing policies for academic appraisal, preparing alternative methods for practical learning.

Likewise, in an online teaching and learning environment, student participation and active involvement, evaluation of performance, absenteeism, and insufficient knowledge on the use of online tools and technologies both by instructors and students are some of the challenges that are significantly felt by institutions, and these must be addressed to ensure teaching and learning through the online environment are taking place effectively. Again, there are limitations in using technological tools for teaching and learning; foremost of them is the efficiency of internet connectivity which disrupts the teaching engagement, restricted access to some of the technological tools, ineffective ways to gauge academic integrity, inability to exhibit the capabilities and skills of learners, etc. are some of the recurring issues that should be seriously viewed by the institutions to resolve in an online mode of education. On the contrary, there is also tremendous scope to explore the concept of online teaching and learning which can be implemented to overcome the challenges and limitations, such as introducing blended learning as an innovative teaching approach, enhancing the skills of instructors and students in utilizing the e-learning tools and techniques, inculcating holistic learning through practical training, and ensuring there are proper guidelines, guidance, and benchmarks available to assess and evaluate the learning from each of the approaches.

3.6.1 Outcomes to the Future of Online Education

The pandemic has knowingly or unknowingly let the education system take a different path of teaching and learning across all educational institutions. This lead the educational system and learners to get familiarized with online education which is now the mainstream of educational systems worldwide. The online education system as part of e-learning involves the integration of ICT, which incurs a certain amount of financial expenditure on educational institutions; it is recommended that large educational institutions may have better scope to imbibe e-learning more effectively into their curriculum than newly established institutions. Studies in this direction are also very important to assess the degree of educational transformation in online teaching and learning. In addition, these objectives can be studied in different contexts categorized at region-level institutions and national-level institutions to draw comparisons between private, public, and university-level academic institutions. Then, the common and institution-specific issues, limitations, challenges, and opportunities that arise in the e-learning mode of teaching and learning among the various level of HEIs can be assessed. The outcome of the study shows that the level of advancement toward online learning because of the COVID-19 pandemic would stay along with traditional classroom teaching and together gradually modernize the educational system.

3.6.2 Future Scope

Undoubtedly, the COVID-19 pandemic has provided opportunities by opening new possibilities to explore in academics. This chapter presents the impact of the COVID-19 pandemic on the educational sector, particularly hindering face-to-face teaching and learning, and it also describes the different methods and approaches HEIs have adopted to overcome the challenges. In addition, the chapter also outlines the various technological initiatives that have taken ingress into the education system, highlighting some of the new trends and developments in the teaching and learning approach. While the chapter mainly attempts to contrast the educational system before and during the COVID-19 pandemic with reference to online or e-learning education, empirical studies are further required to be done in the context of different economies to gauge whether an effective educational transformation has taken place. In this pursuit, studies at local, national, and global levels especially targeting the higher education sector are very important to assess the transformation of the education system in order to compare the extent to which online education as part of e-learning is progressing in different economies and identify the indicators contributing the development.

REFERENCES

1. UNESCO. (2020). *Sustainable Development Goals- Formal Education*. UNESCO. http://uis.unesco.org/node/334633.
2. Searight, E. (2021). 15 online education statistics to help you understand the industry. ESL authority. https://eslauthority.com/blog/online-education-statistics/.
3. Dixon, A. (2019). *The Education Crisis: Being in School Is Not the Same as Learning*. The World Bank. Available From: https://www.worldbank.org/en/news/immersive-story/2019/01/22/pass-or-fail-how-can-the-world-do-its-homework.
4. UNICEF. (2021). Situational analysis on the effects of and responses to covid 19 on the education sector in Asia. Regional Synthesis Report. Available From: https://www.unicef.org/eap/reports/covid-19-education-situation-analysis-asia.
5. Sabine, M., Julian, F., & Rolf, S. (2022). The impact of the covid-19 pandemic on education: International evidence from the responses to educational disruption survey (REDS). UNESCO. Available From: https://unesdoc.unesco.org/ark:/48223/pf0000380398. ISBN: 978-92-3-100502-2.
6. Spielman, A. I., & Sunavala-Dossabhoy G. (2021). Pandemics and education: A historical review. *Journal of Dental Education*, *85*(6), 741–746. https://doi.org/10.1002/Jdd.12615. Epub 2021 Apr 19. PMID: 33876429.
7. Keengwe, J., & Kidd, T. T. (2010). Towards best practices in online learning and teaching in higher education. *Journal of Online Learning and Teaching*, *6*(2), 533–541.
8. Ch, S. K., & Popuri, S. (2013). "Impact Of online education: A study on online learning platforms and EDX. *2013 IEEE International Conference in MOOC, Innovation and Technology In Education (MITE)*, Jaipur (pp. 366–370). https://doi.org/10.1109/MITE.2013.6756369.
9. ICEF Monitor. (2012). 8 countries leading the way in online education. ICEF Monitor. https://monitor.icef.com/2012/06/8-countries-leading-the-way-in-online-education/.
10. Sass, E. (2019). *American Educational History: A Hypertext Timeline*. Retrieved From http://www.eds-resources.com/educationhistorytimeline.html.
11. Tate, B. (2012). A history of education timeline. Retrieved From https://www.tiki-toki.com/timeline/entry/56733/a-history-of-education-timeline/#vars!date=2014-12-05_17:43:25!.
12. Sharma, A., & Alvi, I. (2021). Evaluating pre and post covid 19 learning: An empirical study of learners' perception in higher education. *Education and Information Technologies*, *26*, 7015–7032. https://doi.org/10.1007/s10639-021-10521-3.
13. Donnelly, R., Patrinos, H. A., & Gresham, J. (2021). The impact of covid-19 on education – Recommendations and opportunities for Ukraine. The World Bank. Available From: https://www.worldbank.org/en/news/opinion/2021/04/02/the-impact-of-covid-19-on-education-recommendations-and-opportunities-for-ukraine.
14. Malganova, I. G., Dokhkilgova, D. M., & Saralinova, D. S. (2021). The transformation of the education system during and post COVID-19. *Revista on line de Política e Gestão Educacional, Araraquara*, *25*(1), 589–599. https://doi.org/10.22633/rpge.v25iesp.1.14999.
15. Ahlgren. E., Azevedo, J. P., Bergmann, J., Brossard, M., Chang, G. C., Chakroun, B., Cloutier, M. H., Mizunoya, S., Reuge, N., & And Rogers, H. (2022). The global education crisis is even worse than we thought. Here's what needs to happen. World Economic Forum. Available From: https://www.weforum.org/agenda/2022/01/global-education-crisis-children-students- covid19/.
16. Burgess, S., & Sievertsen, H. H. (2020). Schools, skills, and learning: The impact of COVID-19 on education. https://voxeu.org/article/impact-covid-19-education.

17. Morales, V. J. C., Moreno, A. G., & Rojas, R. M. (2021). The transformation of higher education after the covid disruption: Emerging challenges in an online learning scenario. *Frontiers in Psychology, 12*. https://doi.org/10.3389/fpsyg.2021.616059.
18. UNESCO (2020). Covid-19 educational disruption and response. UNESCO. Available From: https://en.unesco.org/covid19/educationresponse.
19. Ministry of Education (2020). Covid-19: Tertiary student support package, New Zealand government website. https://www.beehive.govt.nz/release/covid-19-tertiary-student-support-package.
20. Department for Education (2020). School funding: Exceptional costs associated with coronavirus (covid-19) for the period March to July 2020- GOV.UK. https://www.gov.uk/government/publications/coronaviruscovid-19-financial-support-for-schools/school-funding-exceptional-costs-associated-with-coronavirus-covid-19-forthe-period-march-to-july-2020.
21. Negash, S., Kartschmit, N., Mikolajczyk, R. T., Watzke, S., Matos Fialho, P. M., Pischke, C. R., Busse, H., Helmer, S. M., Stock, C., Zeeb, H., Wendt, C., Niephaus, Y., & Schmidt-Pokrzywniak, A. (2021). Worsened financial situation during the covid-19 pandemic was associated with depressive symptomatology among university students in Germany: Results of the covid-19 international student well-being study. *Frontiers in Psychiatry, 12*, 743158. https://doi.org/10.3389/Fpsyt.2021.743158.
22. Horváth, D., Ásványi, K., Cosovan, A., Csordás, T., Faludi, J., Galla, D., Komár, Z., Markos-Kujbus, É., & SiVmay, A. E. (2022). Online only: Future outlooks of post- pandemic education based on student experiences of the virtual university. *Society and Economy, 44*(1), 2–21. https://doi.org/10.1556/204.2021.00026.
23. Allen, I. E., & Seaman, J. (2011). Going the distance: Online education in the United States. The Online Learning Consortium. Retrieved From http://Sloanconsortium.Org/Publications/Survey/Going_Distance_2011.
24. Roffe, I. (2004). Innovation and E-Learning: E-Business for an Educational Enterprise. Cardiff, UK: University Of Wales Press.
25. Friedman, J., & And Wood, S. (2022). Discover 15 current online learning trends. Available From: https://www.usnews.com/higher-education/online-education/slideshows/discover-10-current-trends-in-online-education?slide=3.
26. Mariia & Julia. (2021). Online education trends: Elearning predictions for 2022. Available From: https://www.aimprosoft.com/blog/elearning-trends-in-online-education-and-learning/.
27. Wood, J. (2021). Impact report. Serving the world through learning. Available From: https://www.weforum.org/agenda/2022/01/online-learning-courses-reskill-skills-gap/.
28. Bezhovski, Z., & And Poorani, S. (2016). The evolution of e-learning and new trends. *Information and Knowledge Management*, 6(3), 50–57.
29. Verpoorten, D., Huart, J., Detroz, P., & Jérôme, F. (2020). Blended learning in higher education: Faculty perspective through the lens of the planned behaviour theory. In (Ed.), *E-Learning and Digital Education in the Twenty-First Century*. Intechopen. https://doi.org/10.5772/Intechopen.95375.
30. Ordov, K., Madiyarova, A., Ermilov, V., Tovma, N., & Murzagulova, M. (2019). New trends in education as the aspect of digital technologies, *International Journal of Mechanical Engineering and Technology, 10*(2), 1319–1330. http://Www.Iaeme.Com/IJMET/Issues.Asp?Jtype=IJMET&Vtype=10&Itype=2.
31. Lorenza, L., & Carter, D. (2021). Emergency online teaching during COVID-19: A case study of Australian tertiary students in teacher education and creative arts. *International Journal of Educational Research Open, 2*, 100057. https://doi.org/10.1016/j.ijedro.2021.100057.
32. Sunardiyo, S. (2021). Online learning in higher education during the Covid-19 pandemic: A case study in the department of electrical engineering, Universitas Negeri Semarang. The 9th Engineering International Conference, *IOP Conference Series: Earth and Environmental Science*, 700, 012020. https://doi.org/10.1088/1755-1315/700/1/012020.

CHAPTER 4

Factors Affecting E-Learning System in Management Students during COVID-19 in India

Pooja Darda, Om Jee Gupta, and Meenal Pendse

Dr. Vishwanath Karad MIT World Peace University

CONTENTS

DOI: 10.1201/9781003322252-4

4.1 INTRODUCTION

The worldwide impact of the COVID-19 epidemic cannot be overstated. In the wake of COVID-19's impact, several educational institutions have been forced to close for an unknown period. India is not an exception. Since the second week of March, conventional face-to-face teaching and learning have been discontinued. Because of the COVID-19 lockdown, educational institutions have turned to online teaching learning as an alternate means of keeping students, teachers, and staff safe in the event of a public health disaster (Martinez, 2020). The abrupt transitions from traditional face-to-face classroom instruction to online learning have an impact on the students' ability to learn. Most universities, particularly those financed by the state, are not prepared for the online delivery of their courses. Online education is a cognitive burden due to several factors, including inadequate infrastructural facilities, a lack of knowledge and abilities among teachers, students' lack of preparation, and inadequate course design. According to Bhaumik and Priyadarshini (2020), the term for this concept is 'online load'.

A national emergency was declared for most countries at the start of 2020 because of the COVID-19 pandemic. There were several emergency management methods that were put in place across the world due to the pandemic (Zhang et al., 2020). Cities have been put on lockdown, schools have been shut down, and harsh social distancing measures have been implemented by governments around the world, including the US. As a result of the pandemic regime, businesses, scientists, cultural events, management, and educational institutions have moved their operations online (Sulkowski, 2020). There has been a shift in technology and software infrastructure as also social behaviors and communication. According to Kulikowski et al. (2022), COVID-19 had a considerable impact on the main social behaviors and communication channels because of its quick expansion.

As of the 24th of May in the year 2020, there were around 1.72 billion students who were impacted as a direct result of the closure of colleges and universities in response to the pandemic caused by the coronavirus (UNESCO, 2020).

Educational institutions are fast embracing social media and the internet as a means of communication (Leonardi, 2020). More than 99.9% of the world's student population is affected by the closures of 192 educational institutions in 192 nations, according to the UNESCO (UNESCO Report, 2020). When educational institutions in China began closing and moving to online learning, the Chinese government began a strategic initiative called 'Suspending Classes without Stopping Learning' (Zhang et al., 2020). Other nations soon followed suit. In schools and universities around the world, e-learning, online learning, and online learning management systems (LMSs) are changing and altering education delivery techniques dramatically (Hodges et al., 2020). According to Dhawan (2020), academic institutions that had previously been reluctant to adapt their traditional pedagogical approach had little choice but to move toward online education.

Students faced several challenges and constraints because of the transition to the online world, which was a disruptive process. Students abruptly abandoned their campuses, giving themselves very little time to acclimate to the new norm; as a result, greater financial resources were required for learning that was conducted remotely (Aung & Khaing 2015).

Students had not had any official training on technology-based learning, and neither the students nor their lecturers had any prior experience with the technologies that they were experimenting with as they moved to an entirely online environment. It led to an unfavorable educational experience as a result. Students often rely on campus-based resources, like computer laboratories and Wi-Fi hotspots, to meet their needs. Students experienced anxiety because of this unexpected change because the university curriculum was designed for in-person learning, and educational institutions were not prepared for the abrupt change due to a lack of resources to support emergency remote teaching.

In this situation, most governments shut down schools and other facilities where it is impossible to prevent people from entering until further notification (UNESCO, 2020). Since the schools were forced to close as a precautionary measure in reaction to the outbreak on September 20, 2020, around 1.07 billion students had been adversely affected.

Even after the availability of vaccines, higher education institutions experienced several difficulties during the spread of COVID-19. Changes in teaching and learning settings were required to meet the challenge. Schools and universities shifted to virtual classrooms throughout the spread of the pandemic, which was the only option to implement the teaching methodologies and goals (Nuere & de Miguel, 2021; Salloum & Shaalan, 2018). The university environment has changed dramatically, and students now face a slew of new obstacles (Alhassan et al., 2021; MacQuarrie et al., 2021). In the second wave of COVID-19 (Aguilera, 2020; AlQudah et al., 2021), it will be necessary to create a framework for evaluating students' understanding, achievement, and success.

Microsoft Teams, Zoom, and other technologies have emerged as a new way of interacting with students and lecturers. For the successful running of lectures, the students were given adequate rules and recommendations for adapting to this new method of learning (Saxena, 2020). The low attendance rates and lack of learning intentions that have been documented in many studies of students' e-learning situations are concerning, to say the least (Chen et al., 2021). There are a variety of reasons for this, including the difficulties that students and academics alike have when making the switch from traditional to online education at such a quick pace. There are several technological challenges from the student's perspective, such as slow internet speeds, login ID troubles, unheard-of sound, video, and so on (Ekanayake & Weerasinghe, 2020; Andersson, 2008). Even more, students find e-learning environments to be tedious and less participatory. As a result, the study's goal is to better understand the factors impacting Indian management university students' use of e-learning during the COVID-19 pandemic so that academics and other administrators can take remedial action to improve e-learning.

India's Human Resource Development Ministry (MHRD) has recommended higher education institutions continue online teaching and learning considering the lockout that has occurred. Additionally, these online classes (e-learning) are used to keep students in touch with one another, boost their self-esteem, and boost their confidence throughout the COVID-19 pandemic (Kaur et al., 2020). The COVID-19 pandemic necessitated the adoption of digital media by universities, as well. Because of the internet and the World Wide Web, the number of students who can have access to educational resources has risen drastically around the world. Consequently, online education now offers a wealth of information in a variety of media and can facilitate real time and delayed contact between students, teachers, and even among students themselves.

Even though the use of e-learning in schools of higher learning is continuously growing, academic attention is being paid to comprehending the prospective context that lags behind the industry. There are not many previous research studies that have been conducted on the factors that influence e-learning in higher education institutes (for example, Mosunmola et al., 2018; Mei et al., 2018; Tseng et al., 2022), and even those that have been conducted dealt with more incremental transitions from traditional learning to e-learning, creating a gap in the literature on the adoption of e-learning by students, especially in current contexts (Chen et al., 2021). The current study aims to determine the characteristics that influence the adoption of e-learning by Indian management university students during the COVID-19 pandemic (Tandon, 2021). Taking into mind the timeliness and importance of the phenomena, the study's primary focus is to identify these elements.

4.1.1 Overview of E-Learning

Distance education is now more accessible because of recent technological advancements (McBrien et al., 2009; Abbad et al., 2009). Most of the time, words like web-based learning, online learning, blended learning, computer-mediated learning, open learning, and m-learning are used. Any computer connected to a network can be used; therefore, it is possible to do so from any location at any time (Cojocariu et al., 2014).

Learning over the internet is a tool that has the potential to make the teaching and learning process more student-centered, inventive, and adaptable. Learning activities that take place in both online and offline platforms can be accessed by a variety of devices (including mobile phones, laptops, and desktop computers) that are connected to the internet. This is what is meant by the term 'online learning'. Students could independently study on this platform while also interacting with their teachers and with other students (Singh & Thurman, 2019; Mukhopadhyay et al., 2020). Students can structure their learning on the synchronous learning platform by attending live lectures and engaging in real-time communication with their respective lecturers. On the other hand, asynchronous learning platforms are not properly designed, but this method offers the potential for immediate feedback. The information that students need to learn is not delivered through traditional classroom settings but rather through a variety of online and offline learning systems and mediums. Under such a platform, it is not possible to provide immediate feedback or an immediate answer (Littlefield, 2018). Learning in real time provides students with a wealth of options for social interaction (McBrien et al., 2009).

Various learning environments, such as face-to-face learning, distance education, and online education, are available for students to further their education. Most people confuse the phrases 'online learning' and 'e-learning', although there is a distinction between the two. In e-learning, web-based technology tools may be employed in or outside the classroom (Maheshwari & Thomas, 2017; Nichols, 2003). The term 'online learning' refers to learning that takes place entirely online, when students are not in the classroom, and is regarded to be akin to distant learning (Oblinger et al., 2005). Learning outcomes can be achieved more quickly and easily using online delivery modes (Junco et al., 2013).

Technology, user-friendliness, and class activities and assessments all have an impact on the success of online learning (Wijekumar et al., 2006; Shuey, 2002). As a result of the COVID-19 outbreak, many educational institutions have been forced to switch from face-to-face instruction to online instruction. As a result, it is imperative that these institutions learn what factors influence student satisfaction and their future intention to take courses online. They must be ready for anything that may come their way in the years ahead. The purpose of this research work is to find out various determinants that affect the adoption of an e-learning system for management students during COVID-19.

4.1.2 India's E-Learning Education Landscape

Young people make up roughly half of the Indian population (Sharma, 2020). Many e-learning service companies have a new customer base because of this. Additionally, students in India are apprehensive about the new educational platform that utilizes technology (Bhattacharya & Sharma, 2007). The interactive learning environment is a way for students to improve their ability to learn (Yuen & Ma, 2008). The students believe that e-learning has eliminated geographical and budgetary constraints and provides a better-quality education (Ilin, 2020). As evidence, India has a higher percentage of students enrolling in online courses than the United States (Sharma, 2020). Brandfinance.com (2019) reports that India is the second-largest country enrolling in online courses. To begin with, India has a high level of internet usage. India comes in second to China, according to Internetworldstats.com (2019). It is because the number of Indians using smartphones is on the rise. It was found that the other factors included in this study are internet access and cheap data plans, as well as the government's efforts to raise awareness. Through the 'Digital India Campaign', India's government is promoting a digital awareness initiative for school and university students (Nedungadi et al., 2017). As a result, it raises the level of digital literacy and encourages people to pursue online courses (Bharucha, 2018). As a third benefit, the University Grants Commission Act has allowed for an expansion of India's educational landscape, resulting in the addition of several new state-funded technical schools. To keep up with the increasing expansion of educational institutions, more employees and technology are needed. There is also a widening division between teachers and students in terms of their awareness of the latest technical advancements. Educators and experts, particularly in higher education, are in high demand because of this (Nneka Eke, 2010). The National Board of Accreditation research states that many colleges opt to fill this gap by providing online learning solutions. However, the Indian Government has created various schemes like 'Digital India', which give a strong technical foundation. That is why the country's higher education establishments are open to the idea of implementing e-learning (Bliuc et al., 2007; Garcia-Penalvo et al., 2018; Ellis et al., 2009; Reeves, 2000).

4.1.3 E-Learning Is a Necessity, Not a Choice

No one was safe as long as the commands to 'Stay Home and Social Distancing' were being followed. Because of the shutdown, students were unable to leave the building. The fact that India has the largest population in the world presents a significant number of chances for the education sector. Most of the instructions for students at schools, colleges, and other types of institutions now are given online. Lecturers and instructors are constantly putting up online webinars and meetings. Since institutions' doors were shut, there has been a dramatic increase in the demand for online classes and instruction. It exemplifies the large and rapidly expanding demand for online educational opportunities. The use of online instruction as the primary educational approach in many nations has become widespread in recent years. The sole requirements for the online teaching approach are a dependable internet connection and a laptop computer or cellphones, as appropriate. Additionally, students are not required to move from one location to another, which results in a significant reduction in the amount of time spent traveling. Because the student is able to retain more information as a result of their travels, they will not experience feelings of depression. Students always have a pleasant area to focus on when they learn from home since they can choose the setting that is most conducive to their learning. As a result of the significant disruption that COVID-19 has caused to the school year, the importance of learning through virtual means is growing. Because students might be preschoolers or people at higher levels of education, online teaching brings a lot to the table for learning for all of them. A great number of organizations that specialize in digital technology, such as Google Classroom, Zoom, Microsoft Teams, and Blackboard, have been instrumental in the process of transition (Adeoye et al., 2020).

4.1.4 Online Education's Interchangeable Terminology

In online learning, researchers often use the same definitions for a variety of different concepts. Originally, the phrases 'distance education' and 'distance learning' were used interchangeably to denote whether classes were delivered by computer technology and at a distance from the instructors and students, respectively (Belanger & Jordan, 1999). When it comes to distance learning and the use of technology, terms like open distance learning, web-based training (WBT), computer-based training (CBT), technology-based learning, and e-learning have been viewed as interchangeable with the term online learning (Rovai, 2008; Rovai & Downey, 2010, 2009; Sharp, 2009), and online learning is a subset of distance learning (Carliner, 2004). Learning can be classified into four broad categories by Allen and Seaman (2014): traditional (completely in-person), web-assisted, blended/hybrid, and entirely online. Traditional courses are taught face-to-face and do not use any technology at all. A course is considered web-assisted if online technology is used for between 1% and 29% of the class time. Hybrid courses are those that include online learning technology, such as an online discussion board, with face-to-face seminars for between 30% and 80% of the course content. As a result, if more than 79% of the course is conducted online, it is said to be entirely online. Online discussion and web conferencing are the primary methods of education in a fully online course as there is no need for students to meet face-to-face with their instructors.

4.1.5 E-Learning Platforms in India during COVID-19

In a recent study, 90% of schools reported employing pedagogic software tools, 72% used live-streaming videos, and 40% offered connections to progress with online materials (Emami, 2020). There is a growing reliance on electronic channels for theoretical information around the world, including email, educational tools, Skype, Facebook, Telegram, and Google. An approved alternative to classroom activities is offered by educational institutions via internet educational platforms. Many schools make substantial use of the Moodle platform, for example. To enhance learning outcomes and raise student performance in line with international standards, Moodle has created an online community dedicated to promoting blended learning. Moodle can be used to provide educational content, assignments, and assessments, as well as to facilitate one-on-one interactions with students. Students in low-income locations revealed that Moodle provides formative e-assessments that would otherwise be unavailable and accommodates a variety of question styles and ability levels taught by instructors. A favorable performance was typically expected from a student; reactions are given the difficulties of technology and the concerns associated with stress (Passos et al., 2020). A high-quality internet connection and regular maintenance are required to keep those outlets in good working order. Moodle, a personal learning management system (PLS), was utilized by the universities to distribute educational information to students and teachers. There appears to be no evidence that these media sources have a significant impact on students' training and the development of their competence. LinkedIn and Pinterest are some of the other platforms that are commonly used. Video conference technologies, such as Zoom and jitsi, were used at several colleges to provide classroom lectures. Even though they rely largely on a good internet connection, both Zoom and WebEx can be fascinating alternatives to a traditional classroom. (Alshiekhly et al., 2015; Kumar et al., 2021). Student comprehension and retention were excellent. There were no significant changes made to the online platform's assessments, and facilitators rapidly picked up on the new learning strategy. Internet-based discussion on digital platforms provides a more relaxed discussion environment than face-to-face communication education modalities centered on the discussion. While using a robust and portable interface, their performance relies on some prior training (Machado et al., 2020). Like Twitter and other social media platforms like Zoom, it can be used in small groups to stimulate discussion and avoid unnecessary travel during times of social isolation or even social constraint (Das et al., 2020; Ha et al., 2019). It is at this point that new concepts and teaching methods can be united for online courses for primary infection control on the mobile app, which provides current information on infectious illnesses and prevention strategies to students (Wootton et al., 2020). Among other things, Microsoft Teams and Google Meet are often used learning systems, along with Google Classroom and Hangouts. With the option of sharing the presenter's screen with students (lecturer) and allowing for a variety of instructional actions, Google meetings can accommodate up to 250 participants simultaneously. For future reference, all the participants' activities can be documented, logged, and saved to Google Drive. (Martin et al., 2012). During COVID-19, WhatsApp was a vital means of communication and a valuable resource for counselors. We can show greater results with instant multi-media messaging if we combine the reception and reaction timings of WhatsApp with those of regular email in educational platforms (Halpin & Lockwood, 2019). With Facebook, students can discuss subjects more freely and flexibly, with fewer time and location restrictions. The theoretical concept of educational methods can't be taught without it.

Objective of the Chapter

The objectives of the chapter are as follows:

- To determine the factors influencing the effects of the adoption of e-learning systems in management students during COVID-19 in India.
- This study intends to contribute to the growing body of knowledge in the field of literature and tries to incorporate new research variables connected to extrinsic factors into the many models that are currently in use.

Organization of Chapter

The chapter is organized as follows: Section 4.2 elaborates literature review. Section 4.3 discusses the Methodology. Data and Analysis are enlightened in Section 4.4. Section 4.5 entails discussions and implications. Limitations of the study are covered in Section 4.6. Finally, the chapter concludes with future scope in Section 4.7.

4.2 LITERATURE REVIEW

The developments in technology that have taken place recently have made it possible for us to design materials for the internet using a variety of different approaches. When developing online courses, it is essential to consider the interests and perspectives of learners. This will ensure that the courses facilitate learning that is both efficient and fruitful. The preference of the learner is connected to the learner's readiness or willingness to participate in collaborative learning and the elements that influence the learner's preparedness for online learning. In the following section, a concise summary of the key takeaways gained from the analysis of the relevant previous research is provided. In the field of Australian vocational education and training, Warner et al. (1998) presented the idea of preparedness for online learning as a potential factor. Students' preference for online delivery over face-to-face classroom instruction, their confidence in using electronic communication for learning, and their ability to engage in self-directed learning are the three main aspects they emphasized when describing their readiness for online education. McVay (2000, 2001) designed a 13-item scale, which examined students' behavior and attitude as predictors of test performance. To test the validity of the McVay (2000) questionnaire for online readiness, Smith and colleagues did an exploratory study and came up with the two-factor structure, 'Comfort with e-learning' and 'Self-management of learning'. More recent research (Evans, 1999; Smith, 2005) has attempted to put the concept of online learning readiness into action. Among the factors that influenced online learning's readiness were self-directed learning, motivation for learning, learner control, and computer and internet self-efficacy (Guglielmino, 1977; Garrison, 1997; Lin & Hsieh, 2001; McVay, 2000, 2001), and others. User perception is critical to any endeavor to improve online learning. Both positive and negative experiences with online learning have been observed by researchers. Students' opinions of online learning are strongly influenced by their interactions with their instructors, according to several research findings.

The design of the course needs continuous improvement (Swan et al., 2000) to have the ability to interact with course instructors to improve critical thinking and information processing

(Duffy et al., 1998, pp. 51–78; Picciano, 2002; Hay et al., 2004), the rate of interactivity in the online setting (Arbaugh, 2000; Hay et al., 2004), the amount of instructional emphasis on learning through interaction, and the flexibility of online learning (Chizmar & Walbert, 1999). Hence, a good online class needs well-organized course content (Sun & Chen, 2016), well-prepared instructors (Sun & Chen, 2016), modern technology (Sun & Chen, 2016), and feedback and clear instructions (Sun & Chen, 2016; Gilbert, 2015). Several researchers looked at how well online or web-based tutorials work compared to traditional classroom teaching. The kinds of interactions that could happen online are very different from those that could happen in a traditional classroom, and the effect of communicating in one setting or another can have a direct effect on how students and teachers feel. The studies looked at how students and teachers felt about online learning vs. traditional classroom learning (Sorebo et al. 2009). The results were mixed, which means that more research is needed. Some of these areas include analyzing the type and number of online interactions (Moore & Kearsley, 1996), the flexibility and accessibility of web-based instructions (Navarro & Shoemaker, 2000), the skills, motivations, time, and perceptions of the learner and instructor (Yong & Wang, 1996; McIsaac et al., 1999; White, 2003). It was also found that there was no big difference between online learning and in-person classes in how happy students were or how well they did in school (Hara & Kling, 1999). Studies have also shown that an online class can be just as effective as a traditional one if it is set up right (Nguyen, 2015). However, not many research papers had tried to figure out what are the factors that influence Indian students to adopt e-learning platforms. Before the COVID-19 pandemic, it makes sense that only a small number of distance education platforms used online learning. Also, as far as we know, research along these lines has not been done in the field of management education, where online learning projects are even less common, likely because they focus more on practical learning. We try to fill this gap with our study, which focuses only on online learning in management education and uses the literature to help us understand the problem.

Perception of Utility: Knowledge sharing and learning are the primary goals of online learning platforms. As we live in a globalized world, technology has become an essential tool for obtaining knowledge, getting information, and learning. Using and accessing these resources is simple, which makes it easier to share information. Online media and mobile devices play a vital role in the learning process, according to a wide range of studies. Easier access to online learning has increased its adaptability, which has led to better outcomes (Salloum et al., 2019; Erkan & Evans, 2016).

User-Friendliness: An individual's degree of belief that using a specific technology will be effortless is known as user-friendliness (Davis, 1989, p. 320; Davis et al., 1989). User-friendliness is the user's capacity to understand and handle the information system in an efficient and effective manner, without any misunderstanding or difficulty. It has been shown that when people believe that technology skills are simple to use, they are more likely to utilize them. This is in line with the findings of Sun et al. (2008), and Davis (1989). As a result, the following theory is put forth.

For students, e-learning tools are believed to be easier to use when they are perceived to be more effective.

A learner's perception of online learning's usefulness is the degree to which they believe it will help them improve their performance. To demonstrate the efficacy of online learning, learners can save time and money by using a variety of approaches (Erkan & Evans, 2016).

Numerous research has shown that students' attitudes and motivation are favorably impacted by the perception of usefulness, leading to better learning results (Habes et al., 2018; Alhumaid et al., 2020).

Faculty Capability: A learner-centered strategy rather than a teacher-centered one governs online education (Debattista, 2018). When it comes to creating online course materials for higher education, students benefit from a variety of factors, including pedagogical approaches, professional competence, science and technology application level, and the capacity to synthesize and combine various concepts.

When teachers are present and engaged with their students in a positive way, it can have a beneficial impact on their ability to teach and motivate them (Baker et al., 2010). Due to the rise of online education, instructors' workloads, responsibilities, and connections with their students are all being re-examined (Young, 2002). Students and professors can converse freely on this platform, which Harandi (2015) considers being an important part of learning. Student satisfaction with self-paced and online courses is strongly correlated with instructional quality. E-learning may be underutilized by students due to educators' hesitancy to use it, which may have an adverse effect on learning results, as stated by Sorebo et al. (2009). As Yildirim (2000) points out, ineffective or unimpressive technology can cause problems in the learning process; thus, teachers will discourage their students from using it as well as themselves.

Both technology and the deployment of technology have an impact on educational achievements, according to Collis (1991). Also influencing learning results are one's mindset and methods of instruction (Webster & Hackley, 1997). Control of technology by an instructor and time for interaction with students have been found to affect learning outcomes in previous studies (Arbaugh, 2000; Khan, 2005; Leidner & Jarvenpaa, 1993). Timeliness, self-efficacy, technological control, interaction focus, e-learning attitude, student attitude, distributive fairness, procedural fairness, and interaction fairness are all important aspects of a good instructor to consider (Arbaugh, 2002; Chiu et al., 2007; Liaw et al., 2007; Lim et al., 2007; Sun et al., 2008; Webster & Hackley, 1997).

Course Material: Students are more likely to participate and take initiative in class when the subject is interesting and interesting to them (Ashwin & McVitty, 2015; Little & Knihova, 2014; Khamparia & Pandey, 2018; Akyüz & Samsa, 2009). E-learning content encompasses the organization and substance of learning materials' chapters. The e-learning package also offers supplemental materials to help students better comprehend and retain knowledge (Khamparia & Pandey, 2018). Students' analytical and critical thinking as well as problem-solving skills develop because of this component.

Effective educational experiences can be achieved through well-designed courses and curricula (Brophy, 2000). This term refers to the degree to which online course contents are accurate, comprehensive, understandable, and relevant to students' learning goals (Chiu et al., 2007; McKinney et al., 2002). The quality of information is determined by the degree to which it is accurate, timely, complete, relevant, and consistent (DeLone & McLean, 2003). Learning is a multifaceted process that is influenced by a variety of factors, including the quality of the curriculum and the teaching resources used (Sharma & Kitchens, 2004). Researchers Ozkan and Koseler (2009) discovered that a positive correlation existed between student satisfaction and the quality of the material provided. Content relevancy, course quality, and course adaptability are all characteristics of information quality (Arbaugh, 2002; McKinney et al., 2002; Sun et al., 2008)

Course Layout: Students and lecturers can communicate with each other via e-learning forums, which feature a course framework, course creation interface, and testing and assessment methods. Student participation in online classes can be facilitated by a strong course design (Oh et al., 2020). Students' competency and level of understanding are considered when designing the course content, which is tailored to their needs in terms of time and space. (Ahmad et al., 2018; Ricart et al., 2020).

Student Capability: To improve the quality of online learning, social contact between students and lecturers is essential. It is possible to achieve the efficacy of online learning (roughly strong interaction and consistent practice (Jung et al., 2002; Shih et al., 2018). Since the rules and requirements of online learning are more comfortable, proactiveness, self-study ability, and a sense of conformity are also crucial requirements to get greater results in learning. Traditional procedures are more difficult to manage than the (e) process.

Student capability as a concept refers to students' aptitude, abilities, and self-efficiency in using university-provided e-learning tools. It has been widely accepted that computer self-efficacy is a crucial notion in the increasing adoption of technology studies, according to Chien (2012) and Holden and Rada (2011). A person's opinion of their competence to utilize computers in diverse contexts is called computer self-efficacy, according to Thatcher and Perrewé (2002). Both Abbad and Al-Adwan agree that e-learning system users are more likely to be confident in their capacity to master the system on their own and hence more willing to use it. According to Schunk et al. (2014), self-efficacy is the most significant predictor of student performance and motivation.

Peer Influence: It is a term used to describe the motivation, communication, and cooperation among college students, whether or not instructors are present. As defined by Boud et al. (1999), peer learning occurs when students learn from one another without the immediate assistance of a lecturer or their assistants. Learning and engagement in an online debate can be enhanced by 'collaborative interaction between peer students', according to Jung et al. (2002, p.153). According to Paechter et al. (2010), students who study in small groups are more likely to have a deeper understanding of one another, which might spur them on to greater achievement in an online learning environment like Khan Academy. However, Juwah (2006) contends that the mastery of digital tools is vital for improved outcomes in terms of interaction among peers. There is evidence that students' adoption of e-learning is based on the encouragement of their peers and/or teachers, according to studies by Abbad et al. (2009).

Institutional Assistance: Institutional aspects in terms of support, such as administrative support, technological support, university support, and learner characteristics (motivation, self-regulated learning, and self-efficacy), is widely acknowledged as being significant (Yusoff &Yusliza, 2012; Kee et al., 2012). Moore and Kearsley (2011) suggested that the success of online courses is based on improvement, the provision of support resources, and on the encouragement of dialogue. It is anticipated that this will facilitate an improvement in the transactional distance according to the individual learner characteristics and felt needs (Stein et al., 2005).

4.3 METHODOLOGY

The researcher has adopted an exploratory research design to conduct this research work. Moreover, the researcher has used qualitative and quantitative research approaches to conduct this piece of work. First, to find out a broader theme, researchers distributed one open-ended question to 150 management students, i.e., 'What are the factors that affect the adoption of the E-learning system in management students during COVID-19 in India?'.

Each of the students was given 60 minutes to ponder over the question so that they can answer holistically. Out of 150 response sheets, approximately 11 students opted not to answer. Therefore, they were excluded from the study.

The researcher has also undertaken a thorough literature review on the subject matter. Based on the responses received and the existing literature review, the researcher has prepared a questionnaire of 32 statements. The researcher has also framed one negative statement to check the unengaged responses (Podsakoff et al., 2003). Each of the statements was measured using a 7-point Likert scale (1-Strongly disagree, 2–Disagree, 3–Somewhat disagree, 4–Neither agree nor disagree, 5–Somewhat agree, 6–Agree, 7–Strongly agree) (Vagias, 2006).

Thereafter, all 32 statements are placed before four management professors and five EdTech experts. Their opinion was considered and suggested modifications were made to the questionnaire. Using this combination of academia and industry experts, content validity was established. This final questionnaire was distributed using online and offline modes to four prominent business schools. Management graduates are considered as sample units. Questionnaires were distributed to 400 management students, and few of the questionnaires were found either half-filled or unengaged. These unfilled or unengaged questionnaires were omitted from the actual data considered for the study; finally,

333 questionnaires were found to be adequately filled and hence found suitable for the study. Then, the data of (333) was structured, tabulated, and analyzed using SPSS 20.0 software (Zikmund et al., 2013).

The researcher has conducted exploratory factor analysis (EFA). Researchers have used principal component analysis and the Varimax rotation method in SPSS 20.0 to extract factors out of 32 statements. It was found that two statements were not converging to (factor loadings <0.4) any of the factors appropriately and hence were removed from the analysis. Therefore, finally, 30 statements were considered for the study.

Below are the statements considered for the study:

- I have the skills to learn how to use e-learning facilities. (1)
- I am capable of using the different facilities provided to me. (2)
- E-learning is easy and quick to get adapted to the new technology. (3)
- Enhanced motivation and learning style are provided by e-learning. (4)
- The e-learning material is clear and well-structured. (5)
- The e-learning content is full of knowledge. (deleted) (6)
- The e-learning content is skill-oriented and meets future requirements. (7)
- The e-learning content is comprehensive. (8)
- E-learning enhances the student in critical thinking, analysis, and problem-solving. (9)
- The course design is suitable for e-learning. (deleted) (10)
- Learning outcome of the course can be met through e-learning. (11)
- The sufficient classwork and assignments conducted by e-learning. (12)
- Online course curriculum meets modern requirements. (13)
- Peers motivate me for e-learning. (14)
- Peers recommend me for e-learning. (15)
- E-learning facilities become easy with the help of peers. (16)
- I interact with my classmates on the use of e-learning systems. (17)
- Teachers have a lot of experience with e-learning resources and their applications. (18)
- Instructor facilitates classroom community with engaging tone and creativity. (19)
- Faculties are accessible and responsive in e-learning. (20)
- E-learning facilitate instructor in the precise conducting of summative assessments. (21)

- E-learning technologies allow me to execute my learning activities more effectively. (22)
- Using e-learning systems can improve my level of interaction with my professors and classmates. (23)
- Using e-learning technologies can enhance my learning. (24)
- Improved virtual collaboration and communication help me in my future endeavors. (25)
- I may have problems utilizing the university's suggested e-learning system. (26)
- I easily understand the learning system's interface and functionalities. (27)
- E-learning has added flexibility and self-paced learning for me. (28)
- Members of the IT department are always accessible to assist students in need. (29)
- The institute provides online portals to access the textbooks and reference materials. (30)
- The necessary instructions for better use of the e-learning system are available to me. (31)
- E-learning-related training and assistance are provided by the institution. (32)

4.4 DATA ANALYSIS AND FINDINGS

Demographic profile of respondents: The analysis was started after the tabulation of data in an Excel file. After the data cleaning process, the researcher calculated the demographic profile of the respondents. Out of 333 respondents, 183 (54.95%) were male, whereas 150 (45.05%) were female respondents (Table 4.1).

4.4.1 Sampling Adequacy Test

The Kaiser–Meyer–Olkin (KMO) was used to test the sampling adequacy of the data, KMO = 0.923, whereas the KMO value of all the individual statements was also calculated using the anti-image option in SPSS, and it was found to be more than 0.89 except for two statements which later on dropped from the analysis because of low factor loading. This result was well over the required criteria of ≥0.6 (Field, 2013). Then the researcher calculated the Bartlett test of sphericity; $\chi^2 = 9067.82$, $df = 435$, $P < 0.001$. In order to successfully complete the factor analysis, we must have some sort of link between the statements in order for there to be a convergence of components. This means that the correlation between statements was adequately high for principal component analysis.

According to the research that has been done, if the results of the KMO test move in the positive direction (≥0.5), and the results of the Bartlett test are significant ($P \leq 0.001$),

TABLE 4.1 Gender

Gender	No. of Observations
Male	183 (54.95)
Female	150 (45.05)

TABLE 4.2 KMO and Bartlett's Test

Kaiser–Meyer–Olkin Measure of Sampling Adequacy		0.923
Bartlett's Test of Sphericity	Approx. Chi-Square	9067.824
	Df	435
	Sig.	0.000

it indicates that the study will be successful and that the prerequisites for performing EFA have been met. As a result, EFA is regarded as an appropriate method for the study (Hair et al., 2010) (Table 4.2).

4.4.2 Exploratory Factor Analysis

The researcher has used the Varimax rotation method with the Kaiser normalization technique and principal component analysis as an extraction method to conduct EFA.

The number of factors was determined using the criteria of the initial eigenvalue. As per the criteria, if the eigenvalue is more than 1, then we can take that as a factor (Field, 2013). Table 4.3 and Figure 4.1 apparently show that eight factors have an eigenvalue of more than 1, where 30 statements have converged into eight factors.

The researcher has identified eight factors, where each factor explains the percentage of variance:1st factor=41.40%; 2nd factor=50.83%; 3rd factor=58.08%; 4th factor=64.44%; 5th factor=70.09%; 6th factor=74.50%; 7th factor=78.24%; 8th factor =81.75%; cumulatively, all four factors combined together explain 81.75% of the data.

4.4.3 Factor Rotation Analysis

Table 4.4. illustrates that all eight factors are loaded with 30 statements in this study. There is no overlapping of factor loading that shows a particular statement's loading is only considered for a particular extracted factor. All the loadings to the extracted factors are found to be in the range of 0.885–0.714 (Field, 2013). All 30 statements adequately and significantly contribute to their respective construct, i.e., extracted factors. Extracted factors 1 to 6 are contributed by four statements, whereas factors 7 and 8 are contributed by three statements each. The Varimax technique under the rotation method helps in eliminating the probability of dual loadings from a single statement. The rotation method improves the loadings of each statement to the specific extracted factor and minimizes the loading on the remaining other extracted factors. Looking at their properties, we can make some genuine common proposition that carries weight for all the respective items.

4.4.4 Extracted Constructs

The researcher has brought all the statements of respective constructs into one place and given a common proposition to each of the extracted factors in Table 4.5;

TABLE 4.3 Total Variance Explained

	Initial Eigenvalues			Extraction Sums of Squared Loadings			Rotation Sums of Squared Loadings		
Component	**Total**	**% of Variance**	**Cumulative %**	**Total**	**% of Variance**	**Cumulative %**	**Total**	**% of Variance**	**Cumulative %**
1	12.390	41.301	41.301	12.390	41.301	41.301	3.658	12.192	12.192
2	2.860	9.533	50.834	2.860	9.533	50.834	3.593	11.975	24.167
3	2.174	7.246	58.080	2.174	7.246	58.080	3.406	11.354	35.521
4	1.909	6.364	64.445	1.909	6.364	64.445	3.164	10.546	46.067
5	1.694	5.647	70.091	1.694	5.647	70.091	3.025	10.084	56.151
6	1.325	4.417	74.508	1.325	4.417	74.508	2.828	9.428	65.579
7	1.121	3.737	78.245	1.121	3.737	78.245	2.537	8.456	74.035
8	1.053	3.509	81.754	1.053	3.509	81.754	2.316	7.719	81.754
9	0.505	1.684	83.439						
10	0.472	1.572	85.011						
11	0.448	1.492	86.503						
12	0.420	1.400	87.903						
13	0.381	1.270	89.173						
14	0.344	1.145	90.318						
15	0.336	1.119	91.438						
16	0.317	1.058	92.496						
17	0.272	0.906	93.402						
18	0.257	0.856	94.257						
19	0.223	0.742	94.999						
20	0.212	0.706	95.705						
21	0.195	0.650	96.355						
22	0.173	0.577	96.931						
23	0.171	0.569	97.501						
24	0.151	0.502	98.003						
25	0.135	0.450	98.453						
26	0.120	0.399	98.852						
27	0.107	0.357	99.209						
28	0.103	0.344	99.553						
29	0.076	0.255	99.808						
30	0.058	0.192	100.000						

Extraction Method: Principal Component Analysis.

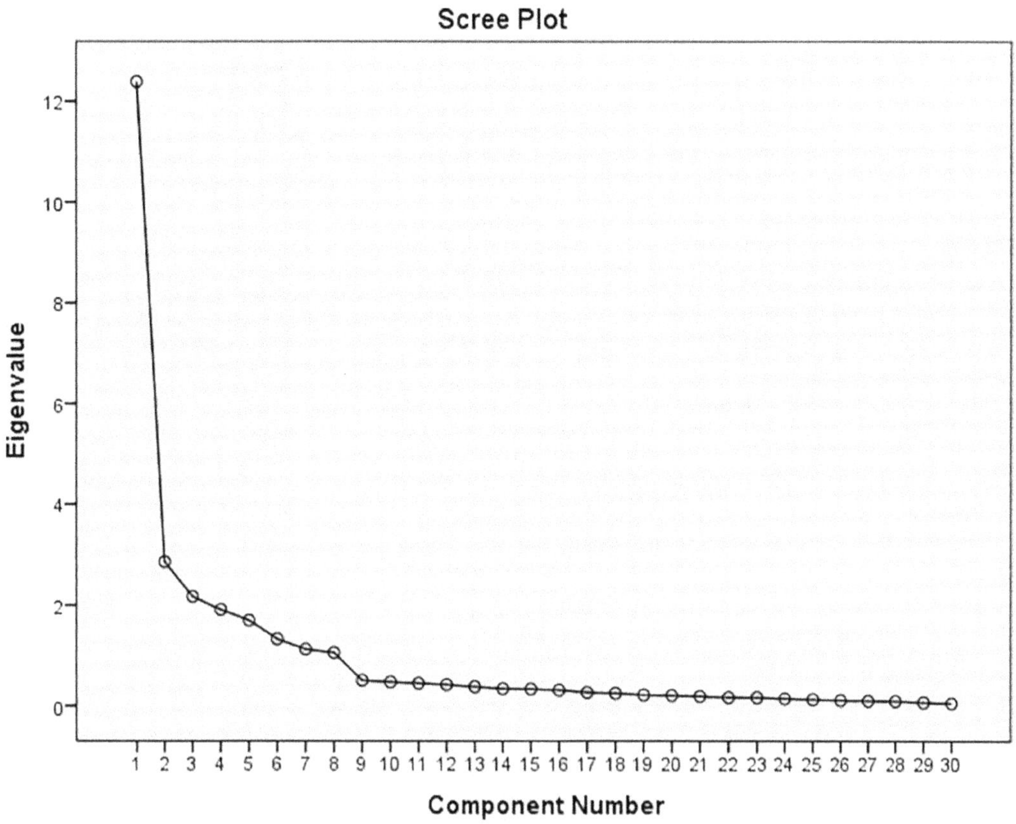

FIGURE 4.1 Screen plot.

thereafter, the researcher has also calculated the internal consistency of each construct using Cronbach alpha. Each construct has shown an adequate level of internal consistency with an alpha value sufficiently above the criteria of 0.70 (Nunnally, 1978; Cronbach, 1951).

Table 4.5 shows that 30 statements have converged into eight factors and based on the ubiquity of respective statements, each factor is given a representative name to that factor. Thereafter, the researcher has drawn managerial implications from each extracted factor for educational institutions and future researchers.

4.5 DISCUSSION AND IMPLICATIONS

This research work illustrates significant implications for management institutions and research scholars. The output of any educational institution depends upon the performance of its students. All the efforts are centered on students. Online educational platforms were challenging for students initially. In this study, it is found that there could be multiple factors that could lead to the adoption of e-learning being much easier for students to acquire. All eight factors have a significant number of implications. Therefore, the researcher has designed the implication for each factor separately so that educational institutions could work on these factors and future researchers could also select these factors for future empirical research work.

TABLE 4.4 Rotated-Component Matrix[a]

	Component							
	1	2	3	4	5	6	7	8
E-learning technologies allow me to execute my learning activities more effectively.	0.874							
Using e-learning technologies can enhance my learning.	0.852							
Using e-learning systems can improve my level of interaction with my professors and classmates.	0.841							
Improved virtual collaboration and communication help me in my future endeavors.	0.831							
Peers recommend me for e-learning.		0.885						
Peers motivate me to e-learning.		0.876						
I interact with my classmates on the use of e-learning systems.		0.860						
E-learning facilities become easy with the help of peers.		0.842						
Faculties are accessible and responsive in e-learning.			0.877					
Instructor facilitates classroom community with engaging tone and creativity.			0.872					
Teachers have a lot of experience with e-learning resources and their application.			0.784					
E-learning facilitates instructors to precisely conduct summative assessments			0.767					
I have the skills to learn how to use e-learning facilities.				0.807				
I am capable of using the different facilities provided to me.				0.794				
Enhanced motivation and learning style provided by e-learning.				0.788				
E-learning is easy and quick to get adapted to the new technology				0.772				
The e-learning material is clear and well-structured.					0.818			
E-learning enhances the student in critical thinking, analysis, and problem-solving.					0.797			
The e-learning content is skill-oriented and meets future requirements.					0.790			
The e-learning content is comprehensive					0.726			
E-learning-related training and assistance are provided by the institution.						0.757		
The institute provides online portals to access textbooks and reference materials						0.745		
Members of the IT department are always accessible to assist students in need.						0.732		
The necessary instructions for better use of the e-learning system are available to me.						0.714		
I easily utilize university-suggested e-learning system (R)							0.820	

(Continued)

TABLE 4.4 (*Continued*) Rotated-Component Matrix[a].

	Component							
	1	2	3	4	5	6	7	8
E-learning has added flexibility and self-paced learning for me.							0.806	
I easily understand the learning system's interface and functionalities.							0.805	
The sufficient classwork and assignments conducted by e-learning								0.804
Online course curriculum meets the modern requirements.								0.775
Learning outcomes of the course can be met through e-learning								0.755

Extraction Method: Principal Component Analysis. Rotation Method: Varimax with Kaiser Normalization.
[a] Rotation converged in six iterations.

Each of the factors has a wide range of managerial consequences for management institutions and research scholars.

Factor 1: Perception of Utility: This factor talks about the perception of utilities that has four components, i.e., 'technology-led learning activities', 'interactions with professors and classmates, 'enhanced learning', and 'virtual collaborations and communication'. The respondents significantly agree on all these four components. Management students agree on this part that technology must play a key important role in the teaching and learning process. This shows their familiarity with technology and its uses. This factor also illustrates that more interaction is possible with professors and even with classmates in an e-learning platform. At the same time, e-learning platforms could also provide enhanced collaboration and communication with the external world easily. All these points clubbed together, we could say that management students' perception of utility is on the higher side; therefore, institutions and future researchers are advised to take note of this variable while framing the e-learning platform for management students.

Factor 2: Peer Influence: This factor talks about peer influence on e-learning platforms. This variable was again measured using four components, i.e., 'motivation', 'recommendation' from peers, 'easy with the help of peers', and 'process of e-learning platform becomes easy'. Major implications in this variable are peers have a lot to 'say' in shaping the mind of students. They draw motivation and recommendations from each other and extend help to understand the e-learning platform. This explicitly shows that peer influence significantly impacts the adoption of e-learning platforms for management students. Therefore, the management institutions are advised to devise mechanisms where students have more online interaction with each other, it could be through some management games, simulations, or any other online activity. More interaction with peers will result in more inclination toward the adoption of an e-learning system.

TABLE 4.5 Reliability and Factor Loadings

Factor No.	Constructs	Cronbach Alpha	Loadings	Statements
1	Perception of utility	0.935	0.874	E-learning technologies allow me to execute my learning activities more effectively. (22)
			0.841	Using e-learning systems can improve my level of interaction with my professors and classmates. (23)
			0.852	Using e-learning technologies can enhance my learning. (24)
			0.831	Improved virtual collaboration and communication help me in my future endeavors. (25)
2	Peer influence	0.937	0.876	Peers motivate me to e-learning. (14)
			0.885	Peers recommend me for e-learning. (15)
			0.842	E-learning facilities become easy with the help of peers. (16)
			0.86	I interact with my classmates on the use of e-learning systems. (17)
3	Faculty capability	0.943	0.784	Teachers have a lot of experience with e-learning resources and their applications. (18)
			0.872	The instructor facilitates classroom community with an engaging tone and creativity. (19)
			0.877	Faculties are accessible and responsive in e-learning. (20)
			0.767	E-learning facilitates instructors to precisely conduct summative assessments (21)
4	Student capability	0.86	0.807	I have the skills to learn how to use e-learning facilities. (1)
			0.794	I am capable of using the different facilities provided to me. (2)
			0.772	E-learning is easy and quick to get adapted to new technology (3)
			0.788	Enhanced motivation and learning style provided by e-learning. (4)
5	Course material	0.965	0.818	The e-learning material is clear and well-structured. (5)
			0.79	The e-learning content is skill-oriented and meets future requirements. (7)
			0.726	The e-learning content is comprehensive (8)
			0.797	E-learning enhances the student in critical thinking, analysis, and problem-solving (9)
6	Institutional assistance	0.87	0.732	Members of the IT department are always accessible to assist students in need. (29)
			0.745	The institute provides online portals to access the textbooks and reference materials (30)
			0.714	The necessary instructions for better use of the e-learning system are available to me. (31)
			0.757	E-learning-related training and assistance are provided by the institution. (32)
7	User-friendliness	0.933	0.82	I easily utilize the university-suggested e-learning system. (26)
			0.805	I easily understand the learning system's interface and functionalities. (27)
			0.806	E-learning has added flexibility and self-paced learning for me. (28)
8	Course layout	0.8	0.755	Learning outcomes of the course can be met through e-learning (11)
			0.804	The sufficient classwork and assignments conducted by e-learning (12)
			0.775	Online course curriculum meets the modern requirements. (13)

Factor 3: Faculty Capability: This factor illustrates the capability of the faculty to implement e-learning platforms in the classroom. This variable was measured using four apparatuses, i.e., 'teacher's experience of e-learning resources', 'tone of the instructor', 'faculty accessibility and responsiveness', and assessment on the e-learning platform. This research work apparently demonstrates that each of these points has significant importance to management students. The tone of the instructor definitely could create a positive vibe, and this will set the path for faculty accessibility and responsiveness on an e-learning platform because, when we teach from the distance, we encounter the disengagement of students on e-learning platforms. These four tools, if institutions work on them, could create a significant amount of difference in the e-learning process. Finally, the assessment of students could also act as a contributor to the student's learning process as the instructor could do multiple layers of assessment of students in the form of constructive feedback. This will help in the overall development of the student. Here, management institutions are also advised to undertake regular faculty development programs so that faculties are familiar with recent developments in the e-learning process.

Factor 4: Student Capability: This factor talks about the capability required at the student level itself. The researcher has used four components here, i.e., 'skills to learn', the capability of using facility', 'ease and quick to adopt new technology', and 'motivation and learning style'. Many times, it is observed that everything is being provided by the institutions but eventually that does not translate into the final outcome for students, and this is because of students in capabilities to assimilate new platforms. Management institutions are advised to devise some mechanism where students' capability could be enhanced on these four parameters. As it is also observed that facilities are in place, learning avenues are available, and newer technologies are adopted by the institutions, but those who lack these parameters, failed to embrace them. Hence, management institutions should also work on creating a bridge for students so that they can cross this journey of skilling to learn, be prompt in using faculties, be forefront in embracing newer technologies, and keep motivation and learning style intact.

Factor 5: Course Material: This factor also illustrates four components of the course material, i.e., 'clear and structured', 'skill-oriented and future requirement', 'comprehensive', and 'critical thinking and problem-solving'. Management institutions and course developers should keep these four components in their mind while preparing course material for management students. They do not require long and lengthy study material. They are much more focused on concise, structured, skill-oriented study material, which meets the future industry requirement. Additionally, they have also emphasized inclusive content so that they could also work on some issues and can come up with some solutions for them. Basically, they are interested in skill-based learning, so they should be industry-ready and meets the requirement of the future.

Factor 6: Institutional Assistance: Institutional assistance is also a significant factor in the adoption of the e-learning platform. Four components were used to measure this factor in this research work, i.e., 'IT department assistance', 'online portals', 'proper instructions for usages of e-learning system', and 'training and assistance'. Comprehensive IT support is required to have a smooth flow of e-learning experience for the students. Right from making the tools, textbooks, or other materials available on the online platform, institutions should provide proper training and necessary support to students in case of any need in accessing and utilizing these online systems.

Factor 7: User-friendliness: User-friendliness is one of the most prominent factors that has emerged in this research work. This factor was assessed using three components, i.e., 'ease in utilizing suggested e-learning system', 'easy interface and functionalities', and 'flexibility and self-paced learning'. Management institutions are advised to consider that e-learning systems should be at ease for all the students and interface and functionality should be easy for management students. The most important point that has emerged in this study is flexibility and self-paced learning. This point alone adds significant value to the e-learning platforms and allows many people to undertake learning at their comfortable time again and again. All these points discussed above should be incorporated by the institutions as these points will add significant value to the adoption of e-learning platforms.

Factor 8: Course Layout: This factor was measured using three components, i.e., 'meeting learning outcome', 'classwork and assignment', and 'curriculum meets modern requirements. Many institutions are advised now to adopt outcome-based learning systems. The layout of the course should be designed to meet course outcomes even on the e-learning platform. There are many ways to conduct classwork and assignments on the e-learning platform. Institutions should understand that shifting to the online platform is opening many unexplored paths of classwork and assignments. Familiarity of students with these platforms is very important to have a smooth transition to an e-learning platform. This research work also illustrates that the curriculum must meet modern industry requirements. Therefore, institutes are also advised to frame the curriculum in such a manner where the industry requirements should be met by the curriculum offered at the e-learning platform.

4.6 LIMITATIONS OF THE STUDY

Due to time constraints, the research was only conducted with management students from India. This is a limitation of the study. In addition, for the sake of simplicity and because we were constrained by time, we decided to restrict the scope of our research to comprehend the viewpoints of students, rather than those of teachers.

In addition, there were just one sort of responders used in the collection of the data and they were the pupils. As a consequence of this, the findings of the study cannot be extrapolated to any other samples. In future research, it may also be beneficial to include the points of view of educators and policymakers to get a greater degree of generalization in the findings. Since the scope of the current research is restricted to only theory classes, it is possible to apply its findings to evaluate students' performance in practical classes. The research has only been conducted on students in India; hence, collecting data from students in other nations may yield more accurate comparative results that may assist in better comprehending students' points of view. This study is limited to evaluating the performance of students; however, in the future, the performance of teachers may be evaluated under situations that are analogous to those currently being evaluated. There is a possibility that the students will have certain difficulties and inconveniences, such as restricted access to the internet or disruptions because of weak signals. Some of the learners might have problems at home, such as disturbances caused by members of the family, which might result in poor academic performance. It is possible to include the aforementioned ideas in the study that will be done in the future.

4.7 CONCLUSION AND FUTURE SCOPE

The use of information technology in education is rising and becoming increasingly commonplace. Students and professors in poor nations have distinct hurdles compared to their counterparts in rich countries, and institutions in these countries must understand what motivates them to use e-learning platforms. The e-learning system's success can be ensured by a better understanding of these problems and taking suitable action. His study found that the implementation of e-learning systems in developing countries should consider eight different dimensions. These include the perception of the utility, peer influence, faculty capability, student capability, course material, user-friendliness, institutional assistance, and course layout.

Universities can use this information to improve the management of their institutions and to develop a high-quality educational system. With the help of this study, e-learning implementations and student satisfaction can be enhanced. People who shift from traditional learning to e-learning require a variety of different resources. This transition process is built on a learning–unlearning–relearning flow that both the teacher and the student must go through. When COVID-19 is over, this study will explain why management students are more likely to use online learning during this time. There are a variety of applications for this information at institutions. It can assist them in managing their institutions and establishing a robust educational system.

People who read this study will be able to utilize it to better their e-learning implementations and make their students even more satisfied with them. The findings of this study can be used to help institutions improve the effectiveness of their e-learning programs and increase student satisfaction even further.

The outcomes of this study allow us to give recommendations to university administrators, developers, designers, and decision-makers to encourage the usage of e-learning.

Furthermore, this study can be considered an addition to the existing literature because it investigates the factors that further influence the instructors' and students' use of e-learning resources that are provided by their universities to enhance their learning processes at any time and from any location, as opposed to the existing literature.

It is recommended that future research extend this study to multiple courses, semesters of study, a focus on learning, and age differences, of which courses, semesters of study, grade achievement, and ranges of age show high or low levels of course satisfaction, for example, education courses versus business courses, first year versus final year of study, low-grade achievement versus high-grade achievement, and younger students versus older students. In addition, it would be beneficial to conduct interviews with either students or teachers or both and triangulate the data as part of a contemporaneous mixed study. Doing so would help to reduce the possibility of bias.

It is also recommended that research of this kind, which is related to the elements of e-learning consumption, can be done at least once every year to evaluate the efficiency of the system that is currently in place because this type of technology is a volatile and rapidly changing trend. The results of this kind of research will provide input that the developer of an e-learning platform can use to provide continual evaluation for improvement. This study is limited to the students of management education at a university; however, it might be expanded to include students from other faculties to determine whether the utilization of an online learning platform is still relevant. The range of future research can be expanded to include the professors, whose points of view will be very distinct from those of the students. The follow-up studies that are related to this research will expand our knowledge of how e-learning platforms are utilized in research universities.

REFERENCES

Abbad, M. M., Morris, D., Al-Ayyoub, A., & Abbad, J. M (2009). Students' decisions to use an eLearning system: A structural equation modelling analysis. *International Journal of Emerging Technologies in Learning, 4*(4), 4–13.

Abbad, M. M., Morris, D., & De Nahlik, C. (2009). Looking under the bonnet: Factors affecting student adoption of e-learning systems in Jordan. *International Review of Research in Open and Distributed Learning, 10*(2), 1–15.

Adeoye, I. A., Adanikin, A. F., & Adanikin, A. (2020). COVID-19 and E-Learning: Nigeria Tertiary Education System Experience. *International Journal of Research and Innovation in Applied Science, 5*(5), 28–31.

Aguilera-Hermida, A. P. (2020). College students' use and acceptance of emergency online learning due to COVID-19. *International Journal of Educational Research Open, 1*, 100011.

Ahmad, N., Quadri, N. N., Qureshi, M. R. N., & Alam, M. M. (2018). Relationship modeling of critical success factors for enhancing sustainability and performance in e-learning. *Sustainability, 10*(12), 4776.

Akyüz, H. İ., & Samsa, S. (2009). The effects of blended learning environment on the critical thinking skills of students. *Procedia-Social and Behavioral Sciences, 1*(1), 1744–1748.

Alhassan, A., Ali, N. A., & Ali, H. I. H. (2021). EFL students' challenges in English-medium business programmes: Perspectives from students and content teachers. *Cogent Education, 8*(1), 1888671.

Alhumaid, K., Ali, S., Waheed, A., Zahid, E., & Habes, M. (2020). Covid-19 & e-learning: Perceptions & attitudes of teachers towards e-learning acceptance in the developing countries. *Multicultural Education, 6*(2), 100–115.

Allen, I. E., & Seaman, J. (2014). *Grade Change: Tracking Online Education in the United States.* Babson Survey Research Group, Boston, USA.

AlQudah, A. A., Salloum, S. A., & Shaalan, K. (2021). The role of technology acceptance in healthcare to mitigate covid-19 outbreak. *Emerging Technologies during the Era of COVID-19 Pandemic*, 223–244. https://doi.org/10.1007/978-3-030-67716-9_14

Alshiekhly, U., Arrar, R., Barngkgei, I., & Dashash, M. (2015). Facebook as a learning environment for teaching medical emergencies in dental practice. *Education for Health, 28*(3), 176.

Andersson, A. (2008). Seven major challenges for e-learning in developing countries: Case study eBIT, Sri Lanka. *International Journal of Education and Development Using ICT, 4*(3), 45–62.

Arbaugh, J. B. (2000). How classroom environment and student engagement affect learning in Internet-based MBA courses. *Business Communication Quarterly, 63*(4), 9–26.

Arbaugh, J. B. (2000). Virtual classroom characteristics and student satisfaction with internet-based MBA courses. *Journal of Management Education, 24*(1), 32–54.

Arbaugh, J. B. (2002). Managing the on-line classroom: A study of technological and behavioral characteristics of web-based MBA courses. *The Journal of High Technology Management Research, 13*(2), 203–223.

Ashwin, P., & McVitty, D. (2015). The meanings of student engagement: Implications for policies and practices. In Adrian Curaj, Liviu Matei Remus Pricopie, and Jamil Salmi PeterScott (Eds.), *The European Higher Education Area.* (pp. 343–359). Springer, Cham.

Aung, T. N., & Khaing, S. S. (2015). Challenges of implementing e-learning in developing countries: A review. In *International Conference on Genetic and Evolutionary Computing* (pp. 405–411). Springer, Cham.

Baker, E. W., Al-Gahtani, S. S., & Hubona, G. S. (2010). Cultural impacts on acceptance and adoption of information technology in a developing country. *Journal of Global Information Management (JGIM), 18*(3), 35–58.

Belanger, F., & Jordan, D. H. (Eds.). (1999). *Evaluation and Implementation of Distance Learning: Technologies, Tools and Techniques: Technologies, Tools and Techniques.* Hershey, PA: Igi Global.

Bharucha, J. (2018). Learning and social software: Exploring the realities in India. *Journal of Information Communication and Ethics in Society, 16*(1), 75–89.

Bhattacharya, I., & Sharma, K. (2007). India in the knowledge economy–an electronic paradigm. *International Journal of Educational Management, 21*(6), 543–568.

Bhaumik, R., & Priyadarshini, A. (2021). Pandemic experiences of distance education learners: Inherent resilience and implications. *Asian Journal of Distance Education, 16*(2), 18–37.

Bliuc, A., Goodyear, P., & Ellis, R. (2007). Research focus and methodological choices in studies into students' experiences of blended learning in higher education. *The Internet and Higher Education, 10*(4), 231–244.

Boud, D., Cohen, R., & Sampson, J. (1999). Peer learning and assessment. *Assessment & Evaluation in Higher Education, 24*, 413–426.

Brandfinance.com. (2019). https://brandfinance.com/wpcontent/uploads/1/global_500_2019_locked_4.pdf.

Brophy, J. (2000). Teaching. Educational practices series-1.

Carliner, S. (2004). *An Overview of Online Learning.* Amherst, MA: HRD Press Inc.

Chen, M., Wang, X., Wang, J., Zuo, C., Tian, J., & Cui, Y. (2021). Factors affecting college students' continuous intention to use online course platform. *SN Computer Science, 2*(2), 1–11.

Chien, T. (2012), Computer self-efficacy and factors influencing e-learning effectiveness. *European Journal of Training and Development, 36*(7), 670–686.

Chiu, C. M., Chiu, C. S., & Chang, H. C. (2007). Examining the integrated influence of fairness and quality on learners' satisfaction and Web-based learning continuance intention. *Information Systems Journal, 17*(3), 271–287.

Chizmar, J. F., & Walbert, M. S. (1999). Web-based learning environments guided by principles of good teaching practice. *The Journal of Economic Education, 30*(3), 248–259.

Cojocariu, V. M., Lazar, I., Nedeff, V., & Lazar, G. (2014). SWOT anlysis of e-learning educational services from the perspective of their beneficiaries. *Procedia-Social and Behavioral Sciences, 116*, 1999–2003.

Collis, B. (1991). Anticipating the impact of multimedia in education: Lessons from literature. *International Journal of Computers in Adult Education and Training, 2*(2), 136–149.

Cronbach, L. J. (1951). Coefficient alpha and the internal structure of tests. *Psychometrika, 16*(3), 297–334. http://doi.org/10.1007/BF02310555.

Das, R., Manaktala, N., Bhatia, T., Agarwal, S., Natarajan, S., Lewis, A. J., & Yellapurkar, S. (2020). Efficiency of mobile video sharing application (WhatsApp®) in live field image transmission for telepathology. *Journal of Medical Systems, 44*(6), 1–9.

Debattista, M. (2018). A comprehensive rubric for instructional design in e-learning. *The International Journal of Information and Learning Technology, 35*(2), 93–104.

DeLone, W. H., & McLean, E. R. (2003). The DeLone and McLean model of information systems success: A ten-year update. *Journal of Management Information Systems, 19*(4), 9–30.

Davis, F. D. (1989). Perceived usefulness, perceived ease of use, and user acceptance of information technology. *MIS Quarterly, 13*(3), 319–340.

Davis, F. D., Bagozzi, R. P., & Warshaw, P. R. (1989). User acceptance of computer technology: A comparison of two theoretical models. *Management Science, 35*(8), 982–1003.

Dhawan, S. (2020). Online learning: A panacea in the time of covid-19 crisis. *Journal of Educational Technology Systems, 49*(1), 5–22.

Duffy, T. M., Dueber, B., & Hawley, C. L. (1998). Critical thinking in a distributed environment: A pedagogical base for the design of conferencing systems. *Electronic Collaborators: Learner-Centered Technologies for Literacy, Apprenticeship, and Discourse, 1*(3), 51–78.

Ekanayake, E. M. H. L., & Weerasinghe, T. (2020). Sustainable engagement of university students in e-learning during the post-pandemic of covid-19: Evidence from faculty of commerce and management studies, University of Kelaniya, Sri Lanka. *Kelaniya Journal of Human Resource Management, 15*(2), 47–73.

Ellis, R., Ginns, P., & Piggott, L. (2009). E-learning in higher education: Some key aspects and their relationship to approaches to study. *Higher Education Research & Development, 28*(3), 303–318.

Emami, E. (2020). COVID-19: Perspective of a dean of dentistry. *JDR Clinical & Translational Research, 5*(3), 211–213.

Erkan, I., & Evans, C. (2016). The influence of eWOM in social media on consumers' purchase intentions: An extended approach to information adoption. *Computers in Human Behavior, 61*, 47–55.

Evans, T. (1999). Flexible delivery and flexible learning: Developing flexible learners? In John Garrick and Viktor Jakupec (Eds.), *Flexible Learning, Human Resource, and Organisational Development*. London: Routledge (pp. 227–240).

Field, A. (2013). *Discovering Statistics Using IBM SPSS statistics*. New Delhi: Sage.

Garcia-Penalvo, F., Fidalgo-Blanco, A., & Sein-Echaluce, M. (2018). An adaptive hybrid MOOC model: Disrupting the MOOC concept in higher education. *Telematics and Informatics, 35*(4), 1018–1030.

Garrison, D. R. (1997). Self-directed learning: Toward a comprehensive model. *Adult Education Quarterly, 48*(1), 18–33.

Gilbert, B. (2015). *Online Learning Revealing the Benefits and Challenges*. Education Masters. Paper 303.

Guglielmino, L. M. (1977). *Development of the Self-Directed Learning Readiness Scale*. University of Georgia, USA.

Ha, N. H., Nayyar, A., Nguyen, D. M., & Liu, C. A. (2019). Enhancing students' soft skills by implementing CDIO-based integration teaching mode. In *The 15th International CDIO Conference. Aarhus University, Denmark* (p. 569).

Habes, M., Salloum, S. A., Alghizzawi, M., & Alshibly, M. S. (2018). The role of modern media technology in improving collaborative learning of students in Jordanian universities. *International Journal of Information Technology and Language Studies, 2*(3), 71–82.

Hair, J., Black, B., Babin, B., Anderson, R., & Tatham, R. (2010). *Multivariate Data Analysis: A Global Perspective*. New Jersey: Pearson Education Inc.

Halpin, P. A., & Lockwood, M. K. K. (2019). The use of twitter and zoom videoconferencing in healthcare professions seminar course benefits students at a commuter college. *Advances in Physiology Education, 43*(2), 246–249.

Hara, N., & Kling, R. (1999). A case study of students' frustrations with a web-based distance education course. *First Monday, 4*(12), 1–21.

Harandi, S. R. (2015). Effects of e-learning on students' motivation. *Procedia-Social and Behavioral Sciences, 181*, 423–430.

Hay, A., Hodgkinson, M., Peltier, J. W., & Drago, W. A. (2004). Interaction and virtual learning. *Strategic Change, 13*(4), 193.

Holden, H., & Rada, R. (2011). Understanding the influence of perceived usability and technology self-efficacy on teachers' technology acceptance. *Journal of Research on Technology in Education, 43*(4), 343–367.

Ilin, V. (2020). The good, the bad and the ugly. A broad look at the adaptation of technology in education. *EDU REVIEW. International Education and Learning Review/Revista Internacional de Educación y Aprendizaje, 8*(2), 103–117.

Internetworldstats.com. (2019). Internet top 20 countries –internet users 2019. https://www.internetworldstats.com/top20.htm.

Junco, R., Elavsky, C. M., & Heiberger, G. (2013). Putting Twitter to the test: Assessing outcomes for student collaboration, engagement and success. *British Journal of Educational Technology, 44*(2), 273–287.

Jung, I., Choi, S., Lim, C., & Leem, J. (2002). Effects of different types of interaction on learning achievement, satisfaction and participation in web-based instruction. *Innovations in Education and Teaching International, 39*(2), 153–162.

Juwah, C. (Ed.). (2006). *Interactions in Online Education: Implications for Theory and Practice* (1st ed.). Routledge, Aberdeen, UK. https://doi.org/10.4324/9780203003435

Kaur, N., Dwivedi, D., Arora, J., & Gandhi, A. (2020). Study of the effectiveness of e-learning to conventional teaching in medical undergraduates amid COVID-19 pandemic. *National Journal of Physiology, Pharmacy and Pharmacology, 10*(7), 563–567.

Kee, N. S., et al. (2012). Towards student-centred learning: Factors contributing to the adoption of e-learn@ USM. *Malaysian Journal of Distance Education, 14*(2), 1–24.

Khamparia, A., & Pandey, B. (2018). Impact of interactive multimedia in e-learning technologies: Role of multimedia in e-learning. In *Digital Multimedia: Concepts, Methodologies, Tools, and Applications* (pp. 1087–1110). IGI Global. https://www.researchgate.net/profile/Aditya-Khamparia/publication/314230554_Impact_of_Interactive_Multimedia_in_E-Learning_Technologies/links/58f58ecc0f7e9b6f82e8ca07/Impact-of-Interactive-Multimedia-in-E-Learning-Technologies.pdf

Khan, B. (2005). Learning features in an open, flexible and distributed environment. *AACE Review (formerly AACE Journal), 13*(2), 137–153.

Kulikowski, K., Przytuła, S., & Sułkowski, Ł. (2022). E-learning? Never again! On the unintended consequences of COVID-19 forced e-learning on academic teacher motivational job characteristics. *Higher Education Quarterly, 76*(1), 174–189.

Kumar, A., Krishnamurthi, R., Bhatia, S., Kaushik, K., Ahuja, N. J., Nayyar, A., & Masud, M. (2021). Blended learning tools and practices: A comprehensive analysis. *IEEE Access, 9*, 85151–85197.

Leidner, D. E., & Jarvenpaa, S. L. (1993). The information age confronts education: Case studies on electronic classrooms. *Information Systems Research, 4*(1), 24–54.

Leonardi, P. (2020). You're going digital-now what. *MIT Sloan Management Review, 61*(2), 28–35.

Liaw, S. S., Huang, H. M., & Chen, G. D. (2007). Surveying instructor and learner attitudes toward e-learning. *Computers & Education, 49*(4), 1066–1080.

Lim, H., Lee, S. G., & Nam, K. (2007). Validating E-learning factors affecting training effectiveness. *International Journal of Information Management, 27*(1), 22–35.

Lin, B., & Hsieh, C. T. (2001). Web-based teaching and learner control: A research review. *Computers & Education, 37*(3–4), 377–386.
Little, B. and Knihova, L. (2014), "Modern trends in learning architecture", *Industrial and Commercial Training*, Vol. *46*(1), 34–38. https://doi.org/10.1108/ICT-07-2013-0046
Littlefield, J. (2018). The difference between synchronous and asynchronous distance learning. Retrieved May, 4, 2020. https://www.thoughtco.com/synchronous-distance-learning-asynchronous-distance-learning-1097959
Machado, R. A., de Souza, N. L., Oliveira, R. M., Júnior, H. M., & Bonan, P. R. F. (2020). Social media and telemedicine for oral diagnosis and counselling in the covid-19 era. *Oral Oncology, 105*, 104685.
MacQuarrie, M., Blinn, N., MacLellan, S., Flynn, M., Meisner, J., Owen, P., & Spencer, B. (2021). Starting from the same spot: Exploring mindfulness meditation and student transitions in the introductory health promotion classroom. *Canadian Journal of Education/Revue canadienne de l'éducation, 44*(1), 256–276.
Maheshwari, G., & Thomas, S. (2017). An analysis of the effectiveness of the constructivist approach in teaching business statistics. *Informing Science, 20*, 83.
Martin, N., Lazalde, O. M., Stokes, C., & Romano, D. (2012). An evaluation of remote communication versus face-to-face in clinical dental education. *British Dental Journal, 212*(6), 277–282.
Martinez-Alvarez, M., Jarde, A., Usuf, E., Brotherton, H., Bittaye, M., Samateh, A. L., ... & Roca, A. (2020). COVID-19 pandemic in west Africa. *The Lancet Global Health, 8*(5), e631–e632.
McBrien, J. L., Cheng, R., & Jones, P. (2009). Virtual spaces: Employing a synchronous online classroom to facilitate student engagement in online learning. *International Review of Research in Open and Distributed Learning, 10*(3), 1–17.
McIsaac, M. S., Blocher, M., Mahesh, V., & Vrasidas, C. (1999). Student interactions and perceptions of online courses. *Educational Media International, 36*(2), 121–131.
McKinney, V., Yoon, K., & Zahedi, F. M. (2002). The measurement of web-customer satisfaction: An expectation and disconfirmation approach. *Information Systems Research, 13*(3), 296–315.
McVay, M. (2000). Developing a web-based distance student orientation to enhance student success in an online bachelor's degree completion program. Unpublished practicum report presented to the Ed. D. Program, Nova Southeastern University, Florida.
McVay, M. (2001). *How to Be a Successful Distance Learning Student: Learning on the Internet*. New York: Prentice Hall, Pearson Custom Pub. USA.
Mei, B., Brown, G. T., & Teo, T. (2018). Toward an understanding of preservice English as a foreign language teachers' acceptance of computer-assisted language learning 2.0 in the People's Republic of China. *Journal of Educational Computing Research, 56*(1), 74–104.
Moore, M. G., & Kearsley, G. G. (1996). *Distance Education: A System View* (No. C10 20). Belmount, California: Wadsworth.
Moore, M. G., & Kearsley, G. (2011). *Distance Education: A Systems View of Online Learning*. Belmount, California: Cengage Learning.
Mosunmola, A., Mayowa, A., Okuboyejo, S., & Adeniji, C. (2018). Adoption and use of mobile learning in higher education: The UTAUT model. In *Proceedings of the 9th International Conference on E-Education, E-Business, E-Management and E-Learning. Association for Computing Machinery, New York* (pp. 20–25).
Mukhopadhyay, M., Pal, S., Nayyar, A., Pramanik, P. K. D., Dasgupta, N., & Choudhury, P. (2020). Facial emotion detection to assess Learner's State of mind in an online learning system. In *Proceedings of the 2020 5th International Conference on Intelligent Information Technology, Hanoi, Vietnam* (pp. 107–115).
Navarro, P., & Shoemaker, J. (2000). Performance and perceptions of distance learners in cyberspace. *American Journal of Distance Education, 14*(2), 15–35.
Nedungadi, P., Mulki, K., & Raman, R. (2018). Improving educational outcomes & reducing absenteeism at remote villages with mobile technology and WhatsAPP: Findings from rural India. *Education and Information Technologies, 23*(1), 113–127.

Nguyen, T. (2015). The effectiveness of online learning: Beyond no significant difference and future horizons. *MERLOT Journal of Online Learning and Teaching, 11*(2), 309–319.

Nichols, M. (2003). A theory for eLearning. *Journal of Educational Technology & Society, 6*(2), 1–10.

Nneka Eke, H. (2010). The perspective of e-learning and libraries in Africa: Challenges and opportunities. *Library Review, 59*(4), 274–290.

Nuere, S., & de Miguel, L. (2021). The digital/technological connection with Covid-19: An unprecedented challenge in university teaching. *Technology, Knowledge and Learning, 26*(4), 931–943.

Nunnally, J. C. (1978). *Psychometric Theory*, New York: McGraw-Hill.

Oblinger, E. D. U. C. A. U. S. E., Oblinger, J., Roberts, G., McNeely, B., Windham, C., Hartman, J., & Kvavik, R. (2005). *Educating the Net Generation* (Vol. 272). New York, USA, Brockport Booksh.

Oh, E. G., Chang, Y., & Park, S. W. (2020). Design review of MOOCs: Application of e-learning design principles. *Journal of Computing in Higher Education, 32*(3), 455–475.

Ozkan, S., & Koseler, R. (2009). Multi-dimensional students' evaluation of e-learning systems in the higher education context: An empirical investigation. *Computers & Education, 53*(4), 1285–1296.

Paechter, M., Marier, B. & Macher, M. (2020). Students' expectations of, and experiences in e-learning: Their relation to learning achievements and course satisfaction. *Computer & Education, 54*(1), 222–229.

Passos, K. K. M., da Silva Leonel, A. C. L., Bonan, P. R. F., de Castro, J. F. L., dos Anjos Pontual, M. L., de Moraes Ramos-Perez, F. M., & da Cruz Perez, D. E. (2020). Quality of information about oral cancer in Brazilian Portuguese available on Google, Youtube, and Instagram. *Medicina oral, patologia oral y cirugia bucal, 25*(3), e346.

Picciano, A. G. (2002). Beyond student perceptions: Issues of interaction, presence, and performance in an online course. *Journal of Asynchronous learning networks, 6*(1), 21–40.

Podsakoff, P. M., MacKenzie, S. B., Lee, J. Y., & Podsakoff, N. P. (2003). Common method biases in behavioral research: A critical review of the literature and recommended remedies. *Journal of Applied Psychology, 88*(5), 879.

Reeves, T. (2000). Alternative assessment approaches for online learning environments in higher education. *Journal of Educational Computing Research, 23*(1), 101–111.

Ricart, S., Villar-Navascués, R. A., Gil-Guirado, S., Hernández-Hernández, M., Rico-Amorós, A. M., & Olcina-Cantos, J. (2020). Could MOOC-takers' behavior discuss the meaning of success-dropout rate? Players, auditors, and spectators in a geographical analysis course about natural risks. *Sustainability, 12*(12), 4878.

Rovai, A. P. (2008). Distance Learning in Higher Education a Programmatic Approach to Planning, Design, Instruction, Evaluation, and Accreditation. New York, NY: Teachers College Press.

Rovai, A. P., & Downey, J. R. (2010). Why some distance education programs fail while others succeed in a global environment. *Internet and Higher Education,* 13(3), 141–147.

Salloum, S. A., Al-Emran, M., Habes, M., Alghizzawi, M., Ghani, M. A., & Shaalan, K. (2019). Understanding the impact of social media practices on e-learning systems acceptance. In *International Conference on Advanced Intelligent Systems and Informatics* (pp. 360–369). Springer, Cham.

Salloum, S. A., & Shaalan, K. (2018). Adoption of e-book for university students. In *International Conference on Advanced Intelligent Systems and Informatics* (pp. 481–494). Springer, Cham.

Saxena, K. (2020). *Coronavirus Accelerates Pace of Digital Education in India.* EDI India, https://ediindia.ac.in/wp-content/uploads/2021/01/Coronovirus-Accelerates-Pace-of-Digital-Education-in-India.pdf

Schunk, D. H., Meece, J. L., & Pintrich, P. R. (2014). *Motivation in Education: Theory, Research, and Applications* (4th ed.). Austin, Texas: Pearson.

Sharma, M. (2020). India' burgeoning youth are the world's future. https://www.livemint.com/Opinion/2WSy5ZGR9ZO3KLDMGiJq2J/Indias-burgeoning-youth-are-the-worlds-future.html.

Sharma, S. K., & Kitchens, F. L. (2004). Web services architecture for e-learning. *Electronic Journal on E-Learning, 2*(1), 203–216.

Sharp, V. F. (2008). *Computer Education for Teachers: Integrating Technology into Classroom Teaching.* John Wiley & Sons. NJ, USA.

Shih, T. K., Gunarathne, W. K. T. M., Ochirbat, A., & Su, H. M. (2018). Grouping peers based on complementary degree and social relationship using genetic algorithm. *ACM Transactions on Internet Technology (TOIT), 19*(1), 1–29.

Shuey, S. (2002). Assessing online learning in higher education. *Journal of Instruction Delivery Systems, 16*(2), 13–18.

Singh, V., & Thurman, A. (2019). How many ways can we define online learning? A systematic literature review of definitions of online learning (1988–2018). *American Journal of Distance Education, 33*(4), 289–306.

Smith, P. J. (2005). Learning preferences and readiness for online learning. *Educational Psychology, 25*(1), 3–12.

Sorebo, O., Halvari, H., Gulli, V. F., & Kristiansen, R. (2009). The role of self-determination theory in explaining teachers' motivation to continue to use e-learning technology. *Computers & Education, 53*(4), 1177–1187.

Stein, D. S., et al. (2005). "Bridging the transactional distance gap in online learning environments." *The American Journal of Distance Education, 19*(2), 105–118.

Sułkowski, Ł. (2020). Covid-19 pandemic; recession, virtual revolution leading to deglobalization? *Journal of Intercultural Management, 12*(1), 1–11.

Sun, A., & Chen, X. (2016). Online education and its effective practice: A research review. *Journal of Information Technology Education, 15*, 157–190

Sun, P. C., Tsai, R. J., Finger, G., Chen, Y. Y., & Yeh, D. (2008). What drives a successful e-Learning? An empirical investigation of the critical factors influencing learner satisfaction. *Computers & Education, 50*(4), 1183–1202.

Swan, K., Shea, P., Fredericksen, E., Pickett, A., Pelz, W., & Maher, G. (2000). Building knowledge building communities: Consistency, contact and communication in the virtual classroom. *Journal of Educational Computing Research, 23*(4), 359–383.

Tandon, U. (2021). Factors influencing adoption of online teaching by school teachers: A study during COVID-19 pandemic. *Journal of Public Affairs, 21*(4), e2503.

Thatcher, J. B., & Perrewć, P. L. (2002). An empirical examination of individual traits as antecedents to computer anxiety and computer self-efficacy. *MIS Quarterly, 26*(4), 381–396.

Tseng, T. H., Lin, S., Wang, Y. S., & Liu, H. X. (2022). Investigating teachers' adoption of MOOCs: The perspective of UTAUT2. *Interactive Learning Environments, 30*(4), 635–650.

UNESCO. (2020). *COVID-19 Educational Disruption and Response.* UNESCO. Retrieved from https://www.unesco.org/en/articles/covid-19-educational-disruption-and-response?TSPD_101_R0=080713870fab200007e9c4ad5e6f30d612c316bf10d35a447a58f1494ea4a06ba0d959e588b9f89a0830453e641430009fdcd3d30c085bdb93e8b725ca7232077bebdc3335d93fb6f2d3727007ec4e42115d194de4cc63f5b95996046f6356a3

Vagias, W. M. (2006). Likert-type scale response anchors. Clemson International Institute for Tourism and Research Development, Department of Parks, Recreation and Tourism Management. Clemson University.

Webster, J., & Hackley, P. (1997). Teaching effectiveness in technology-mediated distance learning. *Academy of Management Journal, 40*(6), 1282–1309.

White, C. (2003). Language learning in distance education. Cambridge: Cambridge University Press (258 p.).

Wijekumar, K., Ferguson, L., & Wagoner, D. (2006). Problems with assessment validity and reliability in web-based distance learning environments and solutions. *Journal of Educational Multimedia and Hypermedia, 15*(2), 199–215.

Wootton, A. R., McCuistian, C., Legnitto Packard, D. A., Gruber, V. A., & Saberi, P. (2020). Overcoming technological challenges: Lessons learned from a telehealth counseling study. *Telemedicine and e-Health*, *26*(10), 1278–1283.

Yildirim, S. (2000). Effects of an educational computing course on preservice and inservice teachers: A discussion and analysis of attitudes and use. *Journal of Research on computing in Education, 32*(4), 479–495.

Yong, Y., & Wang, S. (1996). Faculty perceptions on a new approach to distance learning–Teletechnet. *Journal of Instruction Delivery Systems, 10*(2), 3–5.

Young, J. R. (2002). The 24-hours professor: Online teaching redefines faculty members' schedules, duties, and relationships with students. *Chronicle of Higher Education*, *38*, A31–A33.

Yuen, A. H., & Ma, W. W. (2008). Exploring teacher acceptance of e-learning technology. *Asia-Pacific Journal of Teacher Education, 36*(3), 229–243.

Yusoff, Y. M. (2012). "Self-efficacy, perceived social support, and psychological adjustment in international undergraduate students in a public higher education institution in Malaysia." *Journal of Studies in International Education*, *16*(4), 353–371.

Zhang, W., Wang, Y., Yang, L., & Wang, C. (2020). Suspending classes without stopping learning: China's education emergency management policy in the COVID-19 outbreak. *Journal of Risk and Financial Management*, *13*(3), 55.

Zikmund, W. G., Carr, J. C., & Griffin, M. (2013). *Business Research Methods*. Mason, OH: Cengage Learning.

CHAPTER 5

New Technology Anxiety and Acceptance of Technology

An Appraisal of MS Teams

Shad Ahmad Khan

University of Buraimi

Hesham Magd

Modern College of Business and Science

CONTENTS

DOI: 10.1201/9781003322252-5

5.1 INTRODUCTION AND BACKGROUND OF STUDY

Distance education is not a new phenomenon; there are documentary pieces of evidence that distance education has been practiced for over 300 years. History has taken the name of Caleb Phillips from Boston, USA, who used to offer training in shorthand, in which he used to send weekly lessons via US mail to his trainees (Clark, 2020). Since then, the way and methods of distance learning changed from time to time. Electronic distance learning started with the advent and expansion of radio networks in the 1920s. However, the way in which the internet has boomed distance education, many authors agree that the internet can be given the credit for bringing paradigm change in the arena of distance education. Words like online learning, virtual classes, metaverse, etc. have been possible because of the internet. In modern times, distance learning would be redundant in the absence of online learning and online learning would be meaningless without effective online platforms. The digitalization of education has been an agenda for many countries as part of their sustainability goals. However, the realization of this goal was an issue; one of the reasons for the same was considered non-acceptance of technology. The world is divided into two parts where one section of society is innovating and bringing newer technology every now and then; on the other hand, there is a bigger section of society that is not prepared to accept this new technology. The same was the state of the online learning platform before COVID-19.

COVID-19 acted as a digital lubricant for the education sector. The COVID-19 restrictions led to the quick adoption of digital platforms across all sectors including education. According to Ratten (2020), COVID-19 had an impact on the education systems across the world and had impacted all the areas of education, i.e., teaching, research, and service. In many ways, online learning emerged as an efficient alternative to traditional learning, i.e., face-to-face (Sharma & Gupta, 2021; Khan & Magd, 2021). A good higher education system is considered to be the one that harnesses its returns to the maximum through the creation of knowledge and ensures that society is never short of skills and education at various levels (Cremonini et al., 2014). Higher education Institutions (HEIs) add more dynamism to their system by employing digital platforms (Bhatt & Shiva, 2020). The HEIs play a crucial role in the economy (Ratten, 2017). It contributes to the economy by preparing the future workforce, innovation, and research, overall; it ensures well-being in society. HEIs are viewed as a facilitator of socializing and building communities. Restrictions imposed as part of COVID-19 protocols pushed the HEIs to implement a work-from-home policy that brought in serious changes to the teacher–student relationship. It also brought about change in the ways the teacher–learner behaves in the normal teaching–learning process (Khan & Magd, 2021). The ways HEIs and their stakeholders responded to the pandemic cannot be ignored. The feature of online learning that it can happen anywhere-anytime was harnessed to the optimum level by most of the HEIs. Newer terms became the new normal like virtual meetings, webinars, virtual conferences, online workshops, training, etc. The HEIs through the online platforms not only were engaged in conducting regular classes for their students but also were active in engagement with the community.

During the pandemic, people witnessed the presence of many online platforms; some of them were known to them already like Skype and Hangouts, whereas a few new names they learned during the pandemic like Google Meet, Goto meeting, Goto webinar, CISCO Webex, etc. (Richter et al., 2020; Shrivastava et al., 2021; Khan & Magd, 2021). The platform that gained prominence in the gulf region is MS Teams (Khan & Magd, 2021). The platform developed by Microsoft is provided in the package of Microsoft Office. This enables the users to use outlook email along with other Office 365 applications like Word and Excel. The way the adoption of online learning has happened during the pandemic is going to stay even in the post-pandemic era, and thus, its relevance as a point of study cannot be ignored (Croxton, 2014). An increase in the usage of online learning is forecasted in light of its flexibility, accessibility, and convenience for learners (Nguyen, 2015; Alawamleh et al., 2020). In the body of literature, online education has focused more on the teachers and less space has been given to the learners (Shrivastava et al., 2021).

In the context of distance education in present times, online learning unarguably plays a significant role, and the effectiveness of online learning/teaching platforms does play a crucial role in determining the overall effectiveness of the learning outcomes. It was also noted that distance education often functions in the form of specialized workshops, seminars, webinars, etc. It is worth mentioning that webinars have not found much space in the literature (Khan & Magd, 2021). The researchers believe that for the effectiveness of distance education, the effectiveness of online platforms is important in the present times. It is also noted that most of the studies conducted in the field of online platforms have mainly taken into consideration the perspectives of the teachers, whereas the learners' perspective has been less represented. MS Teams being a new platform has limited representation in the body of literature, and it has gained prominence because of its varied benefits; a study is needed on its use and relevance.

In this study, MS Teams has been analyzed from the perspective of learners, through the technology acceptance model (TAM). Descriptive statistics have been used via SPSS-21 on the data administered from 409 respondents from 15 nations who participated in the online workshop series organized in Oman. The study reveals that despite the popularity of MS Teams, the participants had a high level of new technology anxiety (NTA) in using it. The results can be used as user feedback for MS Teams and can be utilized by educational institutions, and other agencies that are facilitating online education directly or indirectly. On these grounds, the conclusions and direction for future research have been proposed.

The objectives of the chapter are as follows:

- To understand the usage of MS Teams as a potential tool for distance education.
- To study the acceptance of MS Teams through the TAM.
- To assess the impact of NTA in the context of acceptance of MS Teams.

Organization of the Chapter

Section 5.2 of this chapter deals with the literature review, where necessary attempts have been made to understand various concepts associated with this study. The literature review discusses the evolution of distance education over a period of time, distance education, and online learning. Furthermore, the TAM along with the NTA has been discussed followed by a review of MS Teams-related literature that covers its adoption and relevance as well. The establishment of the alternate hypothesis also has been taken care of in this section. After the literature review, Section 5.3 deals with the research methodology and research framework; this section covers the conceptual framework, minimum sample size determination, and demographic profile of the respondents. After research methodology, Section 5.4 deals with the finding and analysis part, where the analysis tools like reliability analysis, mean and median analysis, correlation analysis, and structural equation modeling have been used to draw meaningful conclusions as per the objectives of this chapter. Sections 5.5 and 5.6 deal with the research contribution and limitations of the study, and finally, Section 5.7 concludes the chapter with future scope.

5.2 LITERATURE REVIEW

5.2.1 Evolution of Distance Education Over a Period of Time

Distance education as a concept has evolved through generations. According to Saykili (2018), the evolution of distance learning can be distributed in the form of three generations. It started through mail-based correspondence, then to broadcast, and computer-mediated distance education (Anderson & Simpson, 2012; Clark, 2020; Saykili, 2018). The first generation of distance education is dominated by print technology. During this time, education was made possible beyond the physical boundaries of the HEIs through the proliferation of basic communication systems and postal services (Caruth & Caruth, 2013; Saykili, 2018). Correspondence education was adopted by many institutions aiming for social justice and equal opportunities (Simonson et al., 2015). The first generation of distance education expanded on the principles of inclusion that targeted those who have limited access or sometimes no access to educational resources and institutions (Anderson & Simpson, 2012). During this time, there was a limitation in terms of two-way communication in terms of affordability; thus, the first generation of education was based on behavioral theories of learning such as Holmberg's (1989) guided didactic conversation (Saykili, 2018). Also, distance education research was initiated during this time (Anderson & Simpson, 2012)

The second generation was dominated by broadcast technology in terms of radio and television. Broadcast technology made communication fast and more reachable. However, many researchers view this form of communication as one way on the grounds of limited interaction between teachers and students (Anderson & Simpson, 2012; Saykili, 2018). During this time, few open universities also come into existence around the world (Anderson & Simpson, 2012). It can be noted that distance education requires a medium of technology to reach prospective audiences, and technology has shaped the landscape of distance education. The limitation of the first and second generations led to the birth of the third generation of distance learning, which focused more on two-way communication possibilities, thus strengthening interaction. The terms like video/audio conferencing and synchronous and asynchronous communication were made possible during this time (Saykili, 2018).

The innovation in the field of the internet and the world wide web can be given credit for the communication breakthrough that happened in this generation. The internet played a crucial role in the propagation and adoption of online learning during this period. It is also worth noting that online learning became a prominent part of distance education through the online platforms that enabled synchronous and asynchronous learning possible. As per the investigation of Bozkurt et al. (2015) on research articles published between 2009 and 2015, the variables like community, network, collaboration, and cooperation along with skill-based concepts like critical thinking and problem solving were found as the most used keywords in the conceptual framework of the studies conducted in the arena of distance education. The investigation also reported that delivery methods such as blended learning (a combination of face-to-face and online learning) and mobile learning work on the cognitive elements of the learners. The theories like transactional distance theory, social presence theory, and motivation theories were found to be important constructs for emerging research in distance education (Bozkurt et al., 2015). One of the keywords that have been a focus of education research for more than 20 years is online learning (Shearer et al., 2020). The internet and online learning constitute the major portion of distance learning; in many cases, the terms like blended learning have been used to make distance education more effective and efficient. In the United States of America (USA) a decline has been observed in campus-based registrations, whereas a rise has been observed in distance education-based registrations. Today one out of three students in the US are found to be enrolling in distance education (Seaman et al., 2018; Legon et al., 2019).

5.2.2 Distance Education and Online Learning

Distance education is defined as "planned teaching and learning activities provided through the use of a communication channel within an institutional organization without any time and place limitations" (Moore & Kearsley, 2011, p. 2). The communication channel for the effective implementation odd distance education can be considered an online platform. Online learning being the focus of educational research has found its prominence in the field of distance education as well (Shearer et al., 2020; Singh & Thurman, 2019). Online learning is continuously evolving and has been changing the way it has been offered in the past. It is now far beyond the imitation of face-to-face learning (Shearer et al., 2020). When explored about the need for online learning in the arena of distance education, the need and desire for a personalized learning experience were found to be the most prominent need (Shearer et al., 2020). In order to understand this, the dynamics of online learning are to be mapped with the evolution of technology, which decides the generation and the change in the definition of terminologies (Singh & Thurman, 2019). Online learning in many aspects offers flexibility, accessibility, and affordability to education. The way to access information and communication technology is increasing worldwide, and the narrative for online learning is changing as well (Naidu, 2019). In developed countries like the US, a rise has been observed in distance education registrations, mainly due to the penetrations of online platforms and the internet penetrations (Seaman et al., 2018; Legon et al., 2019; Naidu, 2019). The other contributors have been the availability of devices like laptops, and smartphones along with reliable internet connectivity (Jione et al., 2019).

Wider internet infrastructure ensures that a greater section of the community has access to flexible educational opportunities (Naidu, 2019). As per Wei and Chou (2020), the self-efficacy of students in terms of the internet and computer, along with the motivation for (attitude toward) learning does play a direct, positive effect on their online discussion score and course satisfaction. Online learning acts as a mediator between learning perceptions and online discussion scores. Furthermore, online learning mediates between online learning perceptions and course satisfaction (Wei & Chou, 2020). Thus, online learning brings a variety of opportunities for regular students who are not able to attend physical or face-to-face classes. Furthermore, the learning opportunity also creates a chance for upskilling and re-orientation programs (Naidu, 2019). The aspect that has been now widely practiced is working and studying at the same time. This has a huge implication for educational institutions including the HEIs in terms of possibilities and opportunities. COVID-19 challenged the education fraternity (both teachers and learners) by changing the way teaching was done. All levels of educators were forced to adopt online learning systems and platforms at very short notice (Korkmaz & Toraman, 2020). Many such educators and learners demonstrated fear and anxiety toward the new technology (Bhatt & Shiva, 2020; Khan & Magd, 2021). The online platform companies and the educational institution's management made several efforts to make this adoption process smooth for all the stakeholders (Korkmaz & Toraman, 2020).

Higher education has seen a great transformation through online learning (Stone, 2019). An increase in the number of universities has been seen worldwide that offer degree programs in online, distance modes of study. With the growth of digital communications in modern times, traditional correspondence study has been transformed into online learning (Stone, 2019). There are universities and educational institutions present today that offer entire degree programs online where the students do not have to visit the campus even once. The growth and opportunities brought in by the digital platforms in the field of education are unprecedented and require educators and learners to be abreast with the change in technology (Bhatt & Shiva, 2020; Berge, 2007; Romeo et al., 2013; Khan & Magd, 2021). The dynamism in the digital platform is another aspect that has been changing the way it used to be in the past. The online platforms are customized as per the needs of the educational institutions. The integration of assessment and curriculum tools is one such example (Beller, 2013; Siddiq et al., 2016; Bhatt & Shiva, 2020). The acceptance of digital platforms is based on a variety of factors; to evaluate this type of phenomenon, the TAM has been used and recommended by researchers in the arena of online education (Yaakop, 2015; Bhatt & Shiva, 2020). On these lines, the TAM has been used on MS Teams in the context of this study.

5.2.3 Microsoft Teams

The HEIs are usually prepared for natural disasters and other potential risks, but this preparedness does not include the guidelines for dealing with the COVID-19 type of pandemic (Ratten, 2020). As the COVID-19 restrictions were imposed and the HEIs were forced to switch online, there was an initial struggle for most of the HEIs with regard to the selection of the online platform (Khan & Magd, 2021). MS Teams launched in March 2020 (Mehta et al., 2020) was one such platform that received attention during the time of the pandemic. The features like sharing files, messages, and group video calls with larger gatherings were some of the features MS Teams offered to its users (Bsharat & Behak, 2020). Students' motivation was also found to be on the higher side with the use of MS Teams as it facilitates communication among classmates along with their instructor irrespective of synchronous and asynchronous sessions Singh et.al. (2020). MS Teams supports virtual lessons, meetings, messaging, and live broadcasts; it helped the learners to overcome the traditional method of sharing or submitting documents through hard copy or via email by facilitating sharable files. The platform enables people to work on the same document at the same time by way of having shareable links and necessary access (Mehta et al., 2020; Pal & Vanijja, 2020; Khan & Magd, 2021). The platform also has the feature of recording and sharing live lectures, distributing teaching material, and pre-recorded lectures, enhancing the satisfaction level of students (Sarker et al., 2019). The organization on the other side does not have to pay separately for MS Teams as it comes as a part of the MS Office 365 package; the cost-effectiveness leads to higher acceptability among the decision-makers (Mehta et al., 2020). On the other hand, the user-friendliness of the platform has been questioned as the researchers opine that to use the platform effectively, the users require basic training, without which the maximum utilization of the platform cannot be ensured (Bsharat & Behak, 2020). The poor internet and connectivity infrastructure is also one of the issues that affect MS Teams' performance and create a negative image (Sarker et al., 2019. The platform since its launch date has been evolving, and more improvements have been added considering the use and application of the platform. A study on the need of the users and dissatisfactions, if any, could help the creators of MS Teams to put forward a better platform and overall MS Office package.

5.2.3.1 MS Teams in Terms of Relevance and Adoption

During the pandemic, the biggest challenge encountered by the HEIs was communication as face-to-face classes were restricted during the time of COVID-19 (Khan & Magd, 2021). Online platforms emerged as an alternative among which MS Teams was adapted for lectures and instructions in many Omani HEIs (Khan & Magd, 2021). The platform not only allows the students to interact with their instructors but also with their classmates (Singh et al., 2020). Thus, a virtual replication of peer learning was also possible through MS Teams. This further eases down the work of the instructor who is benefited from the environment of collaborative learning (Khan & Magd, 2021). In the field of medical sciences also, MS Teams is used for illustrating medical concepts, lessons, interdisciplinary meetings, medical consultancy, and meetings (Mehta et al., 2020). The MS Teams platform also enables external entities to attend and join the virtual session; therefore, global collaboration and deliberation are possible, leading to a better level of learning (Mehta et al., 2020; Khan & Magd, 2021).

Another aspect that the HEIs need to consider is the monitoring and quality assessment of the classes being conducted. In face-to-face learning, the supervisors (Dean, Head of the Department, etc.) often make a physical visit to the classroom and assess the teaching–learning process; however, when the teachers and students observe work from home, at times, it becomes difficult to monitor the classes. MS Teams was able to successfully develop and incorporate tools that can take care of other aspects of the teaching–learning process (Khan & Magd, 2021; Pal & Vanijja, 2020). Furthermore, MS Teams facilitates students and teachers to share a file in any format; there are also options where certain file formats can be worked on simultaneously (Pal & Vanijja, 2020). The HEI's faculty can use online platforms to share and exchange teaching materials, pre-recorded lectures, and PowerPoint, which are found very useful by the students (Sarker et al., 2019). This provides access students to an immense amount of information including the course materials. The phenomenon of teamwork, collaboration, and discussion is also possible through this platform (Mehta et al., 2020; Pal & Vanijja, 2020).

MS Teams works on the phenomenon of maximizing the resource's utility. In the United Kingdom, the healthcare industry used MS Teams at a bigger scale in less time with minimal cost in conducting their orientation and sensitization sessions. Medical institutions that lack money could use such online platforms at a reduced cost (Mehta et al., 2020). The activities that saw the highest cost reduction were travel costs, stay costs, face-to-face session infrastructure, and people (Khan & Magd, 2021; Mehta et al., 2020). Educational sessions and data collection are other areas where the execution becomes easier through communication in MS Teams. Furthermore, in developing economies where the population has fewer economic and geographical challenges, MS Teams were found to be effective (Singh et al., 2020). The students do not have to travel for long distances to attend the classes; also, the HEIs get benefited as they can increase the capacity of their classes as there are no physical restrictions like class size, furniture requirement, visibility of the writing board, etc. (Singh et al., 2020). This brings a huge opportunity for educational institutions as well as learners. The pre-recorded lectures also help the students to access the sessions that were missed out due to any emergency or absenteeism (Sarker et al., 2019). The amount of knowledge that was limited in the face-to-face sessions has now become limitless due to platforms like MS Teams (Khan & Magd, 2021).

However, online platform usage largely depends on the users and their orientation toward the use of such platforms. The possibilities that MS Teams offers are immense, and thus, anyone who uses this platform can be benefited to the maximum (Bsharat & Behak, 2020). This phenomenon is true for the otherwise situation as well; that is, if there are good users, they require a good platform to be harnessed to the maximum. Therefore, it is always advised to provide basic training to the users before introducing any platform to them to ensure that they can maximize using it for their educational benefits (Bsharat & Behak, 2020).

The relevance of MS Teams can be seen in the area of blended learning as well. Many HEIs are planning to adopt blended learning so that they can avoid any obstructions to the learning process due to any emergencies (Aguilera-Hermida, 2020). Thus, it can be seen from the above literature that MS Teams complies with the aims and objectives of distance education. The platform provides the facility for file sharing, collaborative work, synchronous and asynchronous sessions, etc. MS Teams not only allows an effective communication platform but also helps in achieving relevant student and learning outcomes.

5.2.4 TAM and Research Hypothesis

The TAM is useful in the adoption of new technology or existing technology. It is a model that is used to analyze the relationship between intentions, beliefs, and attitudes (Khan & Magd, 2021). The model is designed specially to study the factors that would be required in the adoption of computer technology (Singh et al., 2020). The TAM is mainly constituted by five constructs, i.e., perceived usefulness (PUS), perceived ease of use (PES), attitude (ATU), beliefs (BI), and actual use (AU).

PUS is defined as "The degree to which a person believes that using a particular system would enhance his or her job performance" (Davis 1989, p. 320). It is the user's belief in the utility that technology is going to offer. It can be understood by the example of students who study or learn with a belief that it would improve their academic performance. A higher level of study means better chances of academic performance (Farahat, 2012). There have been a good number of researchers (Mohammad AlHamad, 2020; Farooq & College, 2021; Mousa et al., 2021) who provided empirical pieces of evidence of the PUS factors, i.e., image, subjective norms, job relevance, demonstrability, and output quality, and analyzed how the PUS and the factors are significantly related (Henderson & Milman, 2020). Also, the PUS is identified as an important predictor of ATU and BI in online learning. With this argument, hypothesis H1 is formulated as:

H1: PUS has a significant influence on ATU in context of MS Teams' usage.

PUS is "the degree to which a person believes that using a particular system would enhance his or her job performance" (Davis, 1989, p. 320). Perceived ease of use is then defined by the degree of a user's belief that using technology is effortless. In other words, PEOU is how hard or easy learners perceive it (Farahat, 2012). The HEIs require online platforms to disseminate knowledge among students in modern times. They need platforms that are user-friendly and allow easy access to resources (Singh et al., 2020). The user's intention is affected by the user-friendliness of the technology; if the technology is complicated, the chances of its adoption are less (Mohammad AlHamad, 2020). The perceived ease of use not only has an effect on the intention to use technology but also the degree of perceived productivity associated with such platforms in the context of online learning (Mohammad AlHamad, 2020). In the TAM, PUS and perceived ease of use are found to be influencing the next construct, i.e., ATU toward usage (e-learning); it can be defined as "The evaluating effect of the positive or negative feeling of individuals in performing a particular behavior", i.e., "usage of MS Teams in the present context" (Bhatt & Shiva, 2020, p. 74). This leads to the formulation of two hypotheses H2 and H3.

H2: PES has a significant influence on PUS in the context of MS Teams' usage
H3: PES has a significant influence on ATU in the context of MS Teams' usage.

The attitude toward e-learning impacts the behavioral intentions to use (e-learning) in TAM; the behavioral intention is defined as follows: "Behavioral Intention to use is a measure of the strength of one's intention to perform a specific behavior", i.e., usage of MS Teams in the present context (Davis et al., 1989, p. 984). In the context of online learning, Behavioral intention (BI) can be considered as the behavior learners or users would have toward online learning (Farahat, 2012). This can be understood by their decision to use online learning tools and activities. Using technology is considered an implementation tool in online education (Khan & Magd, 2021). However, it is the students' behavior that decides the success of online learning. For an effective realization of online learning objectives, both students and instructors need to utilize and use the learning management system. In the context of the TAM, how easily the online platform can be used and utilized would be the key aspect of online learning success, this might require the HEIs to provide proper training to the users (Henderson & Milman, 2020). The studies of Granić and Marangunić (2019) and Farooq and College (2021) also have found that this relationship between attitudes toward e-learning, BIs, and AU is defined as the AU of technology by a person (Scherer et al., 2018). The hypotheses framed in this context are H6 and H7 as given below:

H6: ATU has a significant influence on BI in the context of MS Teams' usage.
H7: BI has a significant influence on AU in the context of MS Teams' usage.

5.2.4.1 New Technology Anxiety

Technology can affect business operations positively if used in the right manner. If the user is well-versed in the usage of technology (tech-savvy), he/she will be able to simplify and enhance operational activities (Khan & Magd, 2021). On the other hand, the ones who are not that tech-savvy may not be able to take advantage of such technology (Sarker et al., 2019). This applies to the new users of technology as well. Students who are new to online learning may find the online platform difficult to use. The inability to use technology leads to stress and anxiety known as know technology anxiety (Henderson & Milman, 2020). Educators need to put extra effort to explain the learners about the benefits and usage of such platforms and asynchronous discussions. This measure may help to reduce the anxiety related to the new technology. Thus, NTA can be defined as the underlying beliefs and influences resulting from the perceived ease of use of an online system (Bhatt & Shiva, 2020; Venkatesh & Davis, 2000; Venkatesh & Bala, 2008; Demoulin & Djelassi, 2016). This anxiety due to new technology can have a negative influence on the perceived ease of use (Demoulin & Djelassi, 2016) and will also negatively influence technology acceptance (Chen & Chang, 2013). In other words, the users who have a high level of NTA will not be able to adopt the new technology, i.e., MS Teams. Two hypotheses have been framed to address this issue:

H4: NTA has a significant influence on PEU in the context of MS Teams' usage.
H5: NTA has a significant influence on PUS in the context of MS Teams' usage.

5.3 METHODOLOGY AND RESEARCH FRAMEWORK

Based on the background, review of literature, and hypothesis established, the conceptual framework for the present study is proposed (Figure 5.1).

G*Power (version 3.1.9.7) was used to check the minimum number of samples required to achieve the objectives of this study (Faul et al., 2007, 2009). The algorithm of the software suggested a minimum of 134 sample requirements for the statistical efficiency of this study.

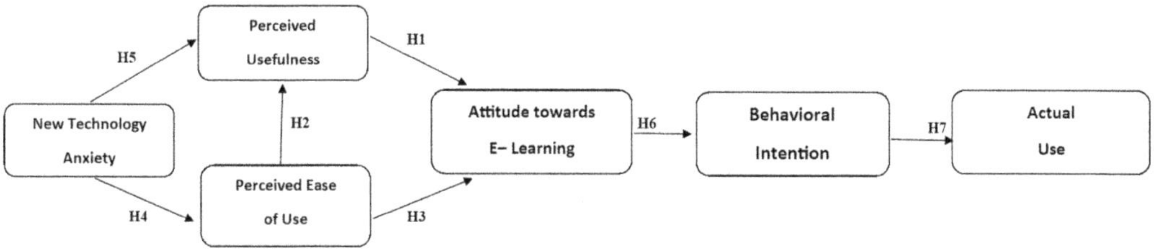

FIGURE 5.1 Proposed conceptual framework.

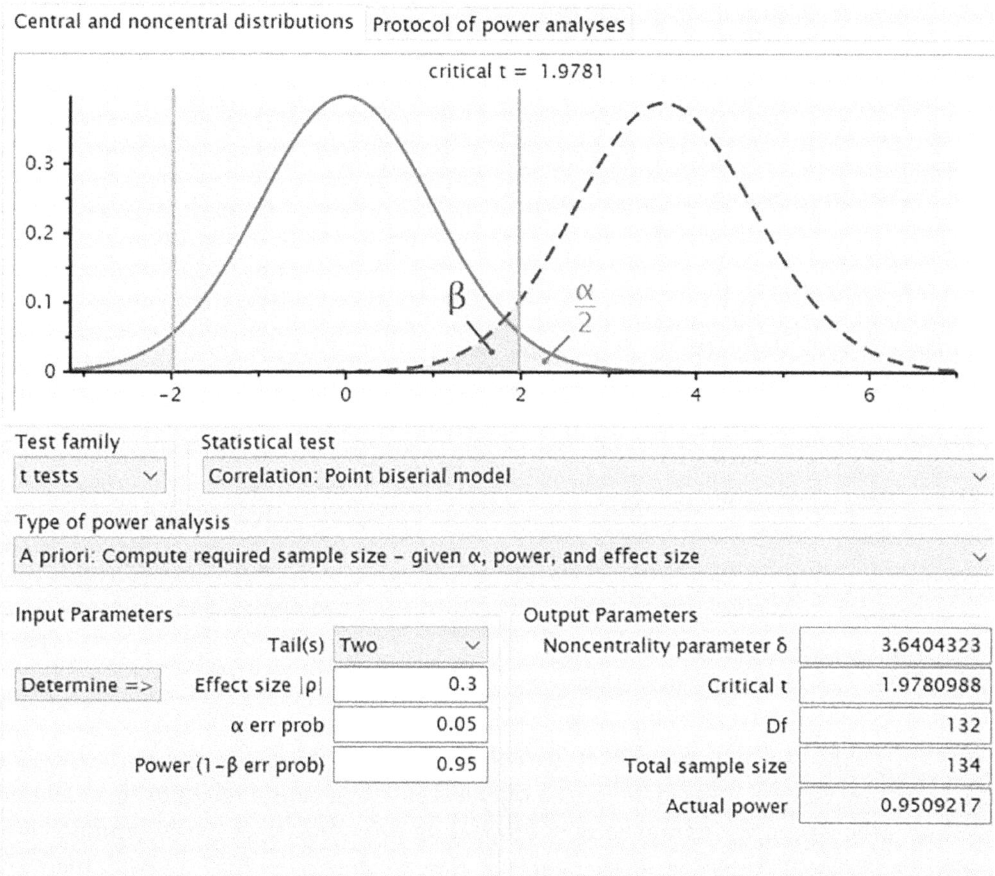

FIGURE 5.2 G* Power minimum sample size analysis. (Source: Authors' Calculations.)

The original sample size administered of 409 was more than the minimum sample size requirement as presented in Figure 5.2.

To test this model and to perform the analysis, a structured questionnaire was distributed to more than 1000 respondents in 15 nations, out of which 409 valid responses were gathered and analyzed for the purpose of this study. The minimum sample size determination was done through the demographic information of the respondents, which can be seen in Table 5.1.

TABLE 5.1 Demographic Profile of the Respondents

Demographic Category	Frequency (*N*)
Category of Respondent	
Faculty/University staff	238
Research scholar	22
Industry representative	6
Student	143
Gender	
Female	172
Male	237
Country of Residence	
Afghanistan	10
Bahrain	15
Bangladesh	60
Bhutan	25
Canada	5
Ethiopia	6
India	118
Indonesia	51
Malaysia	3
Nepal	6
Nigeria	4
Oman	83
Saudi Arabia	12
Tunisia	6
United Arab Emirates	5

N = *409*
Source: Authors' calculations.

For the analysis of the model and the collected data, IBM SPSS version 21 and IBM AMOS version 20 were utilized. Since the model is based on causal relationships and multiple linear regression, structural equation modeling was used as per the recommendation of Hair et al. (2017). Thirty measurement items were categorized into six constructs and were measured on the basis of a seven-point Likert scale where 1 denotes strongly disagree and 7 denotes strongly agree.

The data were collected through the distributed online feedback form, and the respondents were informed about the utilization of their responses for the purpose of research activities like this. To be fair to their privacy and research ethics, an alternate option was given to them to submit their concerns to the email ID of the researchers. The responses were collected as part of the feedback for an international webinar organized in Oman. This webinar was organized using the MS team's platform and was made open to the section of the learning community; there were 1000+ registrations in the webinar, out of which 409 participants filled out this online questionnaire.

5.4 FINDING AND ANALYSIS

5.4.1 Reliability Analysis

For further analysis, it is important that Cronbach's alpha value is more than 0.7 (Hair et al., 2017, 2020) which depicts the reliability value in the present analysis. As presented in Table 5.2, Cronbach's alpha value for all six constructs was found to be above 0.7, with an overall alpha value of 0.966, prescribing the data fit for further analysis.

5.4.2 Mean and Median Analysis

Mean and median analysis is performed to check the orientation of each variable where a value closer to the upper value of the Likert scale (7 in this case) is considered to be on the agreement side, whereas a value closer to the lower side, that is, 1 in the present context is considered to be on the disagreement side. The data are presented in Table 5.3.

As visible from the table above, the lowest mean and median values were obtained from the construct NTA, i.e., new technology usage, where the anxiety was found to be relatively lower; however, the values are still found to be on the agreement side considering the values greater than 4 as agreement and less than 4 as disagreement. For all the other constructs, the mean and median values were found to be oriented toward strong agreement (strongly agree).

5.4.3 Correlation Analysis

Correlation analysis is calculated to understand the inter-relationship between two constructs. This also helps to identify the probability value (*P*-value) and the type of relationship in terms of positive or negative. The correlation analysis for the present study is presented in Table 5.4.

A positive and significant correlation is identified among all the constructs. However, a weak correlation is found between NTA-PUS and NTA-PES; similarly, a relatively weak correlation is found between ATU-NTA, BI-NTA, and AU-NTA. In short, NTA has a weak relationship with all the other TAM constructs. The TAM constructs were found to have a strong correlation among them.

TABLE 5.2 Reliability Analysis

Variable	No. of Items	Cronbach's Alpha
Perceived usefulness (PUS)	5	.949
Perceived ease of use (PES)	6	.960
New technology anxiety (NTA)	4	.912
Attitude toward e-learning (AT)	5	.925
Behavioral intentions	5	.945
Actual use	5	.940
Overall	30	.966

Source: Authors' calculations.

TABLE 5.3 Mean and Median Analysis

Code	Valid	Mean	Median	Std. Deviation
PUS	**Perceived Usefulness**	**6.3604**	**6.8000**	**.94563**
PUS1	Using the MS Teams for this online program improved my understanding of the program conducted	6.350	7.000	1.0302
PUS2	The MS Teams are useful in interacting with the resource persons in a better way	6.311	7.000	1.0864
PUS3	It is useful for the organizers to manage large audiences with less effort	6.411	7.000	.9938
PUS4	It was useful in terms of viewing and accessing the content of the program	6.352	7.000	1.0234
PUS5	The platform is useful in serving the purpose of a webinar and online programs involving large audience	6.379	7.000	1.0505
PES	**Perceived Ease of Use**	**6.3456**	**6.8333**	**1.00482**
PES1	I feel it is easy to get equipped with MS Teams in less time	6.313	7.000	1.1525
PES2	The software is easy to install and get started	6.347	7.000	1.0967
PES3	The MS team's platform is user-friendly	6.323	7.000	1.0794
PES4	It is easy to manage a larger audience with MS Teams	6.421	7.000	1.0118
PES5	I found it easy to adopt and use for myself	6.328	7.000	1.1312
PES6	It is easy to log in, join a session or connect	6.342	7.000	1.1334
NTA	**New Technology Anxiety**	**5.0110**	**5.7500**	**1.95825**
NTA1	I was worried as MS Teams was a new platform for me	5.083	6.000	2.2487
NTA2	I had concerns about whether I will be able to use this platform or not	5.010	6.000	2.2096
NTA3	I feel that another platform like Zoom would have been better for such a program	5.149	6.000	2.0085
NTA4	I wonder would propose to the organizers not to use this platform for future programs	4.802	6.000	2.3237
ATU	**Attitude toward e-learning**	**6.1237**	**6.4000**	**1.10296**
ATU1	I feel MS Teams is a very good platform to conduct such programs	6.171	7.000	1.3155
ATU2	Using this platform has changed my perception of this platform	5.941	6.000	1.4078
ATU3	I feel that it adds value to webinars and online programs	6.213	7.000	1.1616
ATU4	I feel that I have developed an orientation to use this platform for my use as well	6.108	6.000	1.2079
ATU5	It has a positive influence on me on using the Microsoft product	6.186	7.000	1.1756
BI	**Behavioral Intention**	**6.1496**	**6.6000**	**1.13884**
BI1	I will prefer to attend programs/Webinars conducted via MS Teams	6.134	7.000	1.2773
BI2	I would recommend to conduct online programs/ webinars through MS Teams to others	6.164	7.000	1.2967

(*Continued*)

TABLE 5.3 (*Continued*) Mean and Median Analysis

Code	Valid	Mean	Median	Std. Deviation
BI3	I would prefer MS Teams if I have to present something as a resource person or presenter.	6.112	7.000	1.2842
BI4	I feel there are many additional features of MS Teams that need to be explored	6.147	7.000	1.1976
BI5	I will attend the programs organized through MS Teams	6.191	7.000	1.2279
AU	**Actual Use**	**6.1531**	**6.6000**	**1.19490**
AU1	The present version of MS Teams is user-friendly	6.183	7.000	1.2498
AU2	I faced no problem in joining the session	6.078	7.000	1.4207
AU3	I faced no problem in attending sessions via MS Teams	6.115	7.000	1.4173
AU4	I found MS Teams a very dynamic platform for Webinars	6.142	7.000	1.3152
AU5	I liked the appearance of the platform	6.247	7.000	1.2348

Source: Authors' calculations.

TABLE 5.4 Correlation Analysis

	PUS	PES	NTA	ATU	BI	AU
PUS	1					
PES	0.908**	1				
NTA	0.228**	0.157**	1			
ATU	0.825**	0.831**	0.340**	1		
BI	0.819**	0.844**	0.224**	0.914**	1	
AU	0.811**	0.859**	0.183**	0.864**	0.901**	1

** *Correlation is significant at the 0.01 level (2-tailed).*
Source: Authors' calculations.

5.4.4 Structural Equation Modeling for Measurement Model

The structural equation model (SEM) is a model that is a combination of regression analysis and factor analysis (Khan & Magd, 2021). The model suggests whether a proposed model qualifies on the parameters prescribed to be called a good fit or not. For the purpose of this study, a covariance-based SEM was performed using IBM Amos version 20. The decisive variables to ascertain model fit are Chi-square fit statistics/degree of freedom (χ^2 or Cmin/df), goodness-of-fit index (GFI), comparative fix index (CFI), and root-mean-square error of approximation (RMSEA). The χ^2 (cmin/df) statistics should be less than 5.0; the GFI and CFI must be close to 1, but higher than 0.9 is recommended (Hair et al., 2006, 2010) for RMSEA, and less than 0.05 is considered to be ideal; however, a RMSEA less than 0.08 is considered as reasonable (Bentler & Bonett, 1980; Browne & Cudeck, 1993).

5.4.4.1 Path Diagram and Model Fit Indices for PES

As presented in Table 5.5, the model fit indices were found within the acceptable range (Hair et al., 2010) with two co-variances drawn; none of the items were deleted or withdrawn. All the items are found to be defining the construct PES with a factor load above 0.7. This makes the construct PES fully qualified for the conceptual model assessment. The reflective path diagram can be seen in Figure 5.3.

5.4.4.2 Path Diagram and Model Fit Indices for PUS

As presented in Table 5.6, the model fit indices were found within the acceptable range (Hair et al., 2010) with only one covariance drawn between residual e1 and e2; none of the items were deleted or withdrawn. All the items are found to be defining the construct PUS with a factor load above 0.7. This makes the construct PUS fully qualified for the conceptual model assessment. The reflective path diagram can be seen in Figure 5.4.

TABLE 5.5 Model Fit Indices of the Construct (PES)

	Cmin/df	GFI	CFI	RMSEA
Overall model	3.140	0.983	0.994	0.072

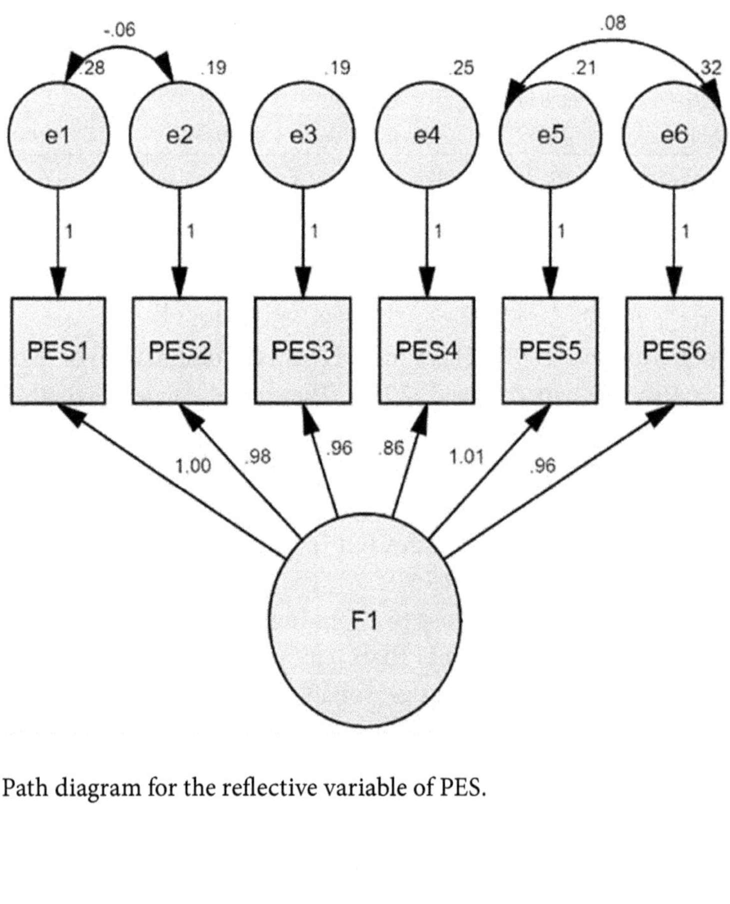

FIGURE 5.3 Path diagram for the reflective variable of PES.

TABLE 5.6 Model Fit Indices of the Construct (PUS)

	Cmin/df	GFI	CFI	RMSEA
Overall model	.961	0.996	1.000	0.000

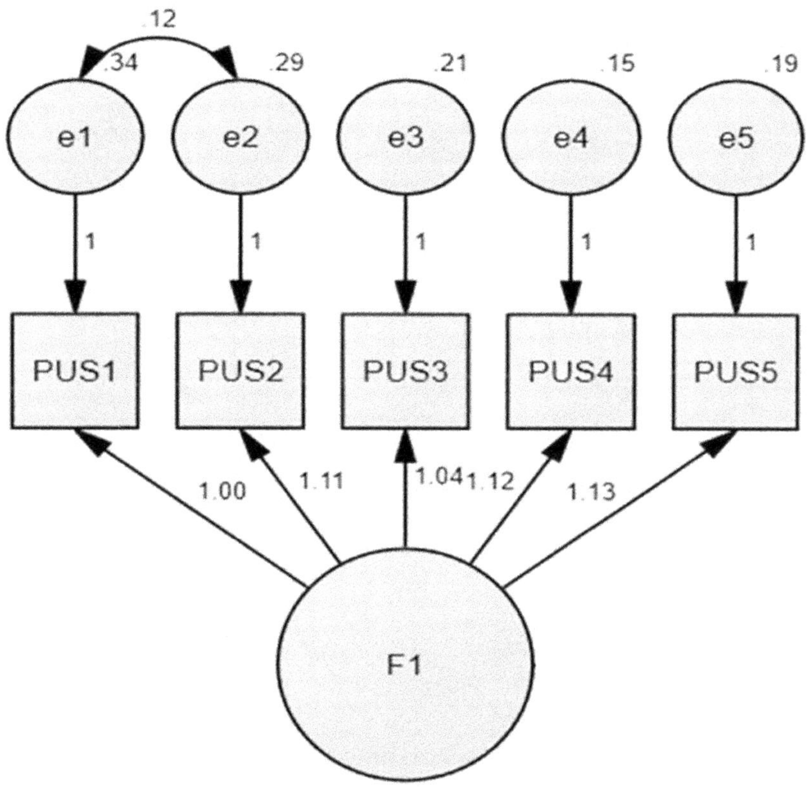

FIGURE 5.4 Path diagram for the reflective variable of PUS.

5.4.4.3 Path Diagram and Model Fit Indices for NTA

As presented in Table 5.7, the model fit indices were found within the acceptable range for the four items of NTA (Hair et al., 2010) with only one covariance drawn between residual e3 and e4 as part of the modification indices; none of the items were deleted or withdrawn. All the items were found to be defining the construct NTA with a factor load above 0.7. This makes the construct NTA fully qualified for the conceptual model assessment. The reflective path diagram can be seen in Figure 5.5.

5.4.4.4 Path Diagram and Model Fit Indices for ATU

As presented in Table 5.8, the model fit indices were found within the acceptable range for the four items of ATU (Hair et al., 2010) with only one covariance drawn between residual e1 and e3 as part of the modification indices; none of the items were deleted or withdrawn.

TABLE 5.7 Model Fit Indices of the Construct (NTA)

	Cmin/df	GFI	CFI	RMSEA
Overall model	.580	0.999	1.000	0.000

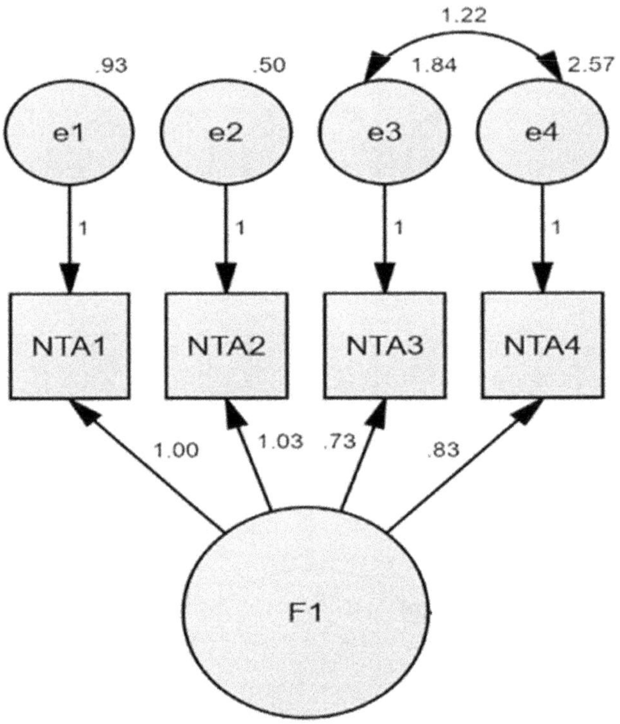

FIGURE 5.5 Path diagram for the reflective variable of NTA.

All five items are found to be defining the construct ATU with a factor load above 0.7. This makes the construct ATU fully qualified for the conceptual model assessment. The reflective path diagram can be seen in Figure 5.6.

5.4.4.5 Path Diagram and Model Fit Indices for BI

As presented in Table 5.9, the model fit indices were found within the acceptable range for the four items of BI (Hair et al., 2010); there was no covariance needed to achieve the prescribed threshold values, and none of the items were deleted or withdrawn. All five items are found to be defining the construct BI with a factor load above 0.7. This makes the construct BI fully qualified for the conceptual model assessment. The reflective path diagram can be seen in Figure 5.7.

5.4.4.6 Path Diagram and Model Fit Indices for AU

As presented in Table 5.10, the model fit indices were found within the acceptable range for the four items of AU (Hair et al., 2010); there was no covariance needed to achieve the

TABLE 5.8 Model Fit Indices of the Construct (ATU)

	Cmin/df	GFI	CFI	RMSEA
Overall model	3.118	0.988	0.995	0.072

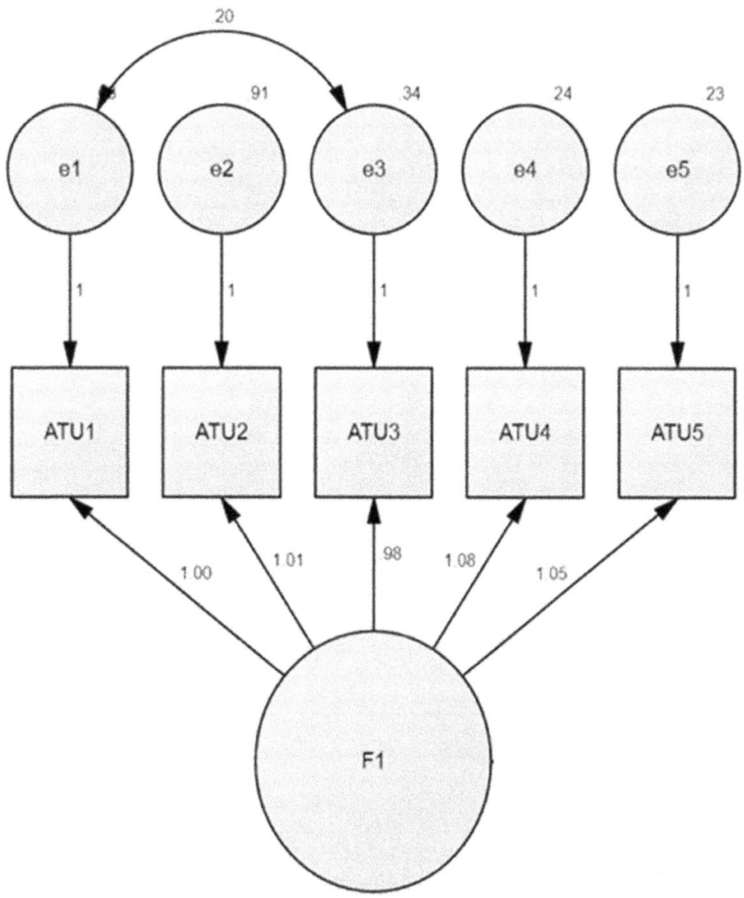

FIGURE 5.6 Path diagram for the reflective variable of ATU.

TABLE 5.9 Model Fit Indices of the Construct (BI)

	Cmin/df	GFI	CFI	RMSEA
Overall model	2.026	0.990	0.998	0.050

prescribed threshold values, and none of the items were deleted or withdrawn. All five items are found to be defining the construct BI with a factor load above 0.7. This makes the construct BI fully qualified for the conceptual model assessment. The reflective path diagram can be seen in Figure 5.8.

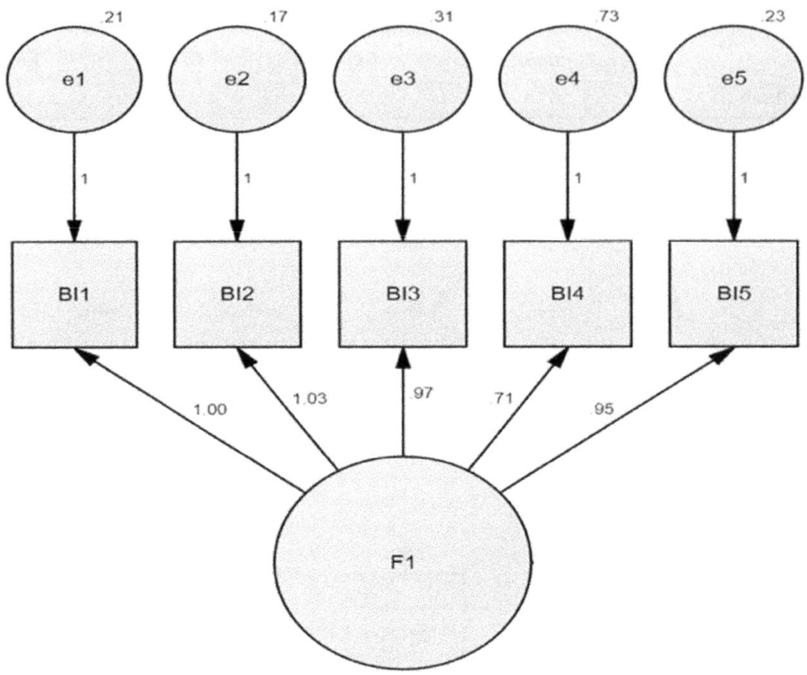

FIGURE 5.7 Path diagram for the reflective variable of BI.

TABLE 5.10 Model Fit Indices of the Construct (AU)

	Cmin/df	GFI	CFI	RMSEA
Overall model	3.450	0.990	0.996	0.077

5.4.4.7 Measurement Model for the Conceptual Framework

After testing all the individual constructs and their factor loads, the conceptual model was tested as shown in Figure 5.9 and presented in Tables 5.11 and 5.12. This analysis was also performed to test the proposed alternate hypothesis.

As visible in Table 5.11, the model fit indices are in the acceptable range (Hair et al., 2010). Hence, the proposed model is acceptable. None of the items were excluded from the final model in terms of factor load (Hair et al., 2010), and all 30 items were retained by creating one covariance. This was obtained by preparing the path diagram as presented in Figure 5.9, and the standardized model value and hypothesis testing can be seen in Table 5.12. Hypotheses H1, H2, H3, H4, H5, H6, and H7 were accepted based on the *P*-value at 0.000. Also, the study was able to give a moderate-to-good fit to the proposed theoretical framework in the context of MS Teams.

5.4.5 Assessment of NTA as a Construct in the TAM Model

An additional assessment of NTA was performed using importance metrics on the final construct, i.e., AU. As presented in Figure 5.10.

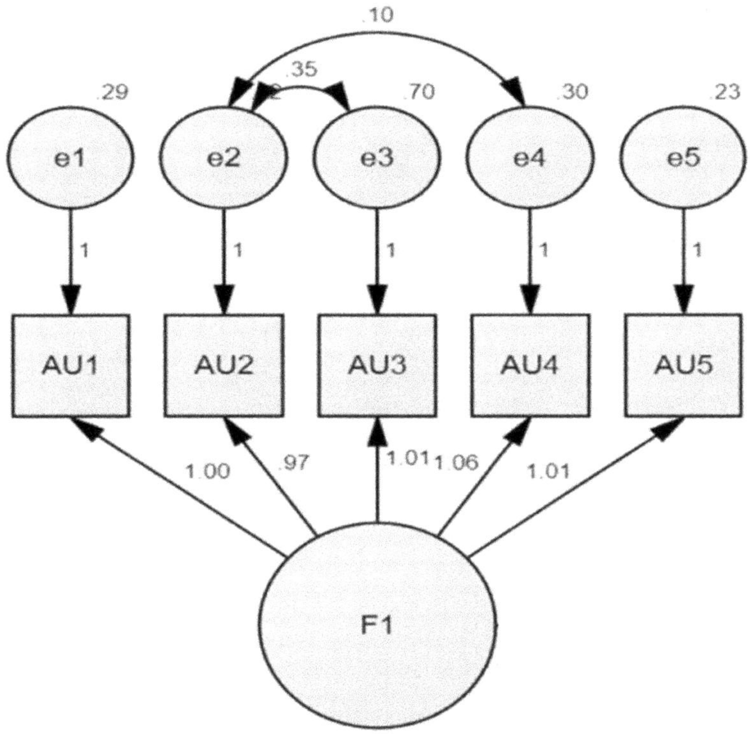

FIGURE 5.8 Path diagram for the reflective variable of AU.

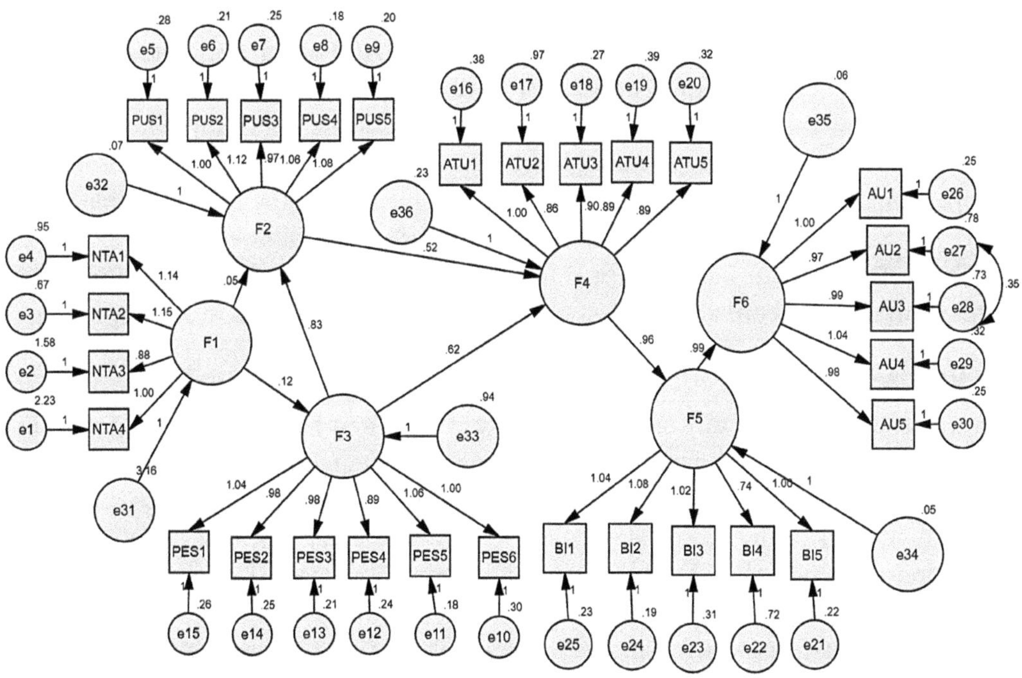

FIGURE 5.9 SEM. Note: F1 = NTA; F2 = PUS; F3 = PES; F4 = ATU; F5 = BI; F6 = AU.

TABLE 5.11 Model Fit Indices of the Structural Model

	Cmin/df	GFI	TLI	CFI	RMSEA
Overall model	4.010	0.907	0.916	0.923	0.086

Note: TLI=Tucker-Lewis index.

TABLE 5.12 Hypotheses Testing Using PLS Structural Model for the Proposed Model

Hypothesis		Path		Estimate	S.E.	C.R.	*P*-value	Decision
H1	ATU	<:	PUS	0.521	0.142	3.669	***	Accepted
H2	PUS	<:	PES	0.828	0.037	22.266	***	Accepted
H3	ATU	<:	PES	0.616	0.126	4.873	***	Accepted
H4	PES	<:	NTA	0.12	0.029	4.124	***	Accepted
H5	PUS	<:	NTA	0.046	0.011	4.192	***	Accepted
H6	BI	<:	ATU	0.956	0.034	28.09	***	Accepted
H7	AU	<:	BI	0.987	0.032	31.07	***	Accepted

*** *Significant at 0.000.*
Source: Authors' calculations.

As seen in Figure 5.10, NTA emerged as the least important construct for the AU. Moreover, it is also found that NTA has a negative impact on AU, which seems to be logically correct, as if the NTA is high among users, they will not adopt a technology to use. To contextualize in the present term, if the NTA for MS Teams is high among users, the users will prefer not to use MS Teams as an online platform. However, among the respondents for the current study, the level of NTA was found to be low; i.e., the respondents did not show a very high level of anxiety, or in other words, they were able to use this platform in an un-stressful manner. The most important construct for the AU was BI which emerged as the most important predicting variable/construct; the second most predictive construct or important construct is PES followed by ATU. This means that the effective implementation of AU, BIs, and perceived ease of use play a dominant role, followed by attitude toward e-learning. The same has been presented in Figure 5.11. With regards to the accuracy level of this analysis was recorded at 75%.

5.5 RESEARCH CONTRIBUTION

This study supports the study of Bhatt and Shiva (2020) and Khan and Magd (2021). The TAM is recognized as a valid model for technology acceptance-related studies in the context of online and distance education. The analysis suggests that BI and perceived ease of use are predictors of the AU of MS Teams. Attitude toward e-learning is also the third most important construct. The learners now do not show the NTA as it was found to be the least important construct followed by PUS. This means that the learners already consider such platforms useful and have sufficient exposure to use them. The learners also have a feeling that such platforms are not that difficult, and with some basic training or exposure, using such platforms will not be a big issue. This can be translated in the

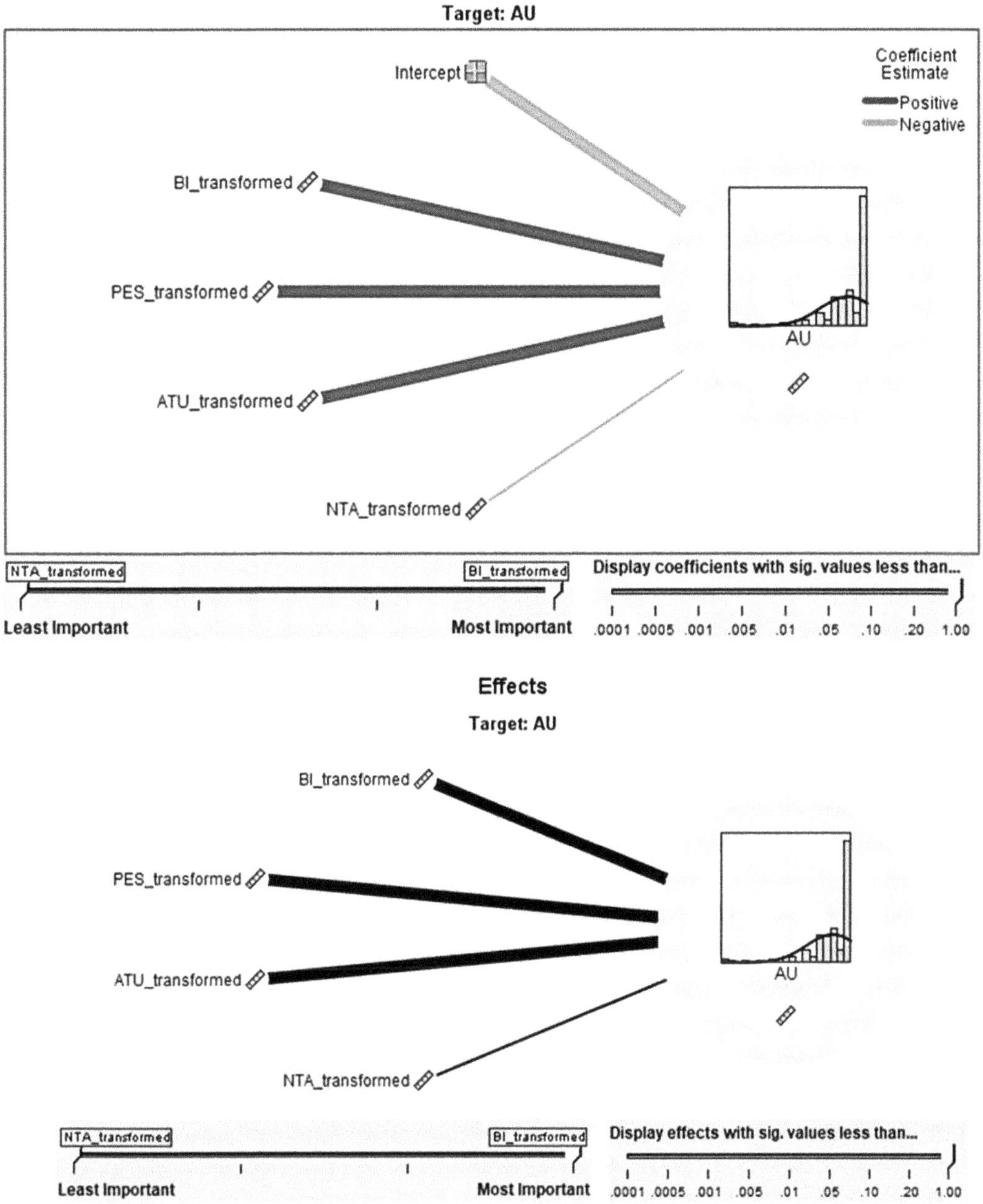

FIGURE 5.10 Importance and effect diagram on AU.

context of MS Teams as the users find it user-friendly and relatively easy to use. It can be also interpreted that the respondents covered in this study were already part of the education system; thus, they might be exposed to such platforms or might have received training to use such online platforms (Ratten, 2020).

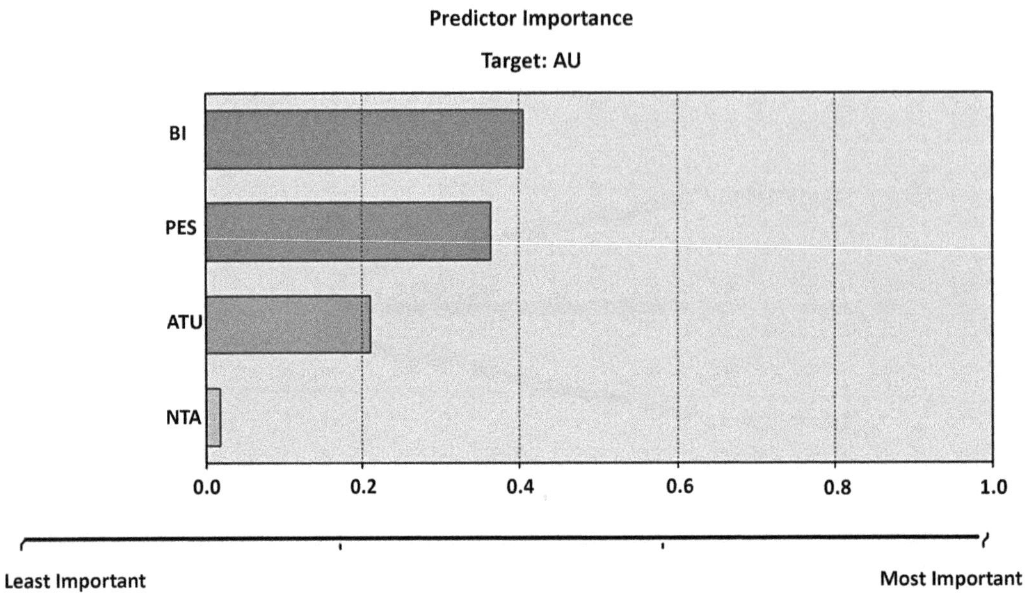

FIGURE 5.11 Importance and effect bar chart on AU.

5.6 LIMITATIONS OF THE STUDY

This study does not take into account full-time distance education learners as the respondents were online participants of an international workshop series organized in Oman. The participants were already either teachers or students in some institute and, thus, might have received formal training in using such platforms or have had enough exposure by attending various online sessions. As in distance education, many learners are expected to get admission, who might not have any previous exposure to dealing with such platforms; the results could be otherwise also. This can be said for continuing education programs or adult education programs that are actually designed for those who had to opt out of the education system due to some reasons. Thus, the results might not be applicable to such a population.

5.7 CONCLUSION & FUTURE SCOPE

Whenever there is a birth of new technology, experiencing apprehension among users is a normal phenomenon. If the anxiety level is high, the use of technology will be less unless and until dedicated efforts are taken to reduce those anxiety levels. In the context of distance education also, the anxiety among the learners is going to determine whether they will make use of the online platforms available to them or not. This is one of the reasons that acceptance of technology has been a point of concern and research among HEIs and education researchers. The digitalization of education that happened during the time of COVID-19 was unprecedented; one of the reasons could be the absence of technology in previous times or epidemic or pandemic. However, an equal concern was the acceptance of technology, as the transition was sudden, and prompt action was to be taken. All the concerns were either not trained or were not prepared for such a transition. A new ecosystem of peer learning emerged that included both learners and instructors, who became part of the process and learned the ways an online platform can be used efficiently and effectively. Under normal circumstances, such instances are expected to witness huge resistance from the side of the users, but the pandemic compelled the users to shift online, and adopt work from home. In a way, all the educational processes were shifted to distance mode. This perhaps could be the reason why the level of NTA was found less in the present study, as the users were aware, had proper devices to use online platforms, and had a positive attitude toward e-learning. Croxton (2014) says that online education will be gaining more prevalence in times to come, and measurement of the effectiveness of existing online platforms becomes essential. Such research is expected to give insight to the platform developers to harness innovative practices and improve their platform by making it more user-friendly and more efficient in terms of online tools. Online education offers a variety of benefits to the users and the institutes as it works on the factors like cost, travel, physical facilities, and convenience and caters to a larger population (Cavanaugh, 2009; Alawamleh et al., 2020; Nguyen, 2015). MS Teams being offered as a part of the MS Office platform is one of the Unique selling propositions that give it an edge over peers like Zoom, Cisco, WebEx, etc. HEIs do not have to consider buying a new platform; rather, they can just take a subscription to MS Office, where MS Teams come as part of the package. Thus, in terms of cost-effectiveness in online learning platforms, MS Teams is clearly a winner. However, concerns have been raised about whether the same is good in terms of acceptance among users.

The TAM model proposed in this study to analyze MS Teams has given a few interesting findings. At an overall level, the acceptance of MS Teams was found to be sufficiently good. All the constructs proposed in the study have found a significant and positive correlation among themselves. The seven hypotheses proposed in the study were also accepted based on the *P*-value, the SEM analysis also suggested that the model was a moderate-to-good fit, which means that a similar model can be used in future research. BI and perceived ease of use were found to be predictors of the AU of MS Teams. NTA, however, emerged as the least important predictor in this study. However, the statistical findings suggest that NTA cannot be ignored as it might become a point of concern in the future if not taken care of properly. Thus, for any online platform, sufficient training and awareness need to be created among learners to reduce their anxiety levels.

As a future scope, pure distance education students can be a part of the respondents in future studies. The way students behave from different streams may also have an impact on the findings of similar studies as students with technical backgrounds like IT might behave differently as compared to the students of liberal arts. As technology is evolving every day, more and more innovation is expected in this field. There could be better platforms or refined platforms emerging in the future, whose acceptance of technology needs to be tested. It would be also suggested to utilize "modified TAM" or "unified theory of acceptance and use of technology," i.e., UTAUT (Venkatesh et al., 2003; Venkatesh & Davis, 2000) in future studies.

REFERENCES

Aguilera-Hermida, A. P. (2020). College students' use and acceptance of emergency online learning due to COVID-19. *International Journal of Educational Research Open*. doi:10.1016/j.ijedro.2020.100011.

Alawamleh, M., Al-Twait, L. M., & Al-Saht, G. R. (2020). The effect of online learning on communication between instructors and students during Covid-19 pandemic. *Asian Education and Development Studies*. doi:10.1108/AEDS-06-2020-0131.

Anderson, B., & Simpson, M. (2012). History and heritage in distance education. 16(2), 1–10. Retrieved from http://files.eric.ed.gov/fulltext/EJ1080085.pdf.

Beller, M. (2013). Technologies in large-scale assessments: New directions, challenges, and opportunities. In M. V. Davier, E. Gonzalez, I. Kirsch, & K. Yamamoto (Eds.), *The Role of International Large-Scale Assessments: Perspectives from Technology, Economy, and Educational Research* (pp. 25–45). Dordrecht: Springer Science+Business Media. doi:10.1007/978-94-007-4629-9_3.

Bentler, P. M., & Bonnet, D. G. (1980). Significance test and goodness of fit in the analysis of covariance structures. *Psychological Bulleting*, 88, 586–606.

Berge, Z. L. (2007). Barriers and the organization's capabilities for distance education. *Distance Learning*, 4(4), 1.

Bhatt, S. & Shiva, A. (2020). Empirical examination of the adoption of zoom software during COVID-19 pandemic: Zoom tam. *Journal of Content, Community & Communication*, 12, 70–88. DOI: 10.31620/JCCC.06.20/08

Bozkurt, A., Akgun-Ozbek, E., Yilmazel, S., Erdogdu, E., Ucar, H., Guler, E., ... Aydin, C. H. (2015). Trends in distance education research: A content analysis of journals 2009–2013. *IRRODL*, 16(1), 330–363. Retrieved from http://files.eric.ed.gov/fulltext/EJ1061066.pdf.

Browne, M. W., & Cudeck, R. (1993). Alternative ways of assessing model fit. In K. A. Bollen, & J. S. Long (Eds.). *Testing Structural Equations Models*. Newbury Park, CA: Sage.

Bsharat, T. R., & Behak, F. (2020). The impact of Microsoft teams' app in enhancing teaching-learning English during the coronavirus (covid-19) from the English teachers' perspectives in Jenin city. *Malaysian Journal of Science, Health & Technology*, 7(1), 102–109.

Caruth, G. D., & Caruth, D. L. (2013). The impact of distance education on higher education: A case study of the United States. 14(4), 121–131. Retrieved from http://files.eric.ed.gov/fulltext/EJ1042587.pdf.

Cavanaugh, C. (2009). *Getting Students More Learning Time Online: Distance Education in Support of Expanded Learning Time in K-12 Schools*. Washington, DC: Center for American Progress, pp. 1–28.

Chen, K., & Chang, M. (2013). User acceptance of 'near field communication' mobile phone service: An investigation based on the 'unified theory of acceptance and use of technology' model. *Service Industries Journal*, 33(6), 609–623.

Clark, J. T. (2020). Distance education. *Clinical Engineering Handbook (Second Edition)* (pp. 410–415). doi:10.1016/B978-0-12-813467-2.00063-8.

Cremonini, L., Westerheijden, D., Benneworth, P., & Dauncey, H. (2014). In the shadow of celebrity? World class university policies and public value in higher education. *Higher Education Policy*, 27(3), 341–361.

Croxton, R.A. (2014). The role of interactivity in student satisfaction and persistence in online learning. *Journal of Online Learning and Teaching*, 10(2), 314.

Davis, F. D. (1989). Perceived usefulness, perceived ease of use, and user acceptance of information technology. *MIS Quarterly*, September, 319–340.

Davis, F. D., Bagozzi, R. P. and Warshaw, P. R. (1989). User acceptance of computer technology: A comparison of two theoretical models. *Management Science*, 35(8), 982–1003.

Demoulin, N. T.M., & Djelassi, S. (2016). An integrated model of self-service technology (SST) usage in a retail context. *International Journal of Retail & Distribution Management*, 44(5), 540–559. doi:10.1108/IJRDM-08-2015-0122.

Farahat, T. (2012). Applying the technology acceptance model to online learning in the Egyptian Universities. *Procedia - Social and Behavioral Sciences*, 64, 95–104. doi:10.1016/j.sbspro.2012.11.012.

Farooq, S., & College, H. (2021). A technology acceptance model for e-learning during covid-19. Empirical insight from Pakistan. *Elementary Education Online*. 20(4), 975–984. doi:10.17051/ilkonline.2021.04.105.

Faul, F., Erdfelder, E., Buchner, A., & Lang, A.- G. (2009). Statistical power analyses using G*Power 3.1: Tests for correlation and regression analyses. *Behavior Research Methods*, 41, 1149–1160.

Faul, F., Erdfelder, E., Lang, A.-G., & Buchner, A. (2007). G*Power 3: A flexible statistical power analysis program for the social, behavioral, and biomedical sciences. *Behavior Research Methods*, 39, 175–191.

Granić, A., & Marangunić, N. (2019). Technology acceptance model in educational context: A systematic literature review. *British Journal of Educational Technology*, 50(5), 2572–2593. doi:10.1111/bjet.12864.

Hair, J. F., Howard, M., & Nitzl, C. (2020). Assessing measurement model quality in PLS-SEM using confirmatory composite analysis. *Journal of Business Research* 109, 101–110.

Hair, J. F., Hult, G. T. M., Ringle, C., & Sarstedt, M. (2017). *A Primer on Partial Least Squares Structural Equation Modeling (PLS-SEM)* (2nd ed.). Thousand Oaks, CA: SAGE Publications.

Hair, J., Black, W., Babin, B., & Anderson, R. (2010) *Multivariate Data Analysis* (7th ed.). New Delhi: Pearson Education.

Hair, J.F., Bush, R., & Ortinau, D. (2006). *Marketing Research*. Boston: McGraw-Hill.

Henderson, J., & Milman, N. B. (2020). The technology acceptance model: Considerations for online educators. *Distance Learning*, 17(3), 104–107.

Holmberg, B. (1989). Theory and Practice of Distance Education. London: Routledge.

Jione, J., Fong, L., & Naidu, S. (2019). *Study of Technology Access and Use by USP Students*. Unpublished internal report, The University of the South Pacific. [Google Scholar]

Khan, S.A., & Magd, H. (2021). Empirical examination of ms teams in conducting webinar: Evidence from international online program conducted in Oman. *Journal Of Content Community and Communication*, 14(8), 159–175. doi:10.31620/JCCC.12.21/13.

Korkmaz, G., & Toraman, Ç. (2020). Are we ready for the post-COVID-19 educational practice? An investigation into what educators think as to online learning. *International Journal of Technology in Education and Science*, 4(4), 293–309.

Legon, R., Gareett, R., & Fredericksen, E. E. (2019). *CHLOE 3: Behind the Numbers: The Changing Landscape of Online and Education (2019)*. Annapolis, MD: Quality Matters. Retrieved from https://www.qualitymatters.org/sites/default/files/research-docs-pdfs/CHLOE-3-Report-2019-Behind-the-Numbers.pdf [Google Scholar].

Mehta, J., Yates, T., Smith, P., Henderson, D., Winteringham, G., & Burns, A. (2020). Rapid implementation of Microsoft Teams in response to COVID-19: One acute healthcare organisation's experience. *BMJ Health Care Informatics*, 27(3), e100209. doi:10.1136/bmjhci-2020-100209.

Mohammad AlHamad, A. Q. (2020). Acceptance of E-learning among university students in UAE: A practical study. *International Journal of Electrical and Computer Engineering*, 10(4), 3660–3671. doi:10.11591/ijece.v10i4.pp3660-3671.

Moore, M. G., & Kearsley, G. (2011). *Distance Education: A Systems View of Online Learning*. Belmont, CA: Wadsworth Cengage Learning.

Mousa, A. H., Mousa, S. H., Aljshamee, M., & Nasir, I. S. (2021). Determinants of customer acceptance of e-banking in Iraq using technology acceptance model. *Telkomnika (Telecommunication Computing Electronics and Control)*, 19(2), 421–431. doi:10.12928/TELKOMNIKA.v19i2.16068.

Naidu, S. (2019). The changing narratives of open, flexible and online learning. *Distance Education*, 40(2), 149–152.

Nguyen, T. (2015), "The effectiveness of online learning: Beyond no significant difference and future horizons", *Merlot Journal of Online Learning and Teaching*, 11(2), 309–319.

Pal, D., & Vanijja, V. (2020). Perceived usability evaluation of Microsoft Teams as an online learning platform during COVID-19 using system usability scale and technology acceptance model in India. *Children and Youth Services Review*, 119. doi:10.1016/j.childyouth.2020.105535.

Ratten, V. (2017). Entrepreneurial universities: The role of communities, people and places. *Journal of Enterprising Communities: People and Places in the Global Economy*, 11(3), 310–315.

Ratten, V. (2020). Coronavirus (Covid-19) and the entrepreneurship education community. *Journal of Enterprising Communities: People and Places in the Global Economy*, 14(5), 753–764. doi:10.1108/JEC-06-2020-0121.

Richter, A., Leyer, M., & Steinhüser, M. (2020). Workers united: Digitally enhancing social connectedness on the shop floor. *International Journal of Information Management*, 52. https://doi.org/10.1016/j.ijinfomgt.2020.102101

Romeo, G., Lloyd, M., & Downes, T. (2013). Teaching teachers for the future: How, what, why, and what next? *Australian Educational Computing*, 27(3), 3–12.

Sarker, M.F.H., Mahmud, R.A., Islam, M.S. and Islam, M.K. (2019). Use of e-learning at higher educational institutions in Bangladesh: Opportunities and challenges. *Journal of Applied Research in Higher Education*, 11(2), 210–223. doi:10.1108/JARHE-06-2018-0099.

Saykili, A. (2018). Distance education: Definitions, generations and key concepts and future directions. *International Journal of Contemporary Educational Research,* 5(1), 2–17.

Scherer R., Siddiq F. & Tondeur J. (2018). The technology acceptance model (TAM): A meta-analytic structural equation modeling approach to explaining teachers' adoption of digital technology in education. *Computers & Education*. doi:10.1016/ j.compedu.2018.09.009.

Seaman, J. E., Allen, I. E., & Seaman, J. (2018). *Grade Increase: Tracking Distance Education in the United States*. Oakland, CA: Babson Survey Research Group. Retrieved from http://onlinelearningsurvey.com/reports/gradeincrease.pdf [Google Scholar].

Sharma, D. & Gupta, S. (2021). Impact of digital media on students' engagement via e-learning: A critical literature review using bibliographic analysis. *Journal of Content, Community & Communication,* 13(7), 27–34. doi:10.31620/JCCC.06.21/04.

Shearer, R. L., Aldemir, T., Hitchcock, J., Resig, J., Driver, J., & Kohler, M. (2020). What students want: A vision of a future online learning experience grounded in distance education theory. *American Journal of Distance Education*, 34(1), 36–52.

Shrivastava, G., Ovais, D. & Arora, N. (2021). Measuring the walls of communication barriers of students in higher education during online classes. *Journal of Content, Community & Communication,* 13(7), 27–34, 263–272. doi:10.31620/JCCC.06.21/22.

Siddiq, F., Scherer, R., & Tondeur, J. (2016). Teachers' emphasis on developing students, digital information and communication skills (TEDDICS): A new construct in 21st century education. *Computers & Education,* 92–93, 1–14. doi:10.1016/j.compedu.2015.10.006.

Simonson, M., Smaldino, S., & Zvacek, S. (2015). *Teaching and Learning at a Distance Foundations of Distance Education*. Charlotte, NC: Information Age Publishing, Inc.

Singh, A., Sharma, S., & Paliwal, M. (2020). Adoption intention and effectiveness of digital collaboration platforms for online learning: The Indian students' perspective. *Interactive Technology and Smart Education*. doi:10.1108/ITSE-05-2020-0070.

Singh, V., & Thurman, A. (2019). How many ways can we define online learning? A systematic literature review of definitions of online learning (1988–2018). *American Journal of Distance Education,* 33(4), 289–306.

Stone, C. (2019). Online learning in Australian higher education: Opportunities, challenges and transformations. *Student Success,* 10(2), 1.

Venkatesh, V., & Bala, H. (2008). Technology acceptance model 3 and a research agenda on interventions. *Journal of Information Technology*, 39, 273–315.

Venkatesh, V., & Davis, F. D. (2000). A theoretical extension of the technology acceptance model: Four longitudinal field studies. *Management Science*, 46(2), 186–204.

Venkatesh, V., Morris, M. G., Davis, G. B., & Davis, F. D. (2003). User acceptance of information technology: Toward a unified view. *MIS Quarterly: Management Information Systems*, 27(3), 425–478. doi:10.2307/30036540.

Wei, H. C., & Chou, C. (2020). Online learning performance and satisfaction: Do perceptions and readiness matter? *Distance Education*, 41(1), 48–69.

Yaakop A.Y. (2015). Understanding students acceptance and adoption of web 2.0 interactive edu-tools. In S. Tang, L. Logonnathan (Eds.). *Taylor's 7th Teaching and Learning Conference 2014 Proceedings*. Singapore: Springer. doi:10.1007/978-981-287-399-6_11.

CHAPTER 6

The Paradigm Shift in Higher Education and Impact of Distance Learning in Era of Industry 4.0 and Society 5.0

R. Thiyagarajan and V. Harish

PSG College of Technology

CONTENTS

DOI: 10.1201/9781003322252-6

6.1 INTRODUCTION

The growth and penetration of technology and tech gadgets are growing day by day, which is evident in many sectors that offer tech-savvy solutions to customers in the comfort of being at their homes or workplace. Since many services are offered online and utilized in the distance mode, the need to visit the service providers physically is close to nil. Entire banking services, electricity payments, income tax filing, corporation tax, property tax, consumer durables, groceries, apparel, vegetables, food, etc. are available online at the touch of a smartphone or the click of the mouse. Similarly, education has also gone online and is available in distance mode.

When the world was moving toward Society 5.0 and the Industry 4.0 stage was at its peak, the COVID-19 pandemic hit hard not only on the economy but also on the education sector which was only tech-supported and not completely tech-driven as other industries. The educational institutions were suddenly closed, and the entire teaching and learning system went through distance mode (Kecojevic et al., 2020). Along with the existing challenges from Industry 4.0 and Society 5.0, higher education across the world had to face the new challenges of distance learning, where the teacher and learner were at different locations. Though tech platforms were available to facilitate distance learning, the stakeholders were not trained enough to handle them initially (Lischer et al., 2021).

Teaching and learning started when human civilization evolved, but the methodology and pedagogy kept changing based on the requirements and the challenges from society and industry. India with its 5000 years of civilization has a very strong foundation in education. The ancient Indian education system dates back to the Vedic period in 1500 BC with a residential learning system called Gurukul (Kashalkar-Karve, 2013). Higher education flourished with ancient universities like Takshashila, Nalanda, and Vikramshila in the medieval period which attracted students from many Asian and European countries. With the entry of the British in the 17th century, the education system started changing to reflect western principles and thoughts (Frederick, 2016).

Teaching and learning in hunting and then in agricultural society were only to master survival skills and produce food for self-consumption. The Industrial Revolution that started around the end of the 18th century in Europe paved the foundation for the demand

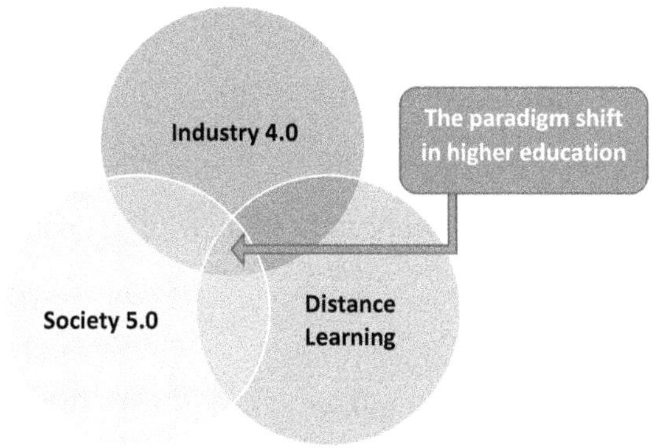

FIGURE 6.1 The paradigm shift in higher education.

for manpower and the needed skill set to perform the job in massive factories (Aquilani et al., 2020). As the number of employees, production, and factory capacity increased, the demand for administration, operations, marketing, logistics, financial planning, and related areas demanded expert human resources at right time in the right place with the right skills.

The paradigm shifts in the higher education sector across the globe saw many transitions from the late 19th century which was more industry-focused to the mid of the 20th century which was more technological and later information-focused (Kumar et al. 2021). The recent trend and needs of Society 5.0 demand education to be smart and human-focused in the 21st century. The recent changeover from offline to online education mainly during the second wave of the pandemic made the stakeholders comfortable with distance learning, though it has some challenges and demerits. Distance learning has grown to be the unavoidable choice in the medium of learning and will continue to grow as also in blended mode along with the traditional way.

Aligning higher education with Industry 4.0, Society 5.0, and blended or distance learning is the need of the hour since the expectations of the stakeholders are mounting and the demand from industry and society also keeps advancing. In order to prepare higher education for anticipated challenges from all corners and ensure the future of learners, this course correction seems very vital. Japan, a country with a more aging population, started this initiative in 2016, and the benefits are better seen since 2020. The need for this future-ready higher education which is the combination of Industry 4.0, Society 5.0, and blended learning is indispensable to make the lives of humans comfortable and peaceful through smart innovation as seen in Figure 6.1.

Objectives of the Chapter

- To understand the transition from Industry 1.0 to Industry 4.0;
- To understand the transition from Society 1.0 to Society 5.0;
- To understand the paradigm shift in education from the industrial era to the smart era;

- To analyze the perception of distance learning since the pandemic;
- To identify the challenges in distance learning;
- To learn from the best practices and case studies across the world;
- And, to study the need of aligning education with Industry 4.0, Society 5.0, and blended learning.

Organization of the Chapter

The rest of the chapter is organized as follows: Section 6.2 of the chapter deals with the transition of the industry from Industry 0 to Industry 4.0 and the expectations of Industry 5.0. Section 6.3 of the chapter deals with the transition of society from Society 1.0 to Society 4.0 and the emergence of Society 5.0. Section 6.4 of the chapter highlights how the COVID-19 pandemic impacted the distance learning mode. Section 6.5 shows the relevant literature review of the chapter title. Section 6.6 explains the paradigm shift in education from the industrial era to the smart era. Section 6.7 discusses the perception of distance learning since the pandemic by stakeholders. Section 6.8 constitutes the challenges in distance learning acknowledged across the three stages. The merits and demerits of distance learning are listed and discussed in section 6.9 and 6.10 respectively. Section 6.11 explains the need of aligning education with Industry 4.0, Society 5.0, and blended learning. The best practices and case studies related to the chapter are discussed in section 6.12 and 6.13, concluding the chapter with future scope.

6.2 TRANSITION FROM INDUSTRY 0 TO 4.0

The transition from Industry 1.0 to 4.0 took around 220 years since the late 18th century; the meaning, core characteristics, types of production, expertise used, focus, and output of each stage are discussed below.

6.2.1 Industry 0 (Before 1.0)

The stage before Industry 1.0 is referred to as Industry 0, where the scope and requirement for industrial products were too low and limited to local or regional markets. This stage existed only to support agriculture and household needs. The focus at this stage was the standard production process with limited resources.

The concept of industry at this stage was the basic level to support the survival and basic needs of human beings like making weapons for protection, utensils for personal needs made out of clay and metals, constructional activities and the materials required for construction, and ornaments using precious stones and metals. It was only on a small scale, mostly focused on agrarian-related sectors, and was also limited to a small region (Figure 6.2).

6.2.2 Industry 1.0

Industrial revolution 1.0 started at the end of the 18th century with the usage of large machinery powered by streams or oil. This stage laid the foundation for the shift from

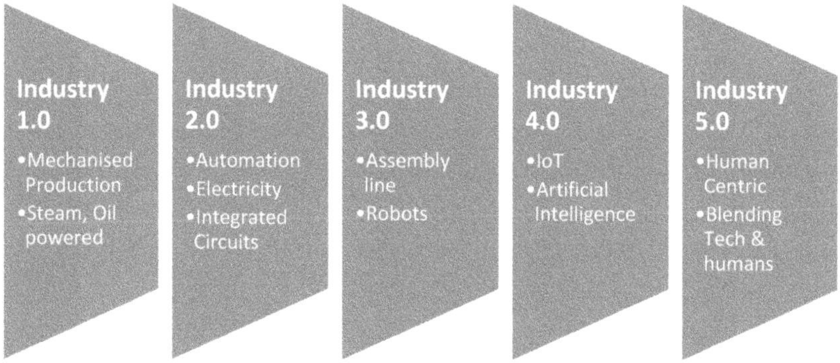

FIGURE 6.2 Stages from Industry 1.0 to 5.0.

TABLE 6.1 Transition from Industry 1.0 to 5.0

Industrial Revolution	Period	Production Type	Expertise/ Powered by	Focus and Output
Industry 0	Till mid-18th century	Handcrafting	Human skills, handmade	Standard production
Industry 1.0	Late 18th century	Mechanized	Stream, oil-powered	Mass production
Industry 2.0	19th century	Automated	Electricity powered	Quality, fast production
Industry 3.0	20th century	Technology supported	Assembly line, robots	Innovation, cost-effective
Industry 4.0	21st century	Technology-driven	IoT, cloud, network, Big Data	Innovation, sustainability
Industry 5.0	Future	Human-centric	AI, Virtual reality	Personalization, inclusive, and smart

agriculture to the industrial economy. Many factories were set up in the western counties and the manpower saw movement from rural to potential urban areas (Carayannis & Morawska-Jancelewicz, 2022). The focus at this stage was mass production, with resources and manpower at the regional level.

This stage saw the growth of industries as potential competitors for the agricultural sector in offering huge employment opportunities. The core advantage of the Industry 1.0 era was that it gave permanent jobs to the workers throughout the year unlike the seasonal nature of the agricultural Industry. The strong criticism was that the working environment was very challenging, and the health of workers was at stake since they were exposed to heat, sound, dust, smoke, and chemicals and had to work for long hours (Table 6.1).

6.2.3 Industry 2.0

The usage of electricity and mass production was the key acceleration in this stage of the Industrial Revolution, and factories started using pieces of machinery powered by electricity. The shift from steam to electricity was the core acceleration for mass production and the speed of industrialization started getting momentum and spread across the world in the late 19th century. The focus at this stage was mass, quality, and fast production.

The industrial sector got a huge acceleration at this stage as the amount of production went higher, and it was also backed by strong demand for the products. New initiatives like selling and marketing activities were started along with exploring new potential markets across the world. The usage of electricity instead of steam and crude oil gave a huge boost to the rapid expansion of industries across all sectors. Industry 2.0 also offered huge employment opportunities and initiated the process of making it more attractive than agriculture by working on reducing and eliminating the hazards in the work environment.

6.2.4 Industry 3.0

Industrial revolution 3.0 started in the 20th century with advanced manufacturing facilities and processes with the support of computers, automation of processes, and assembly lines. Robot usage started in this stage which did challenging jobs with speed, accuracy, and within optimum cost (Lischer, et al., 2021). It started spreading throughout the world. The focus at this stage was innovative and cost-effective production.

This stage saw the transformation from a hazardless environment to a comfortable environment for the employees as the technology was adopted and replaced with humans in the most challenging and risky situations of production. This era also saw the spread of industries across all corners of the world and attracted people to change their perception from traditional agricultural-related jobs to more sophisticated factory jobs in urban areas. The urbanization of society started at a higher level that was never seen before in the previous era.

6.2.5 Industry 4.0

Industry 4.0 is the current trend that supports communication and sharing of key data among the various types of machinery in the same and also in remote locations, powered by advanced technologies like the Internet of Things (IoT), cloud computing, and Big Data (Harish 2019). The interventions of humans got reduced to a great extent in the production process as the technology was empowered to excel (Salimova et al., 2019). The focus at this stage was innovative and sustainable production.

In this stage, Industry 4.0 became the most attractive and highly sought-after career option for job seekers. The growth of information technology companies and the dot.com revolution after 2000 saw a huge transition in the perception of people (Kumar and Nayyar 2020). The growth of the service sector, computerization and digitalization, widespread use of the Internet, mobile phone usage, smartphone usage, social media, and all related developments made a huge impact on the planet. People started enjoying the comfort and ease of life that was made possible by using all technological advancements to support and fulfill the daily challenges of life.

6.2.6 Industry 5.0

Industry 5.0 is the next expected and upcoming trend, which will integrate machines' with highly tech-savvy processes and be more human-centric. The need to be innovative and the demand to produce cost-effective and sustainable products that can be customized to consumer preference is the challenge awaiting in the future. The focus at this stage was smart and personalized production.

From the era of producing and selling in the early stage of Industry 1.0, Industry 5.0 has come a long way to give away future-oriented, human-centric solutions for the problems faced in Society 5.0. The aim is to not just meet, but also to exceed the expectations of people by offering customized, creative, timely, and human-friendly solutions for all their needs. Since the present society is moving from 4.0 (informational era) toward 5.0 (super-smart era), there is a great demand and pressure to upgrade the industry to industry 5.0 to match the level, demands, and quality of Society 5.0.

6.3 TRANSITION FROM SOCIETY 1.0 TO 5.0

Similar to the industrial transformation from 1.0 to 4.0, Japan was the first country to propose a concept in 2016 named Society 5.0 which was a super-smart society for the 21st century with the twin objective of tech-savvy and human-centric society. The challenges faced by humans shall be addressed using Industry 4.0 advancements like IoT, artificial intelligence, Big Data, etc. to give human-centric, innovative, and customized solutions in Society 5.0 (Smuts & Van der Merwe, 2022).

6.3.1 Hunting Society

This traces to the beginning of human evolution where the prime source of survival was hunting. Self-protection and hunting to take care of the basic requirements of food, and clothes were the only objective of this society (Onday, 2019) (Figure 6.3 and Table 6.2).

FIGURE 6.3 Stages from Society 1.0 to 5.0.

TABLE 6.2 Transition from Society 1.0 to 5.0

Society	Period	Society Type	Skills Required
Society 1.0	Since human existence	Hunting society	Skills for hunting and co-existing with nature
Society 2.0	Since agriculture	Agrarian society	Farming, irrigation, and settlements
Society 3.0	18th century	Industrial society	Technical and management skills
Society 4.0	20th century	Informational society	Creativity and tech-savvy
Society 5.0	21st century	Super-smart society	Human-centric and inclusive skills

Source: The evolutionary aspect of the Society 5.0 concept as introduced in the 5th Science and Technology Basic Plan of Japan—Keidanren (2016).

6.3.2 Agrarian Society

Once human beings started agriculture and started to settle down near the river banks, this type called agrarian society started. Understanding the nature and knowing the nuances of agriculture was the prime objective of this society.

6.3.3 Industrial Society

The industrial society began at the end of the 18th century when steam locomotives were used for mass production, and this stage saw the transition from agricultural to industrial society. The demand for technical skills to work in factories started to grow and so the managerial skills to manage the process effectively (Roblek et al., 2020).

6.3.4 Informational Society

This stage saw the use of computers, IoT, AI, and all technical developments to solve the challenges of society. Huge volumes of data were collected, stored, and processed for many purposes (Solanki & Nayyar, 2019). Information Communication Technology (ICT), the Internet, and smartphones had a great impact on people at all levels, and they spread to all corners of the world (Rahmawati et al., 2021). Big Data and data analytics saw a huge demand and contributed more to the well-being of humans.

6.3.5 Super-Smart Society

The future form of society called "Super-smart" or "Society 5.0" is aimed at using the tech advancements of Society 4.0 like IoT, Big Data, AI, cloud, etc. to offer the best solutions for human life (Nayyar et al., 2020). Tech-savvy and human-centric solutions will be given for any problem faced by people in this society (Onday, 2019). The objectives are to leverage technology for the comfort and care of people by offering simple and convenient solutions for their day-to-day activities.

6.4 COVID-19 AND DISTANCE LEARNING

At the end of 2019, the COVID-19 outbreak began. As a precautionary move, all schools and universities in the affected countries were shuttered, forcing the entire education industry to operate remotely. All countries turned to online/distance learning. School and university closures impacted a large number of students in 138 countries (UNESCO, 2020). Closures of educational institutions across the country have already impacted 1,379,344,914 pupils or 80% of all registered students worldwide. A total of 1.37 billion kids were affected, accounting for more than three out of every four children and youth on the planet. Furthermore, nearly 60.2 million instructors have left the classroom (WHO, 2020). As a result of the pandemic, educational institutions were obliged to go online. The pandemic situation contributed more to the digitalization of the educational process and the workplace.

Pupils from lower-income families found this more challenging because they did not have access to the Internet or personal computers or laptops (Kumari, 2021). Students who do not have access to the necessary technology for online learning have limited options for furthering their education. In a short amount of time, the lives of students who were pursuing their school and higher education physically changed online (Deka, 2021). Students who took online programs were frequently forced to move back in with their families, forcing them to leave universities and return to their previous living conditions. This may be far from campus, which has a significant impact on their learning environment (Wieczorek et al., 2021).

6.5 LITERATURE REVIEW

Society 5.0 is the era of integrating humans and technology to address the issues faced by the common man. Researchers saw Society 5.0 as the viewpoint of materialism and existentialism in social science since the situation may turn critical in Society 5.0 if not the problems of the common man are addressed effectively (Rahmawati et al., 2021).

Integrating the concepts of Society 5.0 and Industry 5.0 by the universities can create a mutually beneficial situation to meet all the expectations of the stakeholders in education, society, and industry. As the universities are expected to produce knowledge for developing technology and social innovation, they had to be tuned to meet the requirements of a human-centric society (Carayannis & Morawska-Jancelewicz, 2022). Industry 4.0 and Society 5.0 drive the future of universities in Europe by making human-oriented innovation and achieving sustainability.

How Industry 4.0 and the recent advancements in technology can support the changeover to Society 5.0 and study the role played by open innovation and value co-creation in this challenging transition toward sustainability? Researchers developed a conceptual framework to support the first timing of transforming managers and organizations from Industry 4.0 to Society 5.0 (Aquilani et al., 2020).

In order to have acceptance and smooth adoption of Society 5.0, many organizations need support. Knowledge management concepts can facilitate this transition that integrates cyberspace and physical space. Organizations can leverage knowledge management capacity and learning agility to address the requirements of the new human-centric society (Smuts & Van der Merwe, 2022). The recent developments in technology and the need for switching to sustainability can be used to facilitate the movement to Society 5.0, which demands tech-savvy solutions with a sustainability focus to solve the problems of people. How the incorporation of knowledge management, sustainable development, the Internet, and the emergence of Society 5.0 can benefit society (Roblek et al., 2020).

A paper suggests that European schools offer unity (in having equal resources, opportunities, and possibilities) and not just stop with the uniformity in syllabus and methodology. This level can be achieved in three stages, namely, by enabling a network-centered education system, secondly, cloud computing as the platform, and thirdly, by a new paradigm named "the school on cloud" (Koutsopoulos & Economu, 2016).

A paper studied student behavior and attitude toward online learning from 300 students in Romania in four dimensions, namely, students' personal characteristics, students' requirements, students' knowledge of using online learning platforms, and students' preference for online learning. Though they found it stressful in the beginning, 78% of students felt that distance learning was beneficial for them (Boca, 2021).

India has a long history in teaching and learning as the Gurukul system was in practice since the ancient era. The gurus had a vast accumulation of knowledge that they gathered from their teachers over years. Gurus transcended this wisdom to their students, and this went on and on in the name of "Guru Shishya Parampara" This paper compares the age-old Gurukul system with the recent ones with a special focus on musical education (Kashalkar-Karve, 2013).

A comparative study of the Indian Gurukul system and the western education system found that the former was the most effective one. Since the students had to stay with their gurus in residential learning, the exposure and impact were very high. British rule in India spoiled the Gurukul system and made education less effective by changing the students into customers in learning and gurus into service providers (Frederick, 2016).

Russian industries are aiming at attaining sustainable competitiveness in the era of digitalization and globalization. The research was aimed at studying how a collective approach of competitiveness of industries, sustainable development, and informational society can address the challenges and movement from Industry 4.0 to Society 5.0 (Salimova et al., 2019).

A study was conducted on how Japan pioneered the adoption of Society 5.0 in January 2016 ahead of the world that was in Industry 4.0. Society 5.0 can be explained as a super-smart society that offers a common society infrastructure based on advanced technological developments. Development with a society-focused and tech-savvy is the need of the hour. The transformation from hunting to agrarian to industry to informational society is known, and the next transition will be to a sustainable and human-centric society (Onday, 2019).

Teaching and learning have seen a big paradigm shift from traditional learning to context-aware ubiquitous learning that is offered by tech platforms. The author uses the term "ubiquitous learning" as it is available anywhere at any time. This method replaced the blackboard, the physical load of books, the classroom under a roof, and all infrastructural needs of a school with tech-based hardware and software that support learning (Liu & Hwang, 2010) The author compared the efficiency of online learning and face-to-face learning during the second wave among college students and found that 41% of the students felt that online teaching very effective, 18% of the students felt only face-to-face learning to be effective, and the remaining 41% felt that there was no big difference between the two mediums of teaching (Stevens et al., 2021).

6.6 THE PARADIGM SHIFT IN EDUCATION

Higher education in the 21st century has many challenges and also brought huge opportunities throughout the world. The advancement of technology and communication, the growing concept of a global village, changing lifestyles of people, liberalization of markets demand everything to be of global standard. The education system, more importantly, higher education system, was forced to be updated to reflect global developments in order to compete in the international arena. The higher education system should equip to face the reality in order to teach the students with the needy qualities in order to meet the expectations of Industry 4.0 and Society 5.0 (Figure 6.4).

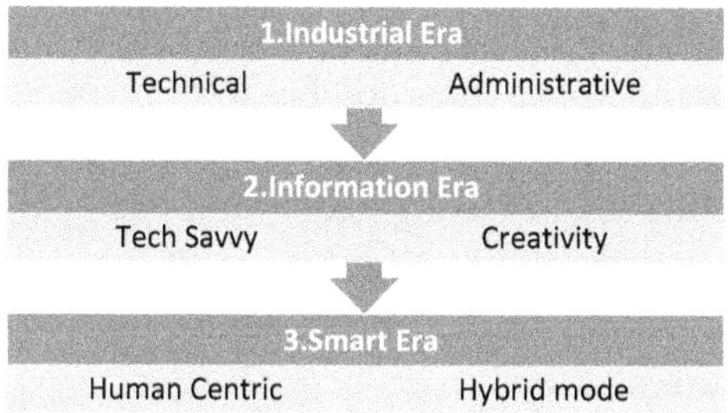

FIGURE 6.4 The paradigm shift in education.

TABLE 6.3 The Paradigm Shift in Education

Stage	Period	Paradigm Shift	Focus
1	1850–1970	Industrial age	Technical and administrative skills
2	1970–2020	Information age	Creativity and tech-savvy
3	2020 onward	Smart age	Human-centric, smart, online, and network

Source: Compiled by the researcher.

6.6.1 Education in the Industrial Era

The higher education system focused on preparing learners with the skills demanded by industries for a long time since the end of the 19th century in India. As the industries were growing, the demand for skilled manpower was also in great need. Higher education institutions were started across the country but in a limited number, so the competition to get in was very high. The Government of India set up the prestigious Indian Institute of Technology, National Institute of Technology (Regional Engineering College formerly), and polytechnic colleges that were exactly training students as per the requirement of the industry through technical educational institutions. The size and demand for the service sector were very limited in those days as the industry was the dominant player (Table 6.3).

6.6.2 Education in the Information Era

Higher education in India witnessed an increasing demand in the information era. After the liberalization of the Indian economy in 1991, the industrial scenario started changing with the entry of multinational companies and the robust growth of Indian players. The effect of liberalization was better felt after 2000, with the giant growth of the service and manufacturing sector making India a competent player in IT, ITES, BSFI, Education, R&D, Tourism, Health care, Automobile, Infrastructure, etc. (Koutsopoulos & Economu, 2016).

This growth provided huge employment opportunities in the above-mentioned private sectors to the workable population in India, especially to the youth workforce. The demand for competent talent also started growing with the rate of the growth of the economy both in quality and quantity to manage and capitalize on the skill of the human resource (Wang, 2005).

6.6.3 Education in the Smart Era

Higher education in India and across the world has to be prepared for this global challenge of preparing students with the skill sets to survive and excel in this smart era of Industry 4.0 and Society 5.0 that aims at integrating people and technology. As this era deals with integrating up-to-date technologies like Big Data, artificial intelligence, the IoT, robots, cloud, etc., the need for educating the students to handle this is a challenge for higher education institutions (Keidanren, 2016).

6.7 DISTANCE LEARNING

As working from home is the new normal for corporate, distance learning has been the new normal for education since March 2020. Though this mode of education was available before the pandemic, the closure of educational institutions forced the world to adopt online mode through distance learning (Mukhopadhyay et al., 2020). The reach of distance learning was at its peak and continues to surge the education space that connects teachers and students in remote areas through a tech platform with numerous advantages and a few challenges as well (Boca, 2021) (Figure 6.5).

6.7.1 Distance Learning in 1st Wave

Distance learning in the first wave of COVID-19 in March 2020 brought a huge cultural shock for the stakeholders of higher education since a majority of them were not prepared or better heard for the transformation (Kleiman et al., 2020). It also brought in huge technological challenges in terms of adapting and using the gadgets that support distance learning (Harish et al., 2020). This stage can be called uncertainty as the stakeholders perceived it to be a sudden compulsion without proper training and breathing space (Table 6.4).

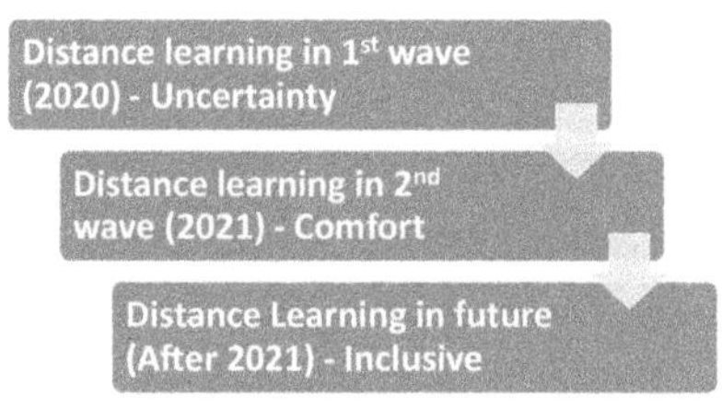

FIGURE 6.5 Stages of distance learning.

TABLE 6.4 Paradigm Shift in Distance Learning

Stage	Perception	Period	Paradigm Shift
1 – 1st wave	Uncertainty	2020	Forced shift from offline to online mode
2 – 2nd wave	Comfort	2021	Planned shift from uncertainty to the comfort of online
3 – Future	Inclusive	After 2021	Both offline and online learning (Inclusive)

Source: Compiled by the researcher.

6.7.2 Distance Learning in the Second Wave

The second stage of distance learning can be named the comfort stage that happened during the second wave of the pandemic in 2021. Most of the stakeholders of higher education were a bit more comfortable with distance learning than the experience they had in stage one. Both the availability of technical gadgets and the level of handling them by instructors and learners increased at this stage (Cicha et al., 2021). All the stakeholders started enjoying the comfort of online learning and started addressing the issues one by one. The perception at this stage had changed from uncertainty to comfort offered by distance learning.

6.7.3 Distance Learning in Future

The need and role of distance learning are growing day by day as a dominant factor in higher education. Considering the advantages, impact, and comfort offered by this mode of learning it is expected to continue and grow exponentially in the future that may be offered as blended or hybrid learning along with traditional or offline learning (Al-Karaki et al., 2021). As this platform is entirely tech-enabled and built on recent technological advancements, this is going to be the future of learning.

6.8 CHALLENGES IN DISTANCE LEARNING

Distance learning was no exception to the challenges to be faced by any new concept or technology. In every stage of its entry, growth, and wide presence, it had to face the challenges that are listed below in Table 6.5.

6.9 MERITS OF DISTANCE LEARNING

Though it might look challenging initially, distance learning has many advantages to offer and transform the teaching–learning process. All that it requires is hands-on training in handling the learning software and supporting electronic gadgets like mobile or laptops (Kireev et al., 2019). Once the students or teachers had mastered the handling of devices and tech platforms, they will start enjoying the benefits and most of them would prefer to

TABLE 6.5 Challenges in Distance Learning

Stage	Challenges
1 – Uncertainty	1. Uncertainty/shock 2. Availability of tech gadgets 3. Training for using an online platform 4. Lack of social interaction
2 – Comfort	1. Mastering the shift in learning 2. Training to be tech-savvy 3. Utilizing allied services 4. Virtual interaction
3 – Inclusive	1. Mastering the shift to online 2. Making distance learning more competitive than offline 3. Updating the pedagogy to match the shift 4. Updating the assessment patterns

Source: Compiled by the researcher.

1	Flexible nature
2	Affordability
3	Time Saving
4	Global Access
5	Wide Talent pool
6	Interactive & wide options
7	Efficient Assessment
8	Reduced Absenteeism
9	Student centered learning

FIGURE 6.6 Merits of distance learning.

go online (Voltmer et al., 2021). Figure 6.6 shows the merits of distance learning, and they are discussed in detail below.

a. **Flexible Nature**
   Distance learning is known for its flexibility as it does not have any locational restrictions. It is accessible anywhere and in case of a recorded session at any time. This flexible nature makes the students study from home similar to the work-from-home option for the new normal in corporate (Liu & Hwang, 2010).

b. **Affordability**
   Distance learning is very affordable as it reduces the cost of learning by minimizing or eliminating the overheads like transportation, maintenance of buildings, outside food, and hard copies of books and materials. Apart from gadgets and the Internet, distance learning reduces most of the costs involved in education.

c. **Time Saving**
   The best advantage of distance learning is that it saves a lot of time in terms of avoiding travel, preparing physical copies of materials, conducting written tests, etc., the instructor can also do multiple assignments in a single day as the time is effectively used.

d. **Global Access**
   A student or teacher is no more confined to a single campus since distance learning offers global access to learning at a very nominal price. The world is open virtually for learning any course offered by any university across the world.

e. **Wide Talent Pool**
   The best faculty or expert in any corner of the world can be available in a short time through distance learning, which provides easy access to the best talent and learning resources in the world.

f. **Interactive and Wide Options**
Distance learning can be very interactive as it has access to videos, podcasts, live demos, animated content, live assignments, and a wide range of tools like online polls, chat options, recordings, and emotional icons available for both students and instructors (Kireev et al., 2019).

g. **Efficient Assessments**
Distance learning offers an efficient and wide variety of assessments as it functions on a tech platform, and the process of conducting assessments is easy for the staff and there is no need to evaluate them as in the conventional method.

h. **Reduced Absenteeism**
Distance learning helps to reduce absenteeism since it does not demand a physical presence, and the flexible nature of distance learning will increase the level of participation in classes.

i. **Student-Centered Learning**
The core advantage of distance learning is that it offers more student-centered learning in contrast with traditional learning which was either teacher- or subject-centered. The tech support and the entire process are designed to keep the interest and comfort of the learner as the first priority (Dumford & Miller, 2018)

6.10 DEMERITS OF DISTANCE LEARNING

Though distance learning has numerous advantages, the demerits cannot be completely ignored. In a situation where the young generation is becoming more addicted to smartphones, other electronic gadgets, and social media, distance learning makes them spend even more time and enable the usage of gadgets in every hand and every corner of the world (Saade, 2017). The core demerits are shown in Figure 6.7 and discussed in detail below.

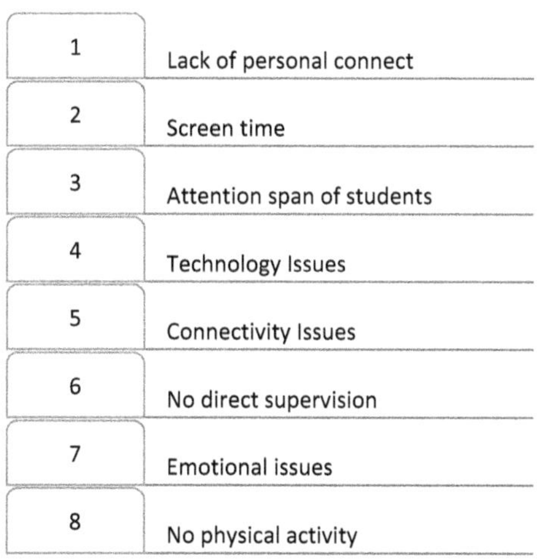

FIGURE 6.7 Demerits of distance learning.

a. **Lack of Personal Connect**
The biggest drawback of distance learning is that it lacks a personal connection between the learner and the teacher. Those comfortable with personal connections and physical classroom setup may not find distance learning appealing (Stevens et al., 2021).

b. **Screen Time**
The screen time spent by students is a great concern in distance learning. Every day many hours need to be spent before laptops or smartphones, which might put the eyes of viewers in trouble (Tus, 2021).

c. **Attention Span of Students**
In distance learning, the attention span of students cannot be known or tested as it varies from one to another student. The teacher has to check this through some kind of interaction or activity in between to get back the attention (Saraswathi et al., 2020).

d. **Technology Issues**
Distance learning is supported by laptops, desktops, tablets, or smartphones; they are prone to any technical issues or physical damages, at any time. The app or software may also malfunction, causing interruptions in learning.

e. **Connectivity Issues**
Internet connectivity is the lifeline of distance learning; in case of disconnection or speed fluctuations, or power cuts, the entire system of distance learning will be affected putting the learner and teacher in a helpless situation.

f. **No Direct Supervision**
Distance learning does not have direct supervision as it happens in traditional classroom learning, virtual supervision may not be that effective. This depends on case to case depending on the student and teacher (Robinson & Hullinger, 2008).

g. **Emotional Issues**
Distance learning has lesser scope to address the emotional issues happening or expected in the teaching–learning process (Salami et al., 2021). A demotivated or isolated student cannot be effectively intervened or timely identified in distance learning.

h. **No Physical Activity**
Learning offline on campus may have more physical activity and long breaks where the mobility of students is more, but in distance learning, the level of mobility is very less or nil. This may affect the health of learners in the future (Fink, 2016).

6.11 ALIGNING EDUCATION WITH INDUSTRY 4.0, SOCIETY 5.0, AND BLENDED LEARNING

In order to meet the growing challenges, the higher education sector across the world has to fine-tune to the requirements of Industry 4.0, Society 5.0, and the new normal in learning. The higher education system needs to be upgraded and also updated to fine-tune

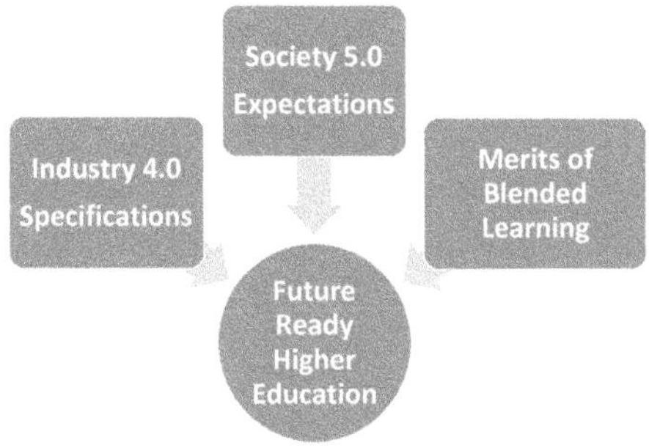

FIGURE 6.8 Future-ready higher education.

the skills required for all the stakeholders. Future-ready higher education is one that can accommodate the specifications of the industry, and requirements of the smart society by exploring the potential of the new normal of hybrid/blended learning (Figure 6.8).

6.12 BEST PRACTICES BY THE JAPAN GOVERNMENT

Japan was the first country to propose the concept of Society 5.0 in 2016 and make it the vision of future education in their country. The former minister for education, culture, sports, science, and technology Mr. Yoshimasa Hayashi said that "We have to give students the skills to both survive that changing society and for them to lead that change," in 2018. He took the initiative to find the feasibility of how the education system can address the needs and values of Society 5.0. A special committee with experts from industry, society, and IT experts in AI, Big Data, and IoT was set up by the ministry in 2017. "In the era of Google, people no longer need to memorize every single fact." Many tasks today are best carried out by computers. Therefore, the emphasis must be on human skills such as communication, leadership, and endurance, as well as curiosity, comprehension, and reading skills, said Mr. Hayashi in 2018. Japan has implemented a new measure where high-school students are segmented into two groups based on their expertise and interest. One group will specialize in humanities and social science, the other will specialize in pure science and mathematics. The universities in Japan are also upgrading their curriculum and approach so as to include the requirements of Society 5.0. Since the country is having a sizable aging population, it mandates them to find solutions for their day-to-day problems and ensure that state-of-the-art technology is put in place to make the lives of its citizens peaceful and comfortable.

6.13 CONCLUSION AND FUTURE SCOPE

The COVID-19 pandemic has forced the entire education system to shift to distance learning. Though it had some concerns initially, the stakeholders have settled down on it as an alternative mode of learning. The concepts of Industry 4.0 and Society 5.0 have proved to be the recent trend across the world demanding that educational institutions more specifically higher education adapt to the specifications and objectives proposed for an inclusive future. Japan's success in adapting to Society 5.0 proves to the world that this is going to be the future to be adopted by other countries of the world.

Policymakers and educationists have the challenging task of redesigning the education system to address the concerns of Society 5.0, meet the expectations of Industry 4.0 and adapt to the new normal of either distance or blended learning. The future-ready education system for the 21st century is expected to empower the learners with skills to adapt and also to lead an inclusive, super-smart society that addresses every problem through state-of-the-art technology and offers a human-centric solution.

6.13.1 Future Scope

Though there were many research studies done over the recent years on Industry 4.0, Society 5.0, and distance learning, there is huge potential for research in this area, especially in the post-pandemic and growing marketplace of distance education. Some of the areas of scope for future research are as follows:

- A detailed investigation of how higher educational institutions had utilized distance learning post-pandemic.
- A comparison between the level of acceptance of distance learning during the first and second wave of pandemic.
- Study and compare the impact of distance learning and traditional learning in higher education.
- Detailed study on the impact of blended learning in comparison with distance learning and/or offline learning.
- An in-depth study on how Industry 5.0 can upgrade to match the needs of Society 5.0.
- Study the paradigm shift of higher education in any domain like management, or engineering to align with Society 5.0.
- Study the success stories and impact after implementing Society 5.0 by any country and Industry 5.0 by any organization.

REFERENCES

Al-Karaki, J. N., Ababneh, N., Hamid, Y., & Gawanmeh, A. (2021). "Evaluating the effectiveness of distance learning in higher education during covid-19 global crisis: UAE educators' perspectives", *Contemporary Educational Technology*, 13(3), ep311.

Aquilani, B., Piccarozzi, M., Abbate, T., & Codini, A. (2020). "The role of open innovation and value co-creation in the challenging transition from industry 4.0 to society 5.0: Toward a theoretical framework", *Sustainability*, 12(21), 8943.

Boca, G. D. (2021). "Factors influencing students' behavior and attitude towards online education during COVID-19", *Sustainability*, 13(13), 7469.

Carayannis, E. G., Morawska-Jancelewicz, J. (2022). "The Futures of Europe: Society 5.0 and Industry 5.0 as Driving Forces of Future Universities". *Journal of the Knowledge Economy*, 13, 3445–3471. https://doi.org/10.1007/s13132-021-00854-2

Cicha, K., Rizun, M., Rutecka, P., & Strzelecki, A. (2021). "Covid-19 and higher education: First-year students' expectations toward distance learning", *Sustainability*, 13, 1889.

Deka, K. (2021). "Covid-19 fallout: The impact on education in India", *India Today*, 11(January), 1–3.

Dumford, A. D., & Miller, A. L. (2018). "Online learning in higher education: Exploring advantages and disadvantages for engagement", *Journal of Computing in Higher Education*, 30, 452–465.

Fink, G. (2016). *Stress: Concepts, Definition and History Change*. Australia: Elsevier Publications.

Frederick, A. (2016). "A comparative study between gurukul system and western system of education", *IRA International Journal of Education and Multidisciplinary Studies*, 3(1), 50–58.

Ha, N. H., Nayyar, A., Nguyen, D. M., & Liu, C. A. (2019). "Enhancing students' soft skills by implementing CDIO-based integration teaching mode". In *The 15th International CDIO Conference*, Aarhus University, Denmark (p. 569).

Harish, V. (2019). "Industry 4.0: Developing sustainable competitive strategies", *Asian Journal of Multidimensional Research*, 8(5), 115–123.

Harish, V., Mansurali, A., & Krishnaveni (2020). *E-Learning – Types, Requirements for Effective E-Learning Teaching, Roadblocks and Future Trends, New Paradigm in Business Education*. New Delhi: National Press Associates. ISBN no: 978-81-944303-8-4.

Kashalkar-Karve, S. (2013). "Comparitive study of Ancient Gurukul system and the new trends of Guru-Shishya Parampara", *American International Journal of Research in Humanities, Arts and Social Sciences*, 2013, 81–84.

Kecojevic, A., Basch, C. H., Sullivan, M., & Davi, N. K. (2020). "The impact of the covid-19 epidemic on the mental health of undergraduate students in New Jersey, cross-sectional study", *PLOS One*, 15(9), e0239696.

Keidanren. (2016, April 19). *"Toward realization of the new economy and society": Reform of the economy and society by the deepening of "Society 5.0"*. https://www.keidanren.or.jp/en/policy/2016/029_outline.pdf

Kireev, B., Zhundibayeva, A., & Aktanova, A. (2019). "Distance learning in higher education institutions: Results of an experiment", *Journal of Social Studies Education Research*, 10(3), 387–403.

Kleiman, E. M., Yeager, A. L., Grove, J. L., Kellerman, J. K., & Kim, J. S. (2020). "Real-time mental health impact of the covid-19 pandemic on college students: Ecological momentary assessment study", *JMIR Mental Health*, 7(12), e24815.

Koutsopoulos, K., & Economu, V. (2016). "School on the cloud: Towards unity not uniformity in education", *British Journal of Education, Society & Behavioural Science*, 16, 1–11.

Kumar, A., Krishnamurthi, R., Bhatia, S., Kaushik, K., Ahuja, N. J., Nayyar, A., & Masud, M. (2021). Blended learning tools and practices: A comprehensive analysis. *IEEE Access*, 9, 85151–85197.

Kumar, A., & Nayyar, A. (2020). "si^3-Industry: A sustainable, intelligent, innovative, internet-of-things industry", In Anand Nayyar and Akshi Kumar (Eds), *A Roadmap to Industry 4.0: Smart Production, Sharp Business and Sustainable Development* (pp. 1–21). Cham: Springer. https://doi.org/10.1007/978-3-030-14544-6

Kumari, S. (2021). "Worries of students regarding online classes", Times of India, 15 July 2021, p. 1.

Lischer, S, et al. (2021). "Remote learning and students' mental health during the Covid-19 pandemic: A mixed-method enquiry", *Prospects*, 1–11.

Liu, G.-Z., & Hwang, G.-J. (2010). "A key step to understanding paradigm shifts in e-learning: Towards context-aware ubiquitous learning", *British Journal of Education Technology*, 41(2), E1–E9.

Mukhopadhyay, M., Pal, S., Nayyar, A., Pramanik, P. K. D., Dasgupta, N., & Choudhury, P. (2020). "Facial emotion detection to assess learner's state of mind in an online learning system", In *Proceedings of the 2020 5th International Conference on Intelligent Information Technology* (pp. 107–115).

Nayyar, A., & Kumar, A. (Eds.). (2020). *A Roadmap to Industry 4.0: Smart Production, Sharp Business and Sustainable Development* (pp. 1–21). Berlin: Springer.

Nayyar, A., Rameshwar, R., & Solanki, A. (2020). *Internet of Things (IoT) and the digital business environment: A Standpoint Inclusive Cyber Space, Cybercrimes, And Cybersecurity. In the Evolution of Business in the Cyber Age* (pp. 111–152). Ontorio, Canada: Apple Academic Press.

Onday, O. (2019). Japan's society 5.0: going beyond industry 4.0. *Business and Economics Journal*, 10(2), 1–6.

Rahmawati, M., Ruslan, A., & Bandarsyah, D. (2021). "The era of society 5.0 as the unification of humans and technology: A literature review on materialism and existentialism", *Journal Sosiologi Dialektika*, 16(2), 151–162.

Robinson, C. C., & Hullinger, H. (2008). "New benchmarks in higher education: Student engagement in online learning", *Journal of Education for Business*, 84(2), 101–109.

Roblek, V., Meško, M., Bach, M. P., Thorpe, O., & Šprajc, P. (2020). "The interaction between internet, sustainable development, and emergence of society 5.0. *Data*, 5(3), 80.

Saade, R. G., Kira, D., Mak, T., & Nebebe, F. (2017). "Anxiety & performance in online learning." *Proceedings of the Informing Science Information Technology Education Conferencs*, Vietnam, pp. 147–157. Santa Rosa, CA: Informing Science Institute.

Salami, M., et al. (2021). "Impact of perceived academic stress and depression on self efficacy beliefs among university students during online learning in Peninsula, Malaysia", *International Journal of Social Learning (IJSL)*, 1(3), 260–269.

Salimova, T., Guskova, N., Krakovskaya, I., & Sirota, E. (2019, March). From industry 4.0 to Society 5.0: Challenges for sustainable competitiveness of Russian industry. In IOP Conference Series: Materials Science and Engineering (Vol. 497, No. 1, p. 012090). IOP Publishing.

Saraswathi, et al. (2020). "Impact of covid-19 outbreak on the mental health status of undergraduate medical students in a covid-19 treating medical college: A prospective longitudinal study", *PeerJ*, 8, e10164.

Smuts, H., & Van der Merwe, A. (2022). "Knowledge management in society 5.0: A sustainability perspective", *Sustainability*, 14(11), 6878.

Solanki, A., & Nayyar, A. (2019). "Green internet of things (G-IoT): ICT technologies, principles, applications, projects, and challenges", In Arun Solanki and Anand Nayyar (Eds), *Handbook of Research on Big Data and the IoT* (pp. 379–405). IGI Global.

Stevens, G. J., et al. (2021). "Online university education is the new normal: But is face-to-face better?" *Interactive Technology and Smart Education*, 18(3), 6–11.

Tus, J. (2021). "Amidst covid-19 pandemic: Depression, anxiety, stress, and academic performance of the students in the new normal of education in the Philippines", *International Engineering Journal for Research & Development*, 6(ICMRD21), 13.

UNESCO. (2020). "1.37 billion students now home as COVID-19 school closures expand, ministers scale up multimedia approaches to ensure learning continuity", United Nations Educational, Scientific, and Cultural Organization. Retrieved June 30, from https://iite.unesco.org/news/1-37-billion-students-now-home-as-covid-19-school-closures-expand/.

Voltmer et al. (2021). "The impact of the COVID-19 pandemic on stress, mental health and coping behavior in German University students–a longitudinal study before and after the onset of the pandemic", *BMC Public Health*, 21(1), 1–15.

Wang, H. (2005). "A qualitative exploration of the social interaction in an online learning community", *International Journal of Technology in Teaching and Learning*, 1(2), 79–88.

Wieczorek et al. (2021). "Class of 2020 in Poland: Students' mental health during the COVID-19 outbreak in an academic setting", *International Journal of Environmental Research and Public Health*, 18(6), 2884, 2.

World Health Organization. (2020). WHO director-general's opening remarks at the media briefing on COVID-19-11. Retrieved June 29th 2022 from https://www.who.int/director-general/speeches/detail/who-director-general-s-opening-remarks-at-the-media-briefing-on-covid-19-11-march-2020.

CHAPTER 7

The Use of Web 3.0 in University E-learning, Quality Assurance, and Knowledge Management

Zahid Hussain

Shaheed Benazir Bhutto University

CONTENTS

DOI: 10.1201/9781003322252-7

7.1 INTRODUCTION

In recent decades, the concept of Web 3.0 but also Web devices and techniques in academic institutions has managed to remain visible and prevalent (Rego, 2011). Sign off: The New York Daily News foresaw the transformation of Web 3.0 technology solutions in 2006. Considering the nature of the delivery of services that academic institutions must provide to their potential students, such educational technological tools facilitated by Web 3.0/Web tools and applications are assured to help the system of education and academic institutions (Rubens et al., 2011). Web 3.0/Web tools have been increasingly being used to endorse enablement as well as research or practice at numerous higher education institutions around the world (Zhang & Adipat, 2005). The introduction of web 3.0 has marked a significant shift in how electronic natives can benefit from such a range of academic software, personalized learning situations, and personal virtual worlds. Such contexts enable digital natives to obtain required details, collaborate, and process data to meet specific needs (Weippl & Ebner, 2008). Change is unavoidable in academic institutions, and an increasing number of them will realize that new teaching tools and procedures are necessary to fulfill the ever-changing requirements of people in accessing and sharing information (Richardson, 2009).

Several researchers (Zhang & Adipat, 2005; McLoughlin & Lee, 2010; Motala & Padayachee, 2018) said that Web 3.0 technologies had the power to change, speed up, create, develop skills, start encouraging and requiring students' learning in education, enhance teaching, and assist higher education institutions in changing. Web 2.0 users not only consume data from the web but also share information with others via the Internet. Numerous famous Web 2.0 dynamic apps exist today, such as blogs, vlogs, mash-ups, Stickers, Google Reader, Web Pages, Torrenting, and so on. The perspectives of various innovators upon the transformation of Web 3.0 vary significantly. Many people believe that technological advancements such as Web 3.0 will change how people use the Internet and open up new opportunities for AI-based apps. Improvements in Internet network speeds, configurable web apps, and computer animation developments, according to other pioneers will be critical in the transformation of the innovative world wide web (WWW) (Moges, 2013). Through such a research methodology, this research examines Web 3.0 as well as Web 2.0 through HEIs with the purpose of improving a comprehensive framework that guides successful adoption. The primary goal of this research is to examine how well the use of social applications (Web 3.0) in higher learning could be combined with conventional initiatives to improve the educational quality of experience as well as provide simple access to academic material and good educational distribution. HEIs influence the degree of information technology trust, willingness, and expertise among teachers and students in order to ensure technology acceptance and advancement in HEIs as well as the issues that hinder the acceptance and use of social applications, information systems, and Internet applications and their apps in HEIs.

Objectives of the Chapter are:

1. Identifying Web 3.0's benefits and characteristics for higher education;
2. Describing the list of Web 3.0 programs modified for distance higher education and their purposes;
3. Elaborating the guidelines for applying Web 3.0 technologies used by applicants to higher education for distance learning;
4. Rationale for implementing Web 3.0 ideas in higher education;
5. Estimating how Web 3.0 technologies affect distance learning results in all levels of learning.

Organization of the Chapter

The chapter is organized as follows: Section 7.2 elaborates literature review regarding Web 1.0, Web 2.0, and Web 3.0. Section 7.3 elaborates on e-learning evaluation with aspects from 1.0 to 3.0 and also discusses the challenges connected to e-learning 3.0. Section 7.4 discusses the research methodology. Section 7.5 highlights the influence of Web 2.0 and Web 3.0 as tutoring agents in HEIs. Sections 7.6 and 7.7 discusses related controversies regarding the adoption of Web 2.0 and Web 3.0 innovation in HEIs and HEI investment in Information and communication technology (ICT) and social apps. Section 7.8 concludes the chapter with future scope.

7.2 LITERATURE REVIEW

7.2.1 Web 1.0

There has been some dispute about whether Web 1.0 would be a distinct phase in evolutionary change (Zhang & Adipat, 2005). "Web 1.0" is indeed a term coined with a valid conclusion to define the beginning stages of the world wide web when the material was stationary and webpages mainly used the HTML programming language. Users had very few opportunities to successfully engage in this context due to the limited resources available, trying to make each other largely customers of information rather than being creative about it. As a result, Web 1.0 is commonly referred to as "learned". Writers of websites will indeed post the information that those who chose to publish will enable users to access. Users might look at and ultimately interact with the writer, even though there was no real correlation between the writer and the reader, or even the reader as well as the subject matter itself (Richardson, 2009).

Web 1.0 is large "an interconnected hypertext file framework network such as the internet" (Rego, 2011), with little to no user involvement. The main focus was solely on the ability to search for information, with site owners and managers aiming to publish certain types of information, especially in a digital space with worldwide visibility. For several businesses, simply being online was a primary objective, and the webpage served as a "business card", going to present the corporation, its objectives, and also its product lines in a descriptive manner to try to contact them directly. Similarly, non-commercial web-pages focused on creating content easily accessible to viewers and offering an organized, effective information source because of whatever aims have been needed, from education to playing games to government.

The aim and vision of Tim Berners-Lee, the creator of the World Wide Web, was always to make the site a cooperative room, in which people separated by geographic location, social, and language issues could come together again and make a contribution to a global communications space (Motala & Padayachee, 2018). As a result, the key aspects of Web 1.0 could be argued to be merely the result of technological limitations as processing and analysis and Internet connections have been extremely limited.

7.2.2 Web 2.0

This situation changed between the 1990s and the first 10 years of the 21st century, as well as the resulting creation of Internet services that become known as Web 2.0 after Tim O' Reilly's coined this term (Rego, 2011). Unless Web 1.0 was "read-only", Web 2.0 is frequently defined as "read-write".

As described in the literature, there are numerous different and more complicated definitions that link the beginnings of Web 2.0 to various perspectives. Motala and Padayachee (2018) defined Web 2.0 in terms of the current innovations, it really has embraced (including blog posts, content sharing, podcasts, and so on), whereas McLoughlin and Lee, 2010, believe that, now, the web service revolution has been fueled by the business industry's need to develop itself over the Internet.

In the most literal sense, Web 2.0 focuses on social interconnectivity, collaborative effort, and information exchange. Previously, subscribers could only access the content. However, the introduction of new Internet techniques, systems, and apps gave the ability to modify that as well as record and share fresh features. From blogs or websites to Wikipedia, YouTube, Facebook pages, and LinkedIn, this advanced generation of the Internet has seen an outburst of networking and sharing of information areas. Computers have always focused on controlling data and knowledge prior to the introduction of Web 2.0, although with Web 2.0, the focus has shifted to methodologies and technologies for the advancement of much more knowledge acquisition generated from social engagement and worldwide involvement (McLoughlin & Lee, 2010).

Web 2.0, according to (Ashworth et al., 2004), is a concept that includes several important principles. Table 7.1 summarizes such principles.

Web 2.0 has a lot of benefits, including the ability to create complicated connections by allowing customers to collaborate with the web's design, resulting in massive quantities of data with several possible applications. This has resulted in the creation of more adaptable, user-centered apps. The Web 2.0 benchmark, on the other hand, seems to have some flaws. As (Ertmer & Ottenbreit-Leftwich, 2010) pointed out, the rise of a much more user-centered Internet results in disorganized, misunderstanding data groupings, as well as an overabundance of useless content. It also causes an increase in security and privacy concerns, with ramifications even now in government policy, as the Wikileaks controversy evidenced.

TABLE 7.1 Key Principles of Web 2.0

Concept	Description
Context created by users	Imaginative techniques for creating individual context, including trying to post one's pictures, text messages, or clips to blog posts, content sharing, social media sites, and so on. This has resulted in such a visible way of life, aided by technological advancements (mobile phones, tablet devices, laptop computers, wireless Internet, and so on).
Use the strength of a crowd to your advantage.	Collaborating and cooperating in groups makes it easier to identify solutions to common problems or concerns. Crowdsourcing is becoming more popular as a way to get beginner content ideas into the public spotlight via social networks and then let people decide on their value or possibilities.
Data on such a massive scale	Big data has become a core component of IT corporations and businesses, including Google, due to the amount of data collected by consumer application areas. One of Web 2.0's most significant challenges is Big Data. Aggregation sites' offerings use the information to make personalized recommendations or adverts.
Participation architecture	The idea is that if enough people participate in such a service, it tends to grow and improve. It is not just a coincidence; this same structure is implemented for using conversations to develop and enhance on its own. Instances included Google or BitTorrent.
Effects of network	The benefit of service increases even though clients are also capable of interacting with it and as many more people participate in it. Clients are indeed a crucial element of the varied features of such offerings as they are only useful and beneficial to the extent they are used. However, it thus presents a problem even though individuals may mobilize around a cheap, sub-standard product since it has a huge number of users.
Openness	Continuing to work with open platforms that use unrestricted data and applications and participating through accessible, cooperative technology have all become extremely prevalent. A good example is Firefox and its numerous extensions.

(Rubens et al. 2011) described three critical issues that are evident here on the Internet these days: conflicting information, copyright obstacles, concerns about confidence, privacy, and safety throughout online spaces. Indeed, the use of social networking sites changes the very rights to private areas as information that was previously restricted to friends and relatives is now available to the entire world (Semweb, 2011). Similarly, Web 2.0 techniques and high-speed Internet connections have made it easier to obtain pirated and unrestricted copies of digital materials. Scholars have indeed stated that the overwhelming volume of data is not the greatest challenge, but rather the biggest opportunity, again for the Internet in the long term (Gray et al., 2010).

7.2.3 Web 3.0

The next era of the Internet recommends a number of solutions to certain problems, including the ability for subscribers to alter data resources on their own. The web will become much more efficient and robust as intelligent robots create content that can know content (rather than just display it), gain knowledge of what subscribers' needs are, and recommend the relevant information that they are looking for (Haigh, 2010). This method, even so, necessitates the introduction of semantic content to the web, and hence, Web 3.0 is also known as the semantic web by incorporating semantic information into a piece of information giving it meaning and significance.

Rather than files, the web is based on datasets (Wood, 2015). This clearly necessitates the implementation of innovative forms of script writing and coding, which was made possible by the growing sophistication of RDF (Resource Description), the linguistic requirement that will control the next creation of the web (Roehrig et al., 2007). Indeed, content interpretation and smart filtration necessitate that computers be given the necessary vocabulary to read, define, and organize data. People will need metadata (Haigh, 2010) and also the markup of web applications to convert those into computer, but also agent-ready artifacts, since they lack the people's capabilities for language (Roehrig et al., 2007).

Such techniques will have to convey logical connections between semantic interpretations and web-based knowledge (Wood, 2015). It will result in a much more organized information system based on rational but also straightforward fundamentals, which, if applied correctly, will enable much more specialized search processes than what Google presently employs (Wastiau et al., 2013). Computer systems will be able to deduce what each component of a search term innately means, including what the participant's motives and wants are.

A collection of conceptual frameworks is among the most crucial components that will have to be applied in order to successfully create a virtual space. (Wood, 2015) explained the characteristics of a truly representative vocabulary for an able-to-share topic or concept. According to a conceptual model, which includes "meanings of courses, interactions, functional areas, as well as other artifacts" (Wood, 2015). In order for computer systems to comprehend related and connected reasons, standardizations should be formed by trying to define conceptual frameworks in diverse spheres. As a result, researchers have argued that the achievement of Web 3.0 would be determined in great part by the proliferation of conceptual frameworks and also the sophistication of domain knowledge design tools (Nwosu & Ogbomo, 2012).

The key aspects of Web 3.0 are summarized in Table 7.2. Though some of these concepts are novel, others have been based on Web 2.0 fundamentals and have evolved over time.

7.3 E-LEARNING EVALUATION

In learning and teaching, e-learning is typically defined as guidance forwarded through a computer. Computer-based coaching, e-learning, online classrooms, and other terms are already used interchangeably with e-learning. Many of these relate to using ICT infrastructure, including all educational programs, if they are done individually or collectively, offline or online, synchronous or asynchronous, on interconnected or self-contained work computers.

In their own attempt to find a system developed for e-learning conditions (HEFCE, 2005, p. 5), the Higher Education Funding Council for England (HEFCE) stated the following in the 2005 E-Learning Strategic Plan:

We discussed whether the need for such a specific idea of e-learning is even necessary, given that it might prevent experimentation and diversification. We still believe that humans must focus their availability of our strategic plan on using technological solutions in educational activities in order to be sufficiently concentrated. Brenton (2008, p. 86) elaborates upon the HEFCE strategic approach by defining e-learning as "that which

TABLE 7.2 Key Aspects of Web 3.0

Concept	Description
Web of Intelligence	Content is defined in a machine-readable format, which allows computers to intelligently organize and sift it, as well as surface or interface searches and enquires. This smart web will also include natural language processing, machine-based teaching, and rationale and will represent a very important role.
Information that is well-organized	Web 2.0's interactive elements and social dimension resulted in an information explosion and disorganized data groupings. Web 3.0 would then allow data to be organized, allowing for more efficient techniques.
Openness	More openness in terms of file types, procedures, and programming interfaces for applications (APIs). It could also relate to increased transparency between subscribers, private details, and private information.
Interoperability	Since Web 3.0 services and applications might very well finally be based on such a simple set of existential fundamentals and language groups, those who would be able to operate proficiently along a multitude of formats as well as devices.
Database on the World	The web will no longer become a collection of individual web pages but instead, a worldwide, huge data system based on structured collected data as well as XML layouts like RDF, OWL, and also SPARQAL.
3D Visualization	With the use of techniques comparable to a new life in virtual space and also emotes and individualized representatives, Web 3.0 might very well rely increasingly on 3D visualization and computation.

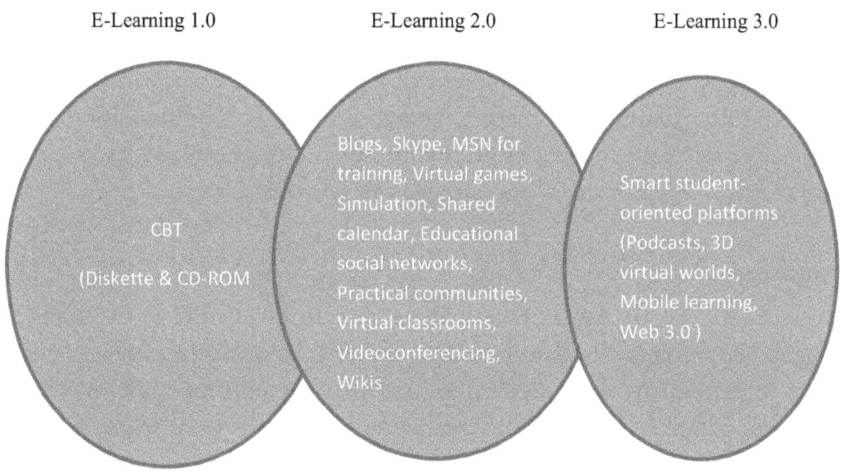

FIGURE 7.1 The evolution of e-learning in terms of definition as well as techniques (Rego, 2011, p. 41).

occurs when students are taught to communicate and information technology" (ICT). It can occur in either online or on-campus classes (it is occasionally referred to as' combined 'or' mixed-mode teaching). This will generally be defined sufficiently to allow users to do it in a variety of ways as either a specialist in their field of study or as an instructor of their students; there is no other way to "do" e-learning. It is not something you are able to "convey." It is something that users give their students a chance of doing (Figure 7.1).

The future learner is completely digitalized. Scholars have labeled educators of a period as "digital natives," "electronic immigrants," "gross-generation," and "other terms" (Prensky, 2002).

7.3.1 E-Learning 1.0

E-learning 1.0 adapted quickly to the emergence of the Web, which made information accessible online, both in terms of building and administering academic digital content. Learning Management Systems (LMS) and Learning Content Management Systems (LCMS) were created to assist with study management as well as curriculum organization for students, and the notion of "learning materials" was born. It is regarded as a much more conventional, patriarchal learning method in which conversation is one-way. The teacher seems to be the supplier of educational tools in a video manner in this simple linear regression model, but, also, he or she may resolve students via multiple communication channels.

7.3.2 E-Learning 2.0

The use of Web 2.0 tools for learning and teaching has been dubbed as "e-learning 2.0" by Stephen Downes (2005), but it has also been dubbed "u-Learning" by Zhang and Adipat (2005). (Motivated by the phrase "pervasive algorithm" or "embedded devices," which refers to computer technology that takes place any place, any time, anyway). While using Wikipedia pages, blog posts, webcasts, as well as other web techniques, Web 2.0 has also changed the classroom through aspects of how it is not just communally and yet cooperatively built. These techniques require energetic content creation, which includes reflective surfaces and discussion, thus needing cooperation and communication (Richardson, 2009). That is a cooperative way of studying in which conversation is two-way; thus, the assumption seems to be that social phenomena are created.

7.3.3 E-Learning 3.0

Educational leaders are already openly using the term e-learning 3.0 in their blogs and online forums. Technological advancements, the rise of cloud technology, collaborative smart sifting, increased and dependable digital storage capabilities, high image resolutions, mega expression gadgets, and a touchscreen interface usher in a new era of e-learning. Among the most important aspects of the third generation of e-learning would be pervasive connectivity to educational resources by the use of smartphones to connect virtually anywhere, at whatever time, from any location.

Digital experts recommend using AI and data mining to create e-learning 3.0 systems that can sort huge amounts of data and offer additional different datasets to aid learners, and achieve a deeper and broader understanding of the learning experience (Rubens et al., 2011). Scholars in higher education assume that "anyway though" might very well help the core idea of "every time, anyplace, and anyone". 3D virtual universes, like virtual worlds and individual emotes, will provide "anyhow" features (Rego, 2011). With the well-known Web 2.0 tools as well as the transition to Web 3.0, researchers are discussing personalized learning environments (PLEs), also referred to as mash-ups. Through today's modern knowledge-based societal structure, character customization should be seen as the crucial method of dealing with the wealth of information. A further e-learning transition might very well happen in various scientific procedures, which will shift from the highest, hypothesized results method to an underside, data-driven strategy. As there is so much available data, large datasets would then give educators and practitioners the opportunity to develop hypotheses.

E-learning 3.0 processes are still not widely available commercially, but scholars are going to propose evidence as well as having to work on rapid prototyping as alternatives. Twine was among the first Internet platforms to use semantic data to instantly as well as smartly organize based on users' particular interests. AHKME (Adaptive Hypermedia Knowledge Management E-Learning Framework), an e-learning information service of Internet-based education requirements that complies with such a Web 3.0 ideology, is yet another example of a successful e-learning 3.0 framework (Rego, 2011).

7.3.3.1 Key Challenges with E-Learning 3.0

According to an evaluation of secondary sources, numerous challenges could arise as a result of the widespread use as well as the adoption of e-learning 3.0 techniques, including:

- Increased risks to security and privacy,
- Web accessibility,
- Users' readiness, whether they are students or teachers,
- Further standardization of e-learning innovations is required, and
- In terms of the growing digital split, there are other social factors to be taken into account.

7.3.3.1.1 Increased Risks to Security and Privacy

Due to the difference in privacy regulations from nation to nation, there seems to be a higher risk of privacy protection inside the mash-ups of a globalized world in hyperdrive. Illegitimate contributions, a lack of http inspections, and inordinate advantages could also pose extra security threats. Researchers worry that ethical issues would be exacerbated as a result of the conceptual web's vast expanse, ambiguity, lack of certainty, and lack of consistency, which might also increase data protection and regulate loss (Alkhateeb et al., 2010).

7.3.3.1.2 Web Accessibility

It will be more difficult to provide ease of access to online content to people with special needs than it will be with some other Internet apps. The Wide Web Consortium (W3C) has begun to take a few steps in the right direction, but more work needs to be done in this area.

7.3.3.1.3 Users' Readiness, Whether They Are Students or Teachers

Some researchers worry about using such cutting-edge technology. "There will still be instructors, let alone students, who really are going to struggle to get web 1.0". As a result, on the one hand, these same questions or queries appear to be with us students who are clearly prepared to be self-directed students. Were our educators willing to accept advanced technologies? As mentioned in Section 7.3, the human element is certainly a factor in the equation; however, as subscribers become more familiar with technology, its impact may diminish.

7.3.3.1.4 Further Standardization of E-Learning Innovations Is Required

E-Learning is regarded as information and material exchange between system applications. A few standardizations, such as the freely distributable curriculum item model (SCORM), the IEEE digital learning guidelines committee (LTSC), the Educational Management Solutions proposal (IMS), and others, need to be improved for the next era of the web. The edu-tech standardization progression has grown in importance, with organizations including the IMS International Education Consortium (IMS, 2011), IEEE (IEEELOM, 202) and many others continuing to work to standardize teaching methods for e-learning apps like communications data, digital repositories, and much more.

7.3.3.1.5 In terms of the Growing Digital Split, There Are Social Factors to Take into Account

Some other concerns, according to scholars, are also widening and also transforming the landscape of the generation gap. "New digital divisions are more likely to be "about to and would not" mental splits, as well as "could indeed and thus can" skill set splits, than "get but have not" social economic divides". As a result, all of the difficulties faced by previous generations will be more dominant, requiring so much focus.

7.4 RESEARCH METHODOLOGY

The research problem is the most important indicator of the type of study method that should be used in the research. Research could be undertaken in one of three ways: qualitative, descriptive, quantitative, or a mixture of the two; even so, neither of these methods is inherently superior to the others. In this case, a mixed-method research design is much more suitable for this study. The research approach and method, in particular, influence the choice of a mixed-method study. This study method is best suited and most effective when working with such a particular research question and achieving a specific research goal. This study's objective of research as well as its research questions cannot be answered with just one research method (Moges, 2013). The benefits of using mixed methods seem to be that they can be used anywhere at any point throughout a research study, can be matched into another framework, can be performed at any stage of the investigative process in just about any proportion, for any tool or technique, and can also be used either with qualitative or quantitative data gathering (Ohei & Lubbe, 2013). In addition, an in-depth review of the literature related to social apps for information technology web devices and services at HEI would be conducted as a phase of this research. A systematic, comprehensive, and informative review of relevant literature would be conducted to lay the perfectly rational foundations for this research. As a result, literary works' data and evidence might very well guide, mentor, clarify, and recognize content that endorses the influence of social devices' strengths and weaknesses as just an educational information technology web service in HEIs. Furthermore, the research will be grounded in both realist and interconnected research philosophies to ground the research. Moreover, it is critical that the study takes a layered or engrained method as it will help the researcher to gain some insight and progress along the journey of data gathering, analysis, and study validity (Moges, 2013). Even so, this method tries to quantify qualitative information in order to fully understand it. This method approach was chosen by scholars because it employs quantitative analysis to demonstrate the findings. This implies that the quantitative data collection approach (seen in Figure 7.2) is deeply embedded or incorporated into the research and serves as

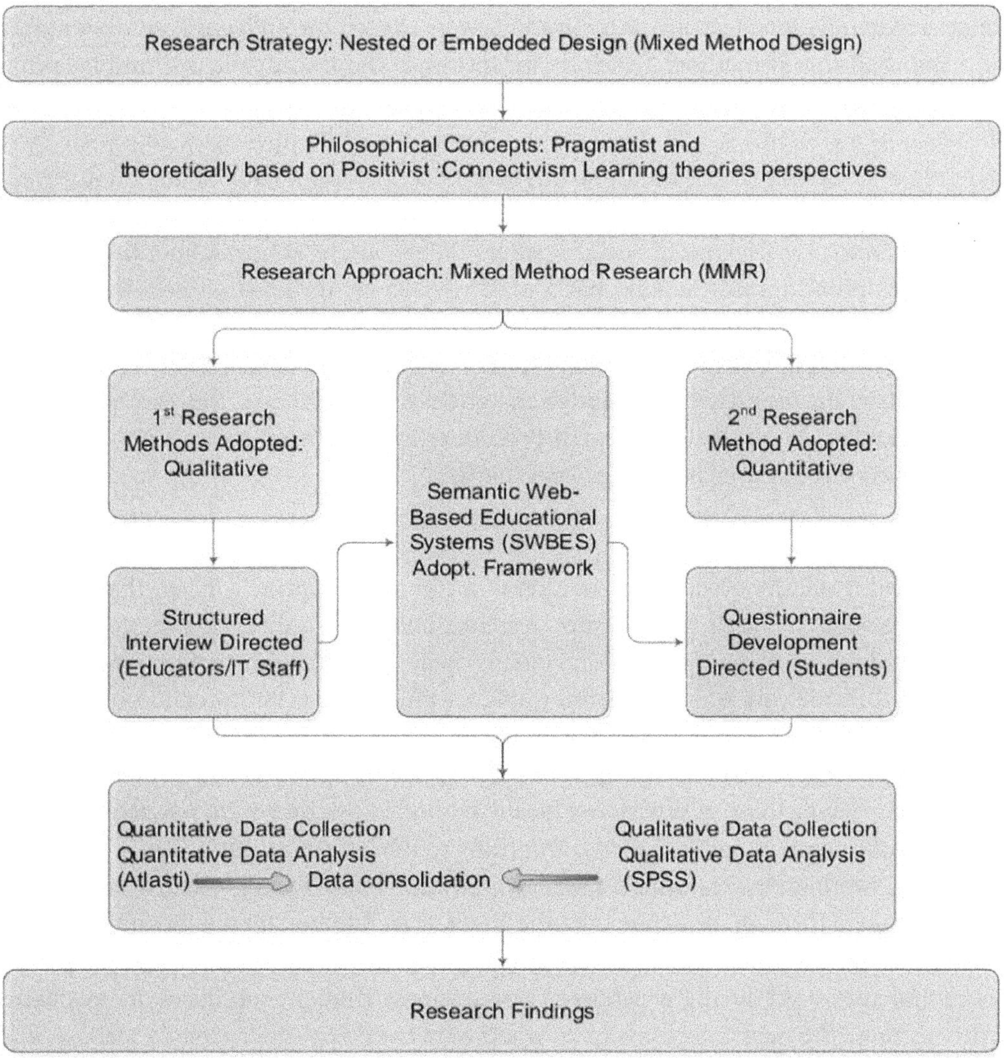

FIGURE 7.2 Research strategy at the highest level.

a complement to the findings. This method is suitable for this research project because it allows the scholars to find the problem, describe its existence, and determine its context, as well as accomplish the fundamental research aims and objectives.

7.4.1 Methods of Data Collection

To gather the necessary data, an organized questionnaire and interview will be used. Interviews will be held with teachers, and questionnaire forms will be distributed to students at three institutions: the Shaheed Benazir Bhutto University, Nawabshah; the Quaid e Awam University of Engineering, Science, and Technology, Nawabshah; and the Peoples University of Medical Sciences for Women, Nawabshah. Teaching staff from various academic fields, such as computer science, engineering, digital marketing, and information technology, who are really engaged in community apps and information technology web applications at Pakistani universities, would be interviewed. Educational staff members from the aforementioned academic institutions were chosen through purposeful sampling and a snowball survey method. To select students from these academic institutions, probability purposive sampling would be used. Each institution will receive five (5) personal interviews as a consequence of the sampling, for a total of 20 interviews across all these academic institutions. Three hundred fifty-five (355) correspondents from the different universities would be implicated. Various analytical tools, like ATLAS, Microsoft Excel, as well as the Statistical Package of Social Sciences (SPSS), are used throughout the data collection. Such programming actions, particularly regarding open coding, selective coding, and data coding, will be used for qualitative research. The next section outlines why methods made by mixing techniques are appropriate, taking into consideration that a mixed method entails the use of both qualitative and quantitative methods. The leading research method (Figure 7.2) details the research procedure and goals for obtaining the relevant data to answer the research issue mentioned earlier.

7.4.2 Conceptual Framework

A theoretical framework is a representation of a complex scientific concept that includes the key constructs and their relationships. Applications show studies and experts in understanding and representing information systems contexts in the information management study field. A framework is used to define guidance philosophies for the conceptual frameworks (McCauley, 2016). A framework is frequently viewed as a road map that ensures the useful research in a scientific investigation is reasonable. Furthermore, this is an organized technique for trying to express why and how a method to discover went into effect, as well as how to fully comprehend its actions, according to (Omona et al. 2010).

The framework of this research is a kind of transition phase theory, which aims to facilitate the scholars through all elements of the study as well as line up the research problem and research questions, so the research comes together well. As a result, the existing techniques and apps used during educational and business studies procedures are evaluated in this section. The question seems to be whether networking sites such as Facebook and information technology website applications can endorse collaborative, engaging, productive, and life-changing learning. This discussion contributes to the development of a comprehensive model for such implementation and acceptance of SS app tools for teaching, allowing students to learn how or wherever they want (Moges, 2013). The section is organized as follows: A comprehensive structure, as described in the previous section, provides an overview of the suggested graphical representation framework. Following that, a brief description of all of the endpoints defined in the suggested framework is provided. The results obtained by analyzing the applicable data for this study are depicted in this proposed methodology evaluation in Figure 7.3. Having followed the structure, each component of the qualitative research would be discussed briefly. This graphical representation structure

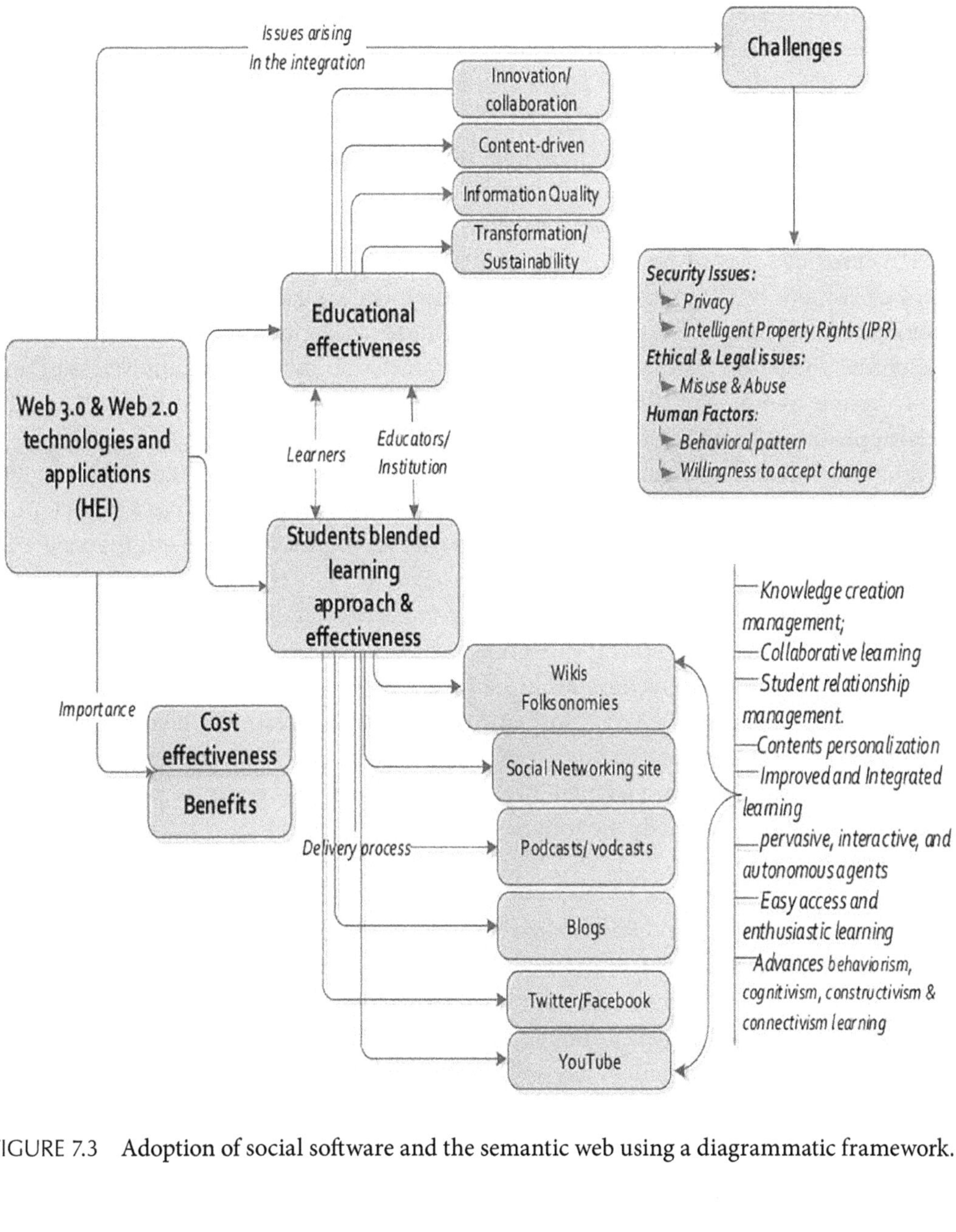

FIGURE 7.3 Adoption of social software and the semantic web using a diagrammatic framework.

in Figure 7.3 appears to broaden the knowledge base and create conceptual frameworks. The advantage of this conceptual methodology, in particular, is that it will expand the understanding of the phenomena being studied. Via graphic explication, it clarifies and reveals the theoretical basis of broader research phenomena. Furthermore, the framework enables web-based information modeling, conceptual analysis, and retrieval of information; system integration; customization; cloud computing; expert systems; and knowledge performance improvement (Omona et al., 2010). The suggested graphical representation framework suggests four particular elements that regulate organizational transformation, each of which is diagrammatically represented by a tree-like structure. The first component of the new framework's graphical model depicts the transitional circumstances and attempts to address the anticipated impact on academics. The importance and enablement of learning via social apps and Internet tutoring agents are the second dimensions. The third and fourth elements are concerned with discourse concerning online networks, Web 2.0, and information technology investments, as also their consequences after adaptation. It enhances the four key endpoints, particularly regarding academic effect, web applications/web information tutoring agent effect, controversial issues, and profitability factors, by requiring the implementation of the semantic internet education system (WBES) provided by this research. The academic effect of the basic structure is divided into four sections by the graphical model: advancement, content divines, quality of information, and transition. 3D virtual globe of Wiki pages; efficient search engines; conceptual online databases; an integrated learning framework; macro/semantic weblogs; and Really Simple Syndication filtration connections are indeed the six components of the social apps/web information tutoring agent underlying cause. These are all the procedures for providing education, and they will be explained briefly. The third element affecting the acceptance of Semantic-Web-Based Educational Systems (SWBES) is indeed the origin of the controversy. That can be based on four: security problems, legal and ethical concerns, human elements, and infrastructure facilities. Privacy as well as intellectual property rights are two of the security considerations. Misappropriation using social application software techniques is an ethical dilemma. Behavior patterns, reactionary tendencies, and approval using these techniques are all examples of human variables. The fourth factor, social software as well as information technology investment opportunities and their implications, were classified into four categories: expenses, advantages, threats from application catastrophes, and adaptability. Even as the study progresses, all the elements of this structure will be briefly mentioned.

7.4.3 Examining the Suggested Structure

7.4.3.1 Semantic-Web-Based Educational Systems

The first part of the discussion focuses on the positive aspects of the framework presented above. Such a framework may assist different users (learners, educators, and the institution itself) to achieve their educational goals. The framework includes a new cohort of SWBESs that are useful for enhancing and improving business processes and that improve the quality-of-service delivery with social software/semantic web technology (Hennessy et al., 2007).

The two dotted double-edged arrows in the framework point toward the educational impact and the social software/semantic web tutoring agent root signifies the dynamic between the learners and educators/institutions in the SWBESs. The learner's role involves the eagerness of learners to collaborate with pedagogical technologies in an attempt to develop their knowledge and to achieve the student's learning goals (Dotsika, 2012). This communication/interaction is attainable through personalized and adaptable educational Web 3.0 content.

The educators'/institution role entails numerous pedagogical events involved in SWBES; this includes collaboration abilities, information quality, content divines and transformation of intuitive teaching and learning, curriculum design, and authoring. Adding to this, the educators play a role in assessing learners' collaborations (problem-solving, assessment, etc.) and helping learners to construct or develop approaches that best suit them. The processes described above are bidirectional and interlinked as learners and educators/institutions are all key components in the SWBES framework (Dotsika, 2012). The next section deals with the transitioning aspect, which includes the educational impact of the adoption of SWBES and ICT into education.

7.4.3.2 Educational Impact

As stated above, the root representing education has four parts, namely, innovation/collaboration, content divineness, information quality, and transformation/sustainability (Dotsika, 2012).

The diagrammatic framework proposes four precise aspects as mentioned in Figure 7.3 that govern institutional transition and is schematically exemplified by a tree structure. The first aspect of the diagrammatic representation of the framework illustrates the conditions for transitioning and addressing the expected educational impact. The second aspect deals with the significance and facilitation of learning through an SW tutoring agent metric. The third and fourth aspects deal with the problems surrounding SW and ICT investments and their effects. The transition involves the integration of SW tools in HEIs, and it complements the five core nodes, namely educational impact, SW or Web 3.0 tutoring agent impact, ontologies, controversies, and SW and ICT investment considerations.

The diagrammatic representation expands on the above and categorizes the educational impact of the basic structure into four leaves, namely, innovation/collaboration, content-drivenness, information quality, and transformation/sustainability. The SW tutoring agent root is categorized into six leaves, namely, the virtual 3D world of wikis/podcasts/vodcasts/folksonomies; intelligent search engines; semantic digital libraries; intelligent tutoring systems; macro/semantic blogging; and RSS filters/mash-up/social networks. These leaves relate to education delivery processes and are briefly discussed in the sections to follow.

The third aspect governing the integration of SW tools in HEIs is the controversy root, which can be categorized into four leaves: security issues, ethical and legal issues, human factors, and ICT infrastructure. The security issues related to privacy and IPR (Intellectual Property Rights). Ethical concerns include the misuse/illegal use of SW application tools. Human factors are divided into behavioral patterns and acceptance/willingness to use these tools. The fourth aspect, which is SW and Information and Communication Technology investments and its implication, is categorized into costs, benefits, risks arising from software crises, and flexibility. All the components of this framework are discussed as the chapter proceeds.

7.4.3.3 Innovation/Collaboration

Innovation is the application of social software, the semantic web, and information technology software tools that support the development of new semantic educational content. It includes the representation of semantically accurate and reliable curriculum design; the facilitation of human intuition; concept searches; procedure setup; and knowledge discovery.

7.4.3.4 Content Divines

This aspect relates to content generation, distribution, repositioning, reusing, retrieval, and distribution. The content generation processes result in improved enactment, with delivery lagging. Progressive automation allows networking to be content and user-directed.

7.4.3.5 The Information Quality Category

The category for information quality is: The accomplishment and profitability of learning are directly impacted by the quality of the information. As shown in Figure 7.3, high-quality information is linked with a range of characteristics, including potential application, evaluation, advantage, speed of delivery, completeness, and capabilities. It is also associated with full representation, including clarity, easiness of ability to comprehend, conciseness, and reliability.

7.4.3.6 Transformation/Sustainability

In this category, the academic transition away from a strong central way of education is linked to information technology and technological devices made possible by social applications and web 2.0, which allow organizational advancements, new functionalities, guidelines, and regulations. The functional requirements of web applications have been enhanced, and information management and knowledge extraction have been revolutionized even before compared to previous web applications. It makes sense to consider the current and prospective effects of online expansion on delivery mechanisms and learning assessments, for example. The smart tutoring and monitoring agent helps students and teachers spend less time sifting through endless amounts of data when it comes to studying evaluation. According to many scholars, semantic information smart searching can help with certain issues in online learning and the management of information (McEneaney, 2011). The impact of Web 3.0 and the technological foundations of Web 2.0 on the learning setting is extensive. Both promote connectivity, promote sustainability, and encourage collaborative intelligence. They can also bring about a revolution that gives a structure a competitive edge (Dotsika, 2012). The effect and enablement of gaining knowledge via social apps and semantic search tutoring agents are covered in the second factor, which is presented in the following section. The influence of the mentoring agent comprised the delivery mechanisms, accompanied by the third and fourth elements.

7.5 THE INFLUENCE OF WEB 3.0 AS WELL AS WEB 2.0 AS TUTORING AGENTS IN HEIS

In the proposed framework, the social software/semantic web tutoring agent branches out into six parts, namely, the virtual 3D world, wikis/podcasts/folksonomies; intelligent search engines; semantic digital libraries; intelligent tutoring systems; macro/semantic blogging; and RSS filters/mash-up/social network.

Figure 7.4 depicts how these tools and applications encourage knowledge creation management; collaborative learning; student management; content personalization; improved and integrated learning; pervasive, interactive, and autonomous agents; easy access and enthusiastic learning; and improved learning theories (behaviorism, cognitivism, constructivism, and connectivism). There are insightful examples of educational institutes that have integrated SW into their administration and facilitation processes.

The illustration shown above demonstrates how all these technologies and techniques support information exchange management, cooperative learning, managing student relationships, personalizing content, improving and integrating teaching, prevalent, engaging, and fully independent agents, ease of access and enthusiasm learning, and improved educational theories (behaviorism, cognitivism, constructivism, and connectivism). The aforementioned literature provided a succinct discussion of the potential advantages of implementing social software and web 2.0 in education.

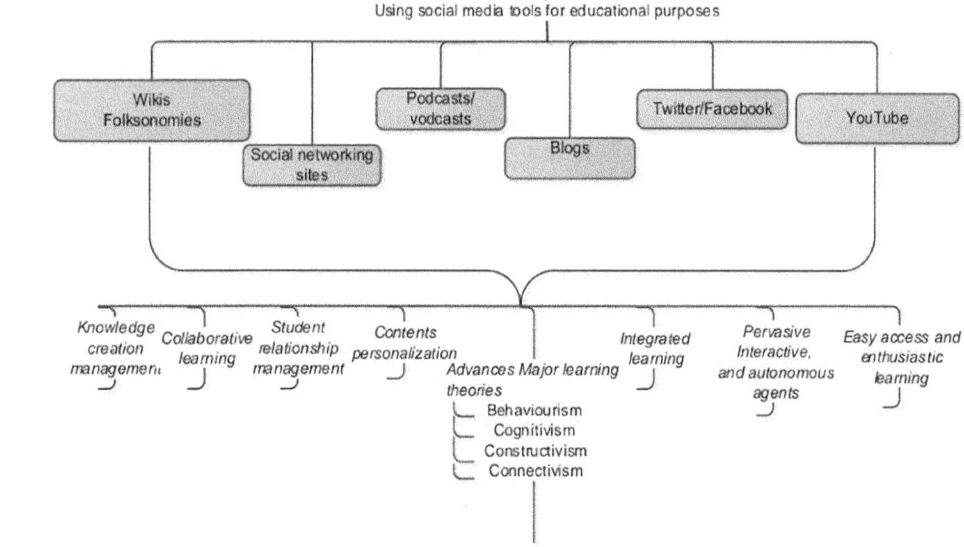

FIGURE 7.4 The tutoring agent's learning delivery processes.

There's been proof that some institutions of higher learning use online networks and Web 2.0 in their facilitation and management procedures. This section presents a brief explanation of Figure 7.4 in terms of web apps and how it corresponds to better theories of learning, particularly internal networks, which is a more suitable theory of learning for such digital natives in such an educational environment.

7.6 CONTROVERSIES ENCOMPASSING THE ADOPTION OF WEB 2.0/WEB 3.0 INNOVATION IN HEIs

The above-mentioned framework increases security issues regarding personal information and intellectual property protection, as well as ethical and practical issues in the event of exploitation and abuse. It also draws attention to potential problems brought on by human elements, like computer anxiety, behavioral patterns, openness to change, and the intention to embrace new models and methods for facilitating business as well as instructional practices. Finally, the framework noted issues with ICT facilities. Numerous authors (Dotsika, 2012) contend that most of the concerns raised by the use of social applications in the public sphere have to be taken into account because they may prevent or delay adoption. The confidentiality and security safety of student engagement with web-based educational devices and services in a public setting is less certain. It is impossible to overstate the persistent uncertainties that widespread online communication raises, particularly those brought on by the application tools made possible by social software. These concerns typically center on the same problems of becoming a victim of predatory unknown outsiders online and increased victimization by peers.

On websites where students rate their teachers, there are cases where students have posted inconvenient short videos or inappropriate comments. According to (Aghaei et al. 2012), improper usage of social applications and semantic web app techniques has been made in the field of education. According to Aghaei's study, students' learning performance suffered as a result of just using social networking websites and all these software techniques frequently. As a result, it would seem that students' grades suffer when they spend so much time on Facebook. Additionally, some students misuse systems like information technology channels or e-learning apps, which lead to failure.

Human activities and ICT facilities are the next topics that will be discussed briefly. High points as to how human behavior patterns and a person's acknowledgment or desire for using technology in teaching may affect the adoption pattern of social apps in HEIs are provided in this section. To understand the underlying principles of adoption studies, the study applies the theories listed below. The theories of reasoned action (TRA), theory of planned behavior (TPB), techniques of user acceptance (TAM), system quality and acceptance model (TRAM), and self-determination theory are further highlighted in this study. Such theories of implementation are defined by (Dotsika's, 2012). research on technology adoption methods and frameworks. There has been a lot of movement in IS studies, as evidenced by the estimate and the below literary works.

7.7 HEI INVESTMENT IN ICT AND SOCIAL APPS

One issue that has received considerable attention is the use of social applications as a teaching tool in educational settings (Abousoliman, 2017). While the case for adopting educational web apps seems strong, it is clear to look more closely at how this involves the implementation of social apps, which raises a number of other issues (Aghaei et al., 2012). Business operations have served as a driving force behind investments in the implementation of social apps and technological tools in institutions of higher learning. It is significant to mention how an organization or institution is interested in adopting technology for a few purposes: perhaps, an organization needs to improve the service quality, invent new products, or modify the existing assistance (growth) to substitute or update facilities and resources that have become outdated (repairs), to cut prices on present or prospective expenditures (price displacement), and to alter the ancient or conventional methods of doing things. Thus, the maintenance and upgrading, growth, cost-effectiveness, and transformational goals of the adaptation of pedagogical innovations in HEIs. Four elements make up investment evaluation, as stated in the proposed methodology: costs, advantages, adaptability, and threats (software crises). Other researchers' discussions of these factors (Haigh, 2010; Chawinga & Zinn, 2016) do not adequately address them. They also noted that when evaluating value for money, it is not just about participating in the effort or having access to information. Firms should develop the unexpected costs of training and making preparations for employees, and the advantages and adaptability that come with time-saving features like enhancing basic maintenance assistance or having to respond to inquiries, the ease of use for students, and the potential for loyalty growth from monitoring all who find it difficult. The focus of educational institutions, according to the authors, should be pushed in the direction of enhancing business effectiveness, reliability, and competitive edge. Numerous problems have surfaced and issues like privacy, prestige, identification, and others based on cultural emergencies have been brought up by academic scholars (beliefs, behavioral patterns). The literature did not go into great detail on the costs, advantages, risks, and flexibility related to an application crisis. Organizations must fund ICT investments in even the most effective way possible to control costs, gain a return on investment (ROI), and transform the company's chance to compete. The researcher is gathering both qualitative and quantitative data in accordance with the guidelines outlined in the aforementioned sections. This merged method of collecting data is known as "mixed methods". In this research, information will be gathered via audible interview sessions from the interviews conducted and analyzed using a straightforward literature review. To create classifications, data coding is used. This categorization of the data into topics is a straightforward but powerful data analysis technique that will enable the discovery of these topics. However, data from educators at all of the chosen academic institutions will be gathered using questionnaires. In order to get a comprehensive basis for future software adoption in HEIs, all methods and quantitative data analysis will be combined.

7.8 CONCLUSION AND FUTURE RESEARCH

This study analyzes how Web 3.0 techniques will impact e-learning 3.0 if a new instructional theory is required given the use of Web 3.0 in instructional technology and also what prospective obstacles and issues may arise with the introduction of e-learning 3.0. The three different eras and the evolution of e-learning (e-learning 1.0, e-learning 2.0, and e-learning 3.0) are connected (Web 1.0, Web 2.0, and Web 3.0). Web 2.0 and its accompanying devices are commonly used and well-liked by users, and their presence in e-learning is prevalent. This chapter, like its predecessor, argues that Web 3.0 techniques will further reshape the field of e-learning once they are well established. The concept of connectivism could perhaps account for the adjustments, and therefore, it does not appear that a new instructional theory is necessary. Nevertheless, as with the introduction and implementation of any innovation, using Web 3.0 as well as e-learning would then raise numerous technical, cultural, lawful, and ethical considerations. In other words, the headline formula can be changed to e-learning 3.0=(e-learning 2.0, Web 2.0, both variables and obstacles), where a feature is used as a prefix.

New tech is not just a facilitating tool for education but somewhat more; it is a force of innovation. Technological improvements will continue to play a crucial role in the development and transformation of e-learning, but they will do so further in the context of interpersonal or social knowledge. As a result, the education system has advanced, including unprecedented technological advancements. Learning and teaching have been significantly impacted by the use of technology in education. Along with the Web, e-learning systems are designed, and technological advances will continue to make it possible to incorporate e-learning theory into practice.

REFERENCES

Abousoliman, O. (2017). *Integrating Social Networks in Teaching in Higher Education*. Biola University.

Aghaei, S., Nemat bakhsh, M. A., & Farsani, H. K. (2012). Evolution of the world wide web: From web 1.0 to web 4.0. *International Journal of Web & Semantic Technology*, 3(1), 1.

Alkhateeb, F., AlMaghayreh, E., Aljawarneh, S., Muhsin, Z., & Nsour, A. (2010). *E-Learning Tools and Technologies in Education: A Perspective. E-learning*. LINC.MIT, USA.

Ashworth, F., Brennan, G., Egan, K., Hamilton, R., & Saenz, O. (2004). Learning theories and higher education. Level3, Issue 2. Retrieved from http://arrow.dit.ie/cgi/ on March 01, 2012.

Brenton, S. (2008). E-learning–an introduction. In *A Handbook for Teaching and Learning in Higher Education*. pp. 103–116. Routledge.

Chawinga, W. D., & Zinn, S. (2016). Use of Web 2.0 by students in the faculty of information science and communications at Mzuzu University, Malawi. *South African Journal of Information Management*, 18(1), 1–12.

Dotsika, F. (2012). The next generation of the web: An organizational perspective. Working Paper, University of Westminster. Series in Business and Management, University of Westminster, London.

Downes, S. (2005). E-learning 2.0. *ELearn*, 2005(10), 1.

Ertmer, P. A., & Ottenbreit-Leftwich, A. T. (2010). Teacher technology change: How knowledge, confidence, beliefs, and culture intersect. *Journal of Research on Technology in Education*, 42(3), 255–284.

Gray, K., Chang, S., & Kennedy, G. (2010). Use of social web technologies by international and domestic undergraduate students: Implications for internationalizing learning and teaching in Australian universities. *Technology, Pedagogy and Education*, 19, 31–46.
HEFCE. (2005). https://www.immagic.com/eLibrary/ARCHIVES/GENERAL/HEFCE_UK/H050300E.pdf
Haigh, C.A. (2010). Reconstructing nursing altruism using a biological evolutionary framework. *Journal of Advanced Nursing*, 66(6), 1401–1408.
Hennessy, S., Wishart, J., Whitelock, D., Deaney, R., Brawn, R., La Velle, L., McFarlane, A., Ruthven, K., & Winter bottom, M. (2007). Pedagogical approaches for technology - integrated science teaching. *Computers & Education*, 48(1), 137–152.
IMS. (2011). https://www.yumpu.com/en/document/view/48656265/2011-ims-annual-report-ims-global-learning-consortium
McCauley, K. (2016). Computer animation in instructional design. In *Society for Information Technology & Teacher Education International Conference* (pp. 2185–2190). Association for the Advancement of Computing in Education (AACE), Waynesville, USA.
McEneaney, J. E. (2011). Web 3.0, litbots, and TPWSGWTAU. *Journal of Adolescent & Adult Literacy*, 54(5), 376–378.
McLoughlin, C., & Lee, M. J. (2010). Personalized and self-regulated learning in the Web 2.0 era: International exemplars of innovative pedagogy using social software. *Australasian Journal of Educational Technology*, 26(1), 28–43.
Moges, B. (2013). The role of information and communication technology (ICT) in enhancing the quality education of Ethiopian universities: A review of literature. *Journal of Education Research and Behavioral Sciences*, 3(8), 246–258.
Motala, I., & Padayachee, I. (2018). Readiness to adopt the internet of things at the University of KwaZulu-Natal. In *ICEL 2018 13th International Conference on e-Learning* (p. 256). Academic Conferences and Publishing Limited.
Mukhopadhyay, M., Pal, S., Nayyar, A., Pramanik, P. K. D., Dasgupta, N., & Choudhury, P. (2020). Facial emotion detection to assess learner's State of mind in an online learning system. In *Proceedings of the 2020 5th International Conference on Intelligent Information Technology* (pp. 107–115).
Nwosu, O., & Ogbomo, E. F. (2012). ICT in education: A catalyst for effective use of information. The official publication of the Pacific Northwest library Association PNLA Quarterly. http:// www. ict in education: as a catalyst for effective use of information Date of access: 28 Jun. 2011.
Ohei, K., & Lubbe, S. (2013). Social differences between information systems students and non information systems students at North West University (Mafikeng Campus). In *Information Technology and Applications (ITA), 2013 International Conference* (pp. 180–184). IEEE.
Omona, W., Van der Weide, T., & Lubega, J. (2010). Using ICT to enhance knowledge management in higher education: A conceptual framework and research agenda. *International Journal of Education and Development using Information and Communication Technology*, 6(4), 83.
Prensky, M. (2002). The motivation of gameplay: The real twenty-first century learning revolution. *On the Horizon*, 10(1), 5–11.
Rego, H. (2011). AHKME eLearning Information System: A 3.0 approach. Unpublished PhD thesis at University of Salamanca.
Richardson, W. (2009). *Blogs, Wikis, Podcasts and other Powerful Web Tools for the Classrooms*. California, USA: Cornwin Press.
Roehrig, G. H., Kruse, R. A., & Kern, A. (2007). Teacher and school characteristics and their influence on curriculum implementation. *Journal of Research in Science Teaching*, 44, 883–907.
Rubens, N., Kaplan, D., & Okamoto, T. (2011). E-learning 3.0: Anyone, anywhere, anytime, and AI. In *International Workshop on Social and Personal Computing for Web- Supported Learning Communities* (SPeL 2011).

Semweb. (2011). Semantic web tutorial. W3Schools. Retrieved from http://www.w3schools.com/semweb/default.asp in March 2012.

Wastiau, P., Blamire, R., Kearney, C., Quittre, V., Van de Gaer, E., & Monseur, C. (2013). The use of ICT in education: A survey of schools in Europe. *European Journal of Education*, 48(1), 11–27.

Weippl, E., & Ebner, M. (2008). Security privacy challenges in e-learning 2.0. I2 C. Bonk et al. (Eds.), *Proceedings of World Conference on E-Learning in Corporate, Government, Healthcare, and Higher Education 2008* (pp. 4001–4007). Chesapeake, VA: AACE.

Wood, L. W. (2015). Faculty perceptions about virtual world technology: Affordances and barriers to adoption. Dissertation, Georgia State University.

Zhang, D. S., & Adipat, B. (2005). Challenges, methodologies, and issues in the usability testing of mobile applications. *International Journal of Human-Computer Interaction*, 18(3), 293–308.

CHAPTER 8

The Influence of Industry 4.0 and 5.0 for Distance Learning Education in Times of Pandemic for a Modern Society

Joseph Bamidele Awotunde
University of Ilorin

Emmanuel Femi Ayo
Olabisi Onabanjo University

Gbemisola Janet Ajamu
Landmark University

Taibat Bolanle Jimoh
Ilorin Zonal Office

Sunday Adeola Ajagbe
Ladoke Akintola University of Technology

DOI: 10.1201/9781003322252-8

CONTENTS

8.1 INTRODUCTION

The first interconnection course, which was conducted via the postal service in Boston, USA, in the 18th century, is where the surprising idea of distance education first emerged [1]. Advancements in technology advancements, particularly in streaming audio and video, have made these courses more accessible and shifted learning to the internet. We are observing an expedited transformation in the learning experience as a result of the current COVID-19 outbreak due to the trend of education toward online learning. Will this harm people's minds in a negative way? [2]. Is mankind ready and knowledgeable enough for such a drastic transformation? Will we go back to traditional in-person teaching, or will virtual distance learning now become the wave of the future? Especially in the fields of medicine, computing, arts, or architecture, it might be difficult for learners to adjust to such a quick change in how they study.

Education distributes possibilities and environmental assets, controls resources efficiently, and establishes institutions for the development of skilled, efficient, and productive labor [3]. This is so mainly because the constructive developmental change in society and the country's life is what progression in education essentially entails [4]. In a number of sectors, including the sociocultural, governmental, industrial, and other spheres of existence. Every change or advancement in national life has an impact on the educational system [5], and every improvement or change in the field of education has an impact on the life of the country [6]. Education must continue in all situations, and the global implementation of absolute lockdown by COVID-19 had a detrimental impact on educational systems [7,8]. However, education activities must continue in any situation, hence the government had made several efforts to reduce this number of students gathering to learn in the same environments [9,10], and various measures have also been taken to make teaching and learning virtual using distance and online methods [11,12].

Since March 2020, teaching and learning activities are conducted by utilizing an online system or various social media platforms. The learning system uses a distance teaching approach rather than face-to-face instruction. Students do not need to travel to schools or institutions to complete their coursework when using the distance learning system. Subsequently, educators use a variety of tools to conduct online teaching and learning activities. The advancement of information and communication technologies (ICTs) cannot avoid using this method of distance learning. Included among the learning resources are the Google Meets app, Zoom, Google Classroom, YouTube, TV, and WhatsApp. The creation of each of these facilities results from the advancement of ICTs.

The distance learning approach does not, however, eliminate the possibility of some learning process issues. Students and instructors from all backgrounds must, of course, have strong internet network connectivity in order to use this distance learning program. However, many locations lack reliable internet connectivity, which is one of the challenges to adequately executing teaching and learning activities. Additionally, a good number of students do not achieve their full potential in terms of learning. During the COVID-19 epidemic, teachers assigned tasks and covered the relevant material.

Fundamentally, the pandemic's effects have the potential to diminish human activity generally. It cannot be contested that many initially believed that the COVID-19 era was a challenging moment for humanity [13]. Without recognizing it, there are numerous benefits to this that we can take away from the current global phenomenon. Everyone may feel the impact because it is genuine [14]. Difficulties cannot, however, be used by the community as an excuse for not fulfilling obligations, particularly in the area of education.

Due to the educated population's drive for personal growth, the rising population, the severe global economic crisis, and the diversity and diffusion of education, education is in more demand and the mandate has increased [15]. In nearly every sector, tremendous developments and transitions are taking place in the 21st century, especially in education, through expanding and advancing industrialization [16]. Individual demands are growing, education is becoming more personalized, and technologies are being used differently, and the movement toward the commercialization of education has advanced due to its use in daily life [17]. The majority of teaching tasks in open and remote learning are stereotyped because they are set by the distance learning courses in addition to technical means [18]. Objectified learning has not just spread outside of regular classrooms but also through the emergence and modern technologies in our daily lives, together with written paper and industrial. Remote learning has become more significant as a critical determinant that industrialization and distance learning had on one another over time [18].

Not just industrial institutions, but also initiating innovations, as well as the labor force, are affected by Industry 4.0 [19]. The tools and materials, as well as the structure and application, have an impact on all businesses. One of these groups that cannot be separated from the ecosystem is education. Both developed and under-developed countries place a high value on education, including distance and online learning in this case, in order to address the demand for individualized education [20,21]. Unlike having to visit a campus or a classroom, distance learning is delivered by employing printed materials or computing infrastructure. The learning environment it provides is available regardless of place, time, or distance. Digital material, including the technological tools it employs, is continually evolving in response to industrialization and the forefront of technological changes.

Given that schools and teachers cannot control all sources of knowledge anymore, with the growth of technology in social interactions, both kids and adults have access to information anytime they want and wherever they choose. Distance learning presents a significant chance to improve society and the dissemination of knowledge via digital means [22,23]. How distance education will be run and administered in this environment in the coming years given the implications of Industry 4.0 will confront us with an issue that has to do with the management style of the organizations that offer distance learning facilities.

Since distance education is an interdisciplinary field dating back hundreds of years to the world, it is applied by both higher education and different institutions [24]. It is impossible to distinguish modern technology from distance education as a system, which is the main driver of these changes, along with globalization and industrialization, due to its inherently unknowable nature. Considering that distance learning is prevalent, the learner receives educational resources through a variety of learning contexts, not in person. Changes are occurring and diversity is growing, from the digital instruments used to impart education, from the design of courses to individual factors, to remote education's organizational framework [25].

In light of the relevance of distance education during the onset of any pandemic internationally, this chapter addresses the paradigm shifts in distance education in the face of a pandemic and deals with the issues given by Industry 4.0 trends. The study determines how Industry 4.0 will impact the remote learning system from the perspective of the industrialization concept. In this scenario, it is crucial to understand how Industry 4.0 will affect the distance education system and how it will direct distance learning in order to create futuristic visions for established structures. Additionally, it is almost certain that institutions of distance learning, which offer educational opportunities to millions of students, will not be impacted by Industry 4.0.

The objectives of the chapter are as follows:

i. To learn what Industry 4.0 and distance learning are predicting for the future;

ii. To understand the applicability of distance learning in the context of Industry 4.0 and Industry 5.0 during the pandemic outbreak;

iii. To explain the technological innovations brought about in distance learning by Industry 4.0 and Industry 5.0 during the pandemic outbreak;

iv. And, to review four major strategies for this distance learning paradigm in the context of Industry 4.0.

Organization of chapter

Section 8.2 presents the various Industry 4.0 during pandemic outbreak technologies used for distance learning using the pandemic outbreak. Section 8.3 discusses the applicability of technologies used by Industry 5.0, and here characteristics of the revolution. Section 8.4 proposed and presents four future strategies for distance learning in Industry 4.0 and Industry 5.0, and finally, Section 8.5 concludes the chapter with future scope.

8.2 INDUSTRY 4.0 AND INDUSTRY 5.0

We have witnessed significant technical advancements during the past 20 years, including new disruptive technologies, both on the hardware and software fronts. Information, communication, and artificial intelligence (AI) are all being combined, along with the merging of many different ideas. These modifications are resulting in the phenomenon known as "digitalization", which is primarily characterized by degradation and divergence [16]. Some convergence phenomena are happening right now, where existing convergence fuels new integration, establishing the groundwork for the eventual emergence of new and improved technologies.

The confluence of the internet and developing technology is what is referred to as Industry 4.0 [26,27]. It fuels an entirely new paradigmatic shift in rapid industrialization [28,29], Industry 5.0 enhances it by emphasizing research and putting innovations at the disposal of the shift to a durable, sustainable, and human-centered industry [30]. The cutting edge of established fields is where this new massive change occurs.

The emerging technologies' fundamental principles are increasingly linked between several fields, in a harmonious, transdisciplinary manner rather than in a multi- or inter-disciplinary way. Simultaneously, the distinctions between several disciplines (as well as many other areas) are blurring; electromagnetic technology, mechatronics, and computer engineering are getting harder to define, as is exactly what computer engineering is. In actuality, the fundamental sciences and engineering are tough to define in modern times. In a nutshell, the phenomena of the convergence of several fields require a detailed discussion of what the future of distance education should entail. "We're going to have problems if we don't transform the way we teach in three decades" [31].

8.2.1 Concept of Industry 4.0 in Distance Learning

The development of the next generation of manufacturing techniques has advanced significantly as a result of the quick development of industrialization and information technology techniques [32,33]. The Industry 4.0 revolution is currently just around the corner. As one of the German government's 10 "Future Projects" outlined in its High-Tech Strategy 2020 Action Plan in 2013, Germany's effort to position itself as a pioneer in defining the structure is said to be represented by the Industry 4.0 initiative [34,35]. The Made-in-China 2025 national plan was launched by the Chinese Executive Council in 2014, which aimed to make China a global industrial force rather than just the world's workforce. The goal of Made-in-China 2025 is to completely modernize China's economy, as well as the industrial sector. Many applications in Industry 4.0 and Made-in-China 2025 demand a synthesis of currently driving innovation, it is fueling the development of Industry 4.0. These technologies come from a variety of fields, notably cyber-physical systems, the Internet of Things (IoT), cloud computing, business process management, service-oriented architecture, industrial information integration, industrial integration, and more terms are used. Utilizing Industry 4.0's maximum capabilities currently faces a significant challenge due to the dearth of formidable technologies. For implementing Industry 4.0, which presents distinct obstacles, formal methods and systems approaches are particularly important.

Industry 4.0 has evolved during the past few years as a viable technology framework for merging and expanding manufacturing operations both within and between organizations. The rapid advancement in ICT has fueled the rise of Industry 4.0. Industry 4.0's achievements and technology advancements will offer a range of realistic solutions to the expanding demands for informatization in manufacturing companies. The fact that an increasing number of businesses globally have investigated the advantages of adopting digital organizations' horizontal processes is confirmation of its potential, and vertical chains, as well as adopting Industry 4.0, are on the verge of becoming the top digital businesses in the intricate industrial landscapes of the future [36].

In the international economy and in the operations of multinational businesses, as we have seen, Industry 4.0 is required to significantly raise the level of industrialization as a whole due to expansion, expertise, and profitability through the digitalization of production [37,38]. Industry 4.0 is widely acknowledged to have a formally proclaimed strategic impact on the advancement of the industrial sector worldwide. Due to the significance of this topic, there is an increasing need for investigation into Industry 4.0 to shed light on the problems, obstacles, opportunities, and answers concerning Industry 4.0's planning, execution, and administration [39,40].

Machines in Industry 4.0 workplaces are backed by sensors and wireless communication. These sensors are linked to a machine that can see the entire manufacturing line, keep track of it, and make choices by itself. To address the COVID-19 outbreak scarcity, Industry 4.0 employs automated production techniques to produce necessary expendable goods. During this situation, it offers a clever supply chain of medical supplies so that victims can access the urgently needed necessary medical supplies [41,42]. Any distance learning component is quickly designed and developed utilizing cutting-edge design technologies [43] and additionally using digital production techniques like 3D printing to create the necessary piece [44].

8.2.1.1 Industry 4.0 Technology-Based Approach for Distance Learning during the COVID-19 Pandemic

The digital revolution of traditional industries is being hastened by rapidly developing technologies like the use of IoT, AI, robotics, 3D printing, and Big Data among others globally.

Because of this, Industry 4.0 puts manufacturing at the forefront while incorporating cutting-edge innovations to enhance its effectiveness, efficiency, and reliability. As a result, Industry 4.0's mass data, intelligent equipment, intelligent materials, and quickly evolving business models all contribute to today's production. The previous industrialization freed us from the constraints of animal power, enabled mass production, and allowed huge numbers of people access to the internet. In ways, we had never envisaged before that Industry 4.0 will allow the merge of the physical, digital, and biological worlds. Industry 4.0 has gained international attention in recent years and depending on a person's age, ethnicity, and level of education, various interpretations are attached to its global advancements. However, a sizable section of the global populace is ignorant of Industry 4.0 and the impending alterations to their way of life. Here, various application of Industry 4.0 technologies is discussed in relation to their applications in distance learning during COVID-19 outbreaks.

8.2.1.2 Applications of Virtual Reality for Distance Learning during the COVID-19 Pandemic

All innovative techniques that encourage the creation of virtual worlds and enable user interaction are referred to as virtual reality [45]. Virtual reality has excelled at enhancing educational abilities in distance learning [46] as well as enhancing students' knowledge. In particular, virtual environments were effective in teaching medical students the basics of anatomy [47,48], radiation oncology [49], endotracheal intubation [50], and other topics, additionally, to encourage medical students to read radiographic images correctly [51]. Virtual patient simulator technologies have evolved significantly during the past 10 years.

It increased students' capacity to gather data during the taking of medical histories in a positive way [52] and enhanced their ability to diagnose problems. When distance education teaching and learning are required, these discoveries offer intriguing insights. Strictly enforced social distancing guidelines have been implemented globally since the COVID-19 outbreak [53], resulting in a disruption of teaching and learning activities and having a decidedly negative significant influence on educational systems around the world. Students who are infected with the virus may in fact spread it if they are undiagnosed, or vice versa, they may contract the virus while in training and spread it [54].

For example, for medical students enrolled in the last year of the degree in medicine and surgery in 2020, the disruption of traditional in-hospital training constitutes a notably significant problem. Who must finish their coursework prior to actually graduating in order to apply for post-graduate residency training and prevent career delays? Consequently, as an alternative to the customary internship at the patient's bedside and to meet the issues of quickly switching the mode of medical education during the COVID-19 epidemic, in order to close these existing gaps, virtual reality was quickly accepted as alternative teaching and learning method school of medicine. Numerous online education strategies have lately been put forth [55]. Reported reports of medical students' enthusiasm for virtual instruction combined with augmented reality (AR) platforms during the COVID-19 pandemic are scarce [56].

Little to nothing about the actual world is present in a completely immersive simulator made possible by virtual reality. It entails the development of events, locations, and things using several media [57]. Additionally, it enables the construction of a virtual laboratory, including a classroom for physics, geography, chemistry, or biology. The cost savings from not having to execute costly activities in practice is just one of the many benefits of this approach and instead creating a virtual reality (VR) simulation of them. This is most appealing to fields like medicine. Safety-related benefits are still another advantage. For instance, a VR chemistry lab enables individuals to safely view a variety of risky events, such as ignition, explosives, and others [58]. Through a single platform, VR intends to connect traditional classrooms and online distance learning. The authors in [59] describe an effort at a pilot virtual technology-based remote learning system. The smartphone or VR viewer (AulaVR), a different prototype that enables students to take face-to-face distance learning courses, is detailed in [60]. The Cave Automatic Virtual Environment is also an intriguing VR application (CAVE). The user is situated in a special space with distinctive representations serving as the walls and flooring. Additionally, there are several applications for teaching culture and heritage [61]. Students embraced the use of VR in higher learning with enthusiasm [62] on a widespread basis.

Building relationships is challenging because of the lack of facial expression engagement with classmates and the educational environment. Online learning innovations can help and hinder. Regrettably, increasingly tools are being created that allow students to falsify answers on online tests or even act as someone else. As a result, many higher education institutions choose to hold exams in a stationary setting, requiring students to physically visit the institution. Additionally, when performing numerous practical exercises is required, such as when welding or conducting chemical investigations, virtual teaching is ineffective. In this instance, the challenge is a lack of resources and equipment at home. Even the most advanced simulators will not be able to replace future doctors or pilots, despite the development of new technologies like VR. Last but also not least, distance learning is insufficient in disciplines requiring technical experience, such as engineering, medicine, etc. [63].

8.2.1.3 Applications of Augmented Reality for Distance Learning during the COVID-19 Pandemic

Applications that use AR accelerate students' learning in virtual classrooms. AR is a unique e-learning innovation [64]. Education systems have been delayed since the COVID-19 outbreak began because of how quickly the sickness spread [65]. A number of authors have emphasized the pandemic as a potent ability to attract new institutional focal points by examining the overall education system and making initiatives toward a sustainable civilization [66–68] and recommended that higher education institutions (HEIs) embrace cutting-edge tactics to enhance distance learning while filling up any gaps that students might experience when engaging in it. ICTs are important in this context, including AR, which can enhance the distance teachable moment and give students the chance to complete practical training. AR provides a dynamic engagement that raises the comprehension level of virtual items in a real-world setting through the use of 3D graphics, films, avatars, and collaborative capabilities to impose computer-simulated material [71]. It enhances direct engagement with the user using an audio–visual smart device such as a smartphone, tablet, or computer to provide the user with information that is relevant to their current situation [69,70]. Despite its uniqueness and technological ingenuity, AR has not received the attention it deserves in HEIs [72].

The educational system is becoming increasingly diversified in the age of cutting-edge technological development. The employment of innovative technologies in the distance learning environment is an experimental significant instrument because it facilitates contact and dialogue between professors and students [69,73]. Additionally, visualization centered on a smart device, such as a smartphone, tablet, or computer, would be crucial in various circumstances. It would cleverly apply to get around the constraints of conventional educational institutions. Similarly, other research looked into how AR applications affected students' desire to learn [72,73]. In comparison with the traditional educational environments for all types of classes, they claim that AR applications can improve students' academic accomplishment and gain experience. The authors in [74] suggested that some students were unable to comprehend their content in traditional online classes. In order to create 3D items during virtual classes, AR is crucial in strengthening the relationship between students and instructors. According to the authors [75], AR can display all the qualities of a virtual presentation of an object in the actual world. AR is one of the appropriate instructional platforms that give students the most efficient strategies for their difficulties. Students can access the simulated reality, take part, and interact with it, thanks to the use of AR applications. Participants engage in real-world interactions with virtual things, improving their spatial ability [76,77].

8.2.1.3.1 AR's Hedonic

The hedonic value of AR conveys the true emotions of virtual things in a physical setting [78,79]. It provides consumers with a real-world experience, which heightens creativity, in terms of hedonic qualities [80]. Hedonic value, according to the authors in [81], is the most crucial component of AR, which greatly arouses the consumers' conscious and subconscious responses, drawing them into a real-time interaction with the virtual goods. Similarly to this, the author in [82] explained how students' encounters with functional and emotional learning have a big impact on their decision to embrace AR-based education. Real knowledge and the desire to learn it are animated by hedonistic objectives, such as characteristics that have a significant impact on students' course participation [83–85]. According to authors in [86], who drew on the research, the hedonic value establishes the users' behavioral intention, which in turn affects behavior intention. It increases the course's effectiveness and enhances understanding of it, which affects the participants' amount of flexibility [87]. Such procedures have been investigated in relation to AR procedures. Technologies based on AR, a breakthrough technology trend, provide a distinct comprehension level [88,89]. It simultaneously offers excitement, self-expression, pleasure, and hedonistic pleasures [90]. AR's hedonic value effectively influences the students' behavioral intention and attitude toward distance learning [91].

8.2.1.3.2 The Utilitarian Value of AR

The reasonable and operational message of the specific object, along with the utilitarian value of AR, has a considerable impact on the behavior of its users [92,93]. It serves as a key indicator of innovation in both new and existing distance learning systems [94,95]. It was found that the students' attitudes toward e-learning are significantly impacted by the accessibility and emulation of AR, which eventually results in perceived behavioral control [96,97]. The practical utility of AR changes the overall presentation of a virtual course, engaging and inspiring students to pursue distance learning [98]. The utilitarian value increases the acceptance of virtual things in a real-world setting, which fosters a friendly perception and a perceived behavioral willingness to use, eventually reinforcing beliefs. Since moving to a distance learning-based educational process, academic institutions around the world have closed their doors to face-to-face instruction due to the COVID-19 outbreak. Students' level of realism concerning the lesson is also affected by the utilitarian valuation model [99]. As a result, it affects the students' reasonable perception of the distance learning system [100].

AR is a helpful learning opportunity for anatomical concepts and can have a good effect on students' academic achievement [101]. In two separate 1-hour sessions, AR was used to teach neuroanatomical [102]. A medial lemniscal pathway connected the somatosensory cortex to the first section. The second component also made it possible to investigate how cerebrospinal fluid moved from the ventricular system to the subarachnoid area. Sixteen first-year physical therapists made up the respondents. Because of this, more than 80% of students desired to incorporate AR into their standard curriculum classes. AR was also implemented in courses for nurses [103]. The forecasting of further stages during interventional procedures was made possible by a specialized application. The Vimedix AR clinical teaching system, developed by CAE Medical, allows students to examine human anatomy in 3D. They can look inside extended scenes of hearts, lungs, or abdomens and watch in real time as the ultrasound beam slices through the internal organs to produce ultrasonic imaging [104]. Authors in [82] provided a compelling point of view. The piece is laid out as a discussion with medical practitioners. One of them claimed that AR technology aids students in understanding spatial information. However, they do have certain drawbacks, such as some kids experiencing slight headaches, eyestrain, or motion sickness. In conclusion, AR can be a useful tool for improving one's perspective of how to behave in real-life scenarios, increasing the appeal of intellectual material, and improving students' effective learning results [105].

8.2.1.3.3 Applications of Mixed Reality for Distance Learning during the COVID-19 Pandemic

In collaboration with the Cleveland Clinic (Cleveland, Ohio), Case Western Reserve University School of Medicine (Cleveland, OH, USA) offered the first available commercial medical course based on Mixed Reality (MR) innovation in 2016 [106]. The "HoloAnatomy" program is based on a Microsoft HoloLens-based 3D holographic human anatomy education. With the help of this application, users may rotate, examine, and virtually examine an entire body as well as individual pieces in order to observe and comprehend the architecture, mechanisms, and functions. It was discussed by authors in [107] how HoloAnatomy compares to conventional cadaveric dissection. Additionally, this program was 1 of the 50 finalists for the 2017 Digital Edge Award. In response, Pearson suggested an educational resource for nursing staff and related medical institutions [108]. The virtual "Next Surgery" tool for scheduling procedures with Microsoft HoloLens was created by Brazilian business Digital Pages. Holographic simulations are used by the software to design surgical treatments. A HoloLens application to aid in the teaching of aesthetic medicine was presented by the French start-up Next Motion. The application directs the surgeon through the full surgical operation while highlighting the patient's body parts that need to be addressed and those that should be disregarded for safety reasons. The program deploys a real 3D hologram customized for the patient using facial recognition technology and records his activities to preserve the consistency of the image modification. This enables the surgeon to see the capillaries, vessels, and facial muscles, as well as the Botox intravenous fluids in aesthetic surgery. Two Australian higher education physiology and anatomy classrooms were examined [109], focusing on how kids see things. It was discovered that classes supplemented by AR are more well-liked and efficient than other types of distance learning, However, because AR devices are rather pricey, they come at a greater cost. MR is employed at primary, secondary, and high schools in addition to universities. For instance, an application with two Solar System and Plant System modules was created in a local elementary school in Singapore [1].

8.2.1.3.4 Applications of Mixed Reality for Distance Learning during the COVID-19 Pandemic

The educational process continued even after schools all around the world were shut down in early 2020 as a result of the COVID-19 outbreak [110]. Students can continue their academic endeavors while studying from a secure distance, thanks to digital innovations and e-learning principles [111]. The IoT is presently one of the digitalization methods that is growing the fastest, and among the most effective teaching techniques is online learning [112]. Electronic sensors and technological innovations let teachers focus on innovative teaching and learning approaches while minimizing time waste in modern society. The virus outbreak has put a stop to the teaching strategy, resulting in subpar exam outcomes and harming students' prospects for employment after graduation. Countries have introduced modern approaches to their learning environments [113].

Due to the recent rapid development of new educational technology and the appearance of a transformation in this sector, one of the most significant and lasting aspects of the educational process nowadays is modern technology [114]. These techniques have great potential for developing teaching strategies and learning objectives [115]. E-learning, on the other hand, makes use of contemporary technology to access instructional resources that are accessible through the internet [116]. E-learning is no longer just about virtual classes and online lectures, due to the quick advancement of ICT, and generated content. Early forms of e-learning have been touched by the IoT, becoming a more engaging and collaborative method [117]. Globally, the IoT network architecture is altering how kids learn and fostering an infrastructure that speeds up communication between students and teachers.

The IoT is a sizable platform made up of a variety of items with electronics, computers, sensors, and network infrastructure that provides a satisfactory experience in incorporating vital services globally [118]. Data is continuously exchanged between manufacturers, clients, and other computing equipment to achieve this [119]. The IoT is a globally dispersed system because anything in it can interact with other things via the network infrastructure [120]. Connecting everything on the Earth in order to enable universal internet access is the IoT's main objective. By the end of 2021, there will be about 25 billion connected devices through the IoT, predicts Gartner [121]. The effectiveness and excellence of this infrastructure that it enables are determined by the numerous IoT-based devices. There are many benefits of integrating IoT with teaching methodologies [122]. The IoT promotes online education. In the upcoming years, most colleges will implement this technology to improve e-learning and instruction [123]. The IoT will help the growth of contemporary knowledge and the establishment of more electronic establishments [124]

The authors in [125] provided descriptions of sensing devices, ZigBee/GPRS wireless network systems, database innovation, decentralized programs over web applications, and embedded technology. This technique connects numerous test terminals, servers, and customers over three interdependent connections, which encourages the quick exchange of information and makes it easier to carry out remote learning. The IoT can enhance both teachers' instructional strategies and students' educational achievement, according to learning results. As a result, it is believed that there is a problem area in contemporary educational systems that requires special attention.

The authors in [126] created a common distance learning architecture that utilizes the IoT to integrate several e-learning apps. This innovation strategy results in higher academic achievement, thus providing a cost-effective education while strengthening the bond between the learners and instructors. This training system delivers an appropriate and ideal teaching model using GSM and GPS technologies. In order to be successful, students will gain from implementing this method in school buses equipped with computer educational resources. This strategy has a serious flaw in that it pays insufficient attention to detail.

Additionally, an IoT-based technique for setting up online virtual labs was been proposed by the author in [127], a necessary condition for any educational system to be effective and successful in the world market. Due to the lower costs and improved performance of modern technological modules made possible by this technique, it enables the creation of a number of web services, including information processing and internet-based services. This method explains how to use the IoT in a practical situation by coupling Arduino software to the solutions to meet the needs of web service. Sadly, this method offers little privacy.

Furthermore, the authors in [128] proposed the greatest and anticipated impacts of the IoT on university education, including statistical analysis and theoretical study. In the context of the e-learning framework, it evaluates the most significant and pertinent intellectual and analytical outputs of IoT, especially when it comes to organizational and learning factors. This work stands out from the rest of the field due to factors including interactionists, collaboration, and research chances. Nevertheless, this technique has some drawbacks, one of which is that it downplays IoT potential in the field of education.

The key benefits of IoT in e-learning and the significance and significance of using IoT in this respect are indicated by authors in [129] discussion of IoT-based e-learning in several areas. Current problems and available remedies are presented and examined. The findings suggest that IoT would eventually be a core component of emerging e-learning techniques. Institutions will gain a lot from IoT-based interactive learning paradigms if the right infrastructure is in place and used effectively. It goes without saying that the proper structures for colleges intending to use IoT must also be thoroughly examined, and thoroughly examined. This strategy lacks consideration for effective interventions, which is seen as one of its inconsistencies and primary shortcomings.

The VR-based Education eXpansion (VREX) teaching platform, which combines online and offline models, is another service provided by the authors in [130] to enhance the post-secondary curriculum and instructional environment. VR is the essential component of this platform. By giving students an emotional understanding of the theoretical concepts that are challenging for teachers to explain, this project seeks to increase the effectiveness of teaching and learning in an experiential setting. It has been applied to enhance the curriculum's efficacy in an experiential setting. In this technique, the slides are turned into VR environments, enabling students to see and learn in a completely virtual but real setting. Students can take part in the interactive learning process with VREX's distributed model from any location at any time. It is disappointing that this technique does not permit the convergence of AR and MR.

Last but not the least, authors in [131] developed a concept that utilized ubiquitous, VR powered by spherical videos (SVVR) and contrasted their layout with that of a group of map guides who utilized tablets to learn academic library procedures. Similar assessments of background experience, responsibilities, exhibit a positive, and cognitive functioning was made by both groups. The findings demonstrate that the wearable VR-based model can increase innovation and complexity levels while enhancing learning. SVVR-based educational systems have a good chance of altering how students learn, enhancing the quality of conventional teaching, and improving students' topic comprehension. Using modern technologies, the length of time needed to execute activities and outcome effective implementation are a few of the problems that this model does not recognize.

8.2.2 Concept of Industry 5.0 in Distance Learning

The Japanese government has ushered in a new age known as "Industry 5.0", which calls on people to be capable of more than just mastering science and technology, yet always be competent to accomplish their human duties [132]. This suggests that people are expected to be cutting-edge in their approach to solving new social issues and to make use of their intelligence, innovation, and economy for the sake of sustainability welfare [133,134]. Due to the failure of Industry 4.0, which encouraged instant satisfaction and made people dissatisfied, Industry 5.0 came into existence, emotionally unstable, and hasty gratification fulfillment [135]. This exacerbates difficulties in daily life because it weakens a person's ability to fight for what they want until Japan finally had the notion to create an Industry 5.0 period in order to strike a balance between technology and social problem-solving and accomplish sustainable development goals.

The building of psychosocial adjustment to deal with rapid developments is one of the objectives of sustainable development, like the existence of the "Industry 5.0" period. In fact, it is also believed that developing resilience is a crucial action that must be taken to aid people in coping with shifting circumstances or harsh environmental conditions, one of which is academic. An individual's personal capacity to successfully adjust to challenges, interruptions, or distress is referred to as resilience, be able to create goals that are specific and attainable, engage with others successfully, feel at ease around them, and be able to manage difficulties as they arise [136]. This justification demonstrates that resilience is not just connected to a person's capacity for adapting to hardship but also to being able to resolve issues and set precise objectives. Similarly, academic definitions of resilience include a person's capacity to endure challenging life experiences, rise above challenges, be able to conquer them, and successfully adjust to academic expectations [137]. Therefore, intellectual resilience can be defined as a person's capacity to recover, find answers, set objectives, adjust, and feel at ease with academic pressures and other challenges.

AI, the IoT, and other key applications were made possible by Industry 4.0, and other cutting-edge technology. Smart IoT is enabled by Industry 5.0, followed by infrastructure and automation systems. This revolution aims to increase manufacturing efficiency and do away with human needs. With the help of collaborating robots, industrialization will improve. Humans and machines converge during this revolution. People must visualize what they desire, and pieces of machinery will carry out the necessary duty effortlessly [138]. To increase performance and productivity, the global medical corporate consistently focuses on the creation and implementation of new technologies. IoT and Big Data are examples of sophisticated technologies that are included in Industry 5.0. To increase automation and effectiveness in healthcare, this transformation adds a personalized human connection. With the help of Industry 5.0 innovations, automated machines demonstrate the capacity for direct customer communication [139].

These technologies accomplish challenging jobs and group activities with an understanding akin to general intelligence. This transformation unites humans and machines and creates infinite potential in distance learning and education. School administrators can produce their creative work more effectively and efficiently [140]. Industry 5.0 technology brings a clever fix for difficult operations and therapies. These technologies help enterprises create a better hospital administration system and improve patient care by reducing treatment-related errors [141]. Additionally, this allows for quicker and more accurate assessment with less difficulty during the whole psychotherapy. An easy illustration is a fact that robots can now be used to routinely do patient scanning and assist in the treatment of COVID-19 patients. All repetitious duties can be mechanized by technology to increase the effectiveness of the therapeutic process. Computerized controls can be used to time the arrival of necessary medical supplies during a prescription [142].

Nowadays, students can typically learn the most crucial information about a subject via resources besides the traditional way of teaching. For instance, MOOCs and collaborative learning using various online teaching like using social media platforms such as Zoom, Telegram, WhatsApp, etc. Through this, you can complete the degree without attending any formal lectures and pass the exams. Students can now learn independently without necessarily needing to be in the classroom, thanks to distance learning education. Mozilla currently issues emblems to recognize such learning [143]. It is just another practical method for allowing learners to advance at their own speed; it maintains their interest and motivation to continue. For pupils who possess the insignia (which supports intelligence) and for a subgroup of subjects that do not require intense practical experiments, universities might offer course waivers or initiatives as doing so would not interfere with achieving their learning objectives and where some of the course material crosses over into other courses. Although not everyone may benefit, many competent students, whose population is continually increasing, will benefit from it.

Furthermore, autonomous learning allows students who excel in particular subjects to participate immediately and understand more efficiently so that the course can be finished more quickly. The authors in [144] highlight advanced learning initiatives, such as fast-track degrees. They contend that these courses are not created to replace the conventional degree already offered, but to present students with various choices. They emphasize that augmentation benefits "abler students, i.e., those who are mature enough, motivated, and committed to deal with the increased effort". Responding to the requirements of both companies and students is one of the advantages, especially for someone who has a pressing need to finish sooner. Additionally, some students' speedier graduation times allow them to enter the employment market at a younger and more youthful age, thereby improving their economic contribution. Furthermore, more students will be able to enroll in the current programs as a result, so addressing the issue of "massification" by satisfying the ever-rising demand for higher education. The authors in [144] provide a full account of the measures implemented in this context at various institutions in the United States, the United Kingdom, Australia, and Norway.

8.2.2.1 The New Society and the New Systems

The nature of modern higher education is continually shifting. More and more classroom interactions between instructors and students are taking place online, implementing the virtual classroom model [145,146]. There will be other modifications as Society 5.0 and Economy 5.0 emerge. The need to combine the physical and digital worlds, the rise in information consumption, the time constraints, and higher education will result from the necessity to balance expert work and education. Future citizens must be prepared to perform in the field of resource distribution being made to undergo additional adaptive changes. According to the wider definition of "Society 5.0", there is a shift away from the emphasis on technologies and toward human behavior, which, according to "a system that tightly connects cyberspace and real space", can "combine wealth and prosperity with alleviating social concerns" [147,148]. The idea of Economy 5.0 is also introduced, and it relates to cooperation in the fields of technology, entrepreneurship, and intense competition so that societies and individuals might develop special means of providing value to institutional systems.

The term "Industry 5.0" applies to the connections between people and intelligent production systems, with an emphasis on the human factor (or "human touch") in industry. It hence transcends the creation of products and services for financial gain, well-being of employees, consideration for precious resources, and building an industry that can withstand unexpected crises like COVID-19. This interdependence is facilitated by version

5.0 of industry, which takes a comprehensive perspective on the world with people and machines. AI-based models and cognitive technologies are allocated a significant role in this reorganization, and it will make it possible for human–machine interactions to be more deeply engaged [149]. Companies will be able to accomplish the depth of personalization like never before attributable to human–intelligent-robot partnerships. Significantly, the idea of Industry 5.0 opposes the idea that automation should take precedence over humanity. Technology is increasingly being used to assist people in doing jobs [150]. In contrast to the Industry 4.0 idea, the progression of occurrences is as follows: persons, procedures, and innovation are only in third place. Even the most cutting-edge technology should not be superior to people.

The new Industry 4.0's new products are a concoction of computer programs, working embedded innovations, sensors/actuators, and telecommunication technologies [147]. These intricate systems function in more intricate and loosely regulated contexts, operate with a high degree of autonomy, interact with the internet and its services, often have survival components, and engage individuals in the loop. Problems with system applications must be addressed by such systems if they are to accomplish the specified standards of protection, confidentiality, and transparency, and required failure mechanisms [151].

Due to technological breakthroughs in the Industry 4.0 revolution era, systems are evolving at an extraordinary speed, and the properties of the socioeconomic divisions change along with society. Social eras are groups of people that were born around the same time and whose experiences and perspectives are comparable. The Lost Generation, the Greatest Generation, the Silent Generation, Baby Boomers, Generation X, Millennials (sometimes known as Generation Y), and Generation Z are the first generations featured in Wikipedia's list of generations in the Western world [16]. Due to its distinctive traits, Gen Z is also known as Gen-C.

Some claim that Gen-C refers to an attitude rather than a specific age group, and Gen C members range in age from 15 to 75. One thing that all of its members have in common is that they are extremely tech-savvy and digital natives that always keep interconnected. They also communicate, are automated, emphasize content, value connection, and click. It is believed that this group will shape the future, and they are the main considerations in markets. To say that they are responsible for the majority of innovations would not be an error.

Environmental consciousness is another key trait of Generation Z, and social conscience would influence how the future generation of cyber-physical systems are designed. For example, a Pew Research Center survey [152] reveals that younger generations, including Millennials (81%) and Gen Xers (75%), are more inclined to concur that global warming is occurring compared to Silents (63%) and Baby Boomers (69%). In a similar vein, Gen Zers' opinions on climate change are essentially the same as those of Millennials and do not differ significantly from those of Gen Xers. The perspectives of the two newer generations are very different from those of their previous generations and encourage governmental action to address issues. This has an impact on the recent school walkouts against climate change, such as Fridays for Future, which are coordinated by thousands of freshmen high school students [16].

The present Industry 4.0 plan, according to the author in [153], is to ensure protection of the environment, but sadly, environmental protection is not a major emphasis of Industry 4.0; neither has it concentrated on developing technology to increase the Earth's environmental sustainability, even though many various AI algorithms have been employed to conduct investigations from a sustainability standpoint [154] in recent time. Future systems are anticipated to close this gap, thanks to Industry 5.0 and services that use data and technological breakthroughs along with a social and environmental focus.

8.3 DISTANCE LEARNING AND DIGITAL TRANSFORMATION

Diversification innovations strategies produce new business opportunities based on digital information. Future value creation is anticipated to come from cutting-edge information-based technologies like the IoT, AI, robotics, and automation. Organizations from all industries around the world have been interested in "digital transformation" (DT) [155]. It makes an effort to gauge how much an organization can gain from the usage of technological advancement. However, it is also viewed as an evolutionary change by which IT is becoming a crucial component of daily life, influencing every aspect including both individuals and the organization itself [156].

One of the main issues that all organizations today face is incorporating and experimenting with new digital technology. The repercussions of DT affect every industry, particularly higher education. The term "digital transformation" refers to advancement from the perspective of incorporating not only technology but also individuals. An organization's vision, strategy, organizational structure, operations, and culture must all be reinvented. Organizations may encounter both opportunities and threats from AI, so they must devise their own skills to efficiently navigate the transformation to the future they want. There are numerous definitions of DT in the literature that emphasize the importance of contextual elements, the organization's size or location, as well as how deeply rooted it is in a region's culture [25,157–162].

In order to benefit from developing technology and its quick uptake in human endeavors, organizations need to redefine themselves and change every aspect of their operations. Due to this, DT necessitates a shift in perspective and calls for technological innovation and changing institutional culture to ensure DT development [157]; thus, universities can also benefit from this. The authors in [156] provide a methodology to evaluate the level of technological preparedness in institutions.

According to [163], DT has given colleges a chance to strengthen both students and academics. Learning can be enhanced to master a variety of pertinent subjects through an independently online-based learning program or methodology is allowed to penetrate the workforce. In particular, it addresses (i) educational materials, (ii) domain competence, and (iii) other capabilities. It has given every link in the competitive landscape with digital yet internet-based solutions; universities can thus be an exception in this case. Not only has this method led to the creation of novel and creative company concepts across the globe, but universities are also becoming more innovative. Universities can expand and transcend their typical virtual limits, thanks to the use and implementation of digital technology, controlling the delivery methodology, the entire investment chain, and the course repertoire of an institution [164]. Universities are currently adopting technology in response to a radical transformation, a complicated concept where technologies are concerned, and a networked setting that facilitates digital learning [157].

8.3.1 Distance Learning Technological Exemplification

Intelligent education, sometimes known as distance learning, includes new instructional settings whose significance centers on how the learner and teacher use freely accessible, cutting-edge technologies [165]. This requires not just the gear and software that are accessible but also how they are combined during in-person or distance learning. The Nigeria Ministry of Education sees distance learning as a self-directed, adaptive, motivated, resource-enriched, and technology-embedded learning paradigm. These characteristics imply that distance learning broadens both students' and teachers' methodologies, competencies, material, and educational environments [166].

The personality learning feature gives you the freedom to learn whenever and wherever you want, but it also has the opportunity to lengthen your period of adjustment. By guaranteeing empirical evidence, the motivational component of intelligent learning enhances educational techniques and classroom collaboration between learners and lecturers [161]. The responsive mechanism in turn expands education opportunities by providing personalized and customized learning. By making a variety of educational materials freely available, the multiplicity of platforms also expands the reach of educational content. Furthermore, ICT enables endless access to educational resources by providing local and international communications infrastructure (Figure 8.1).

The generally known information is essential for the dynamic emergence and evolution of machine learning techniques in this environment. These technologies including the processing of enormous data sets (Big Data), 3D visualization, machine learning (ML), AI, VR, AR, and 3D visualizations are examples of ICT-related developments [167–169]. Based on these technologies and using well-defined mobile applications under the cloud computing architecture, educational solutions will influence how individuals learn (in a more general sense), and intellectual participation in education, for around 100 years. With regard to widely used distance learning infrastructure (LMS class systems), their implementation in a cloud paradigm is now essentially the norm, this makes a variety of integration opportunities possible. The fully virtual teaching support platforms Blackboard Learn Ultra or OpenLMS are built into the SaaS (Software as a Service) or PaaS (Platform as a Service) cloud paradigm [170]. Such a venue offers practically limitless integration opportunities with virtual learning environments, business effectiveness of the control, technologies for analysis, 3D printing, or frameworks for commercial simulation models (e.g., Marketplace). Universities now have access to virtual modeling and prototyping, which was previously only available to large corporations. Additionally, the digitalization of educational procedures makes it simpler to create dual learning, which combines educational and commercial components. As a result, the line separating the institution from the business in the virtual space is becoming hazier.

Education has always involved teaching–learning process contact in which knowledge is passed from one to the other. This method of teaching will not alter, although much of the

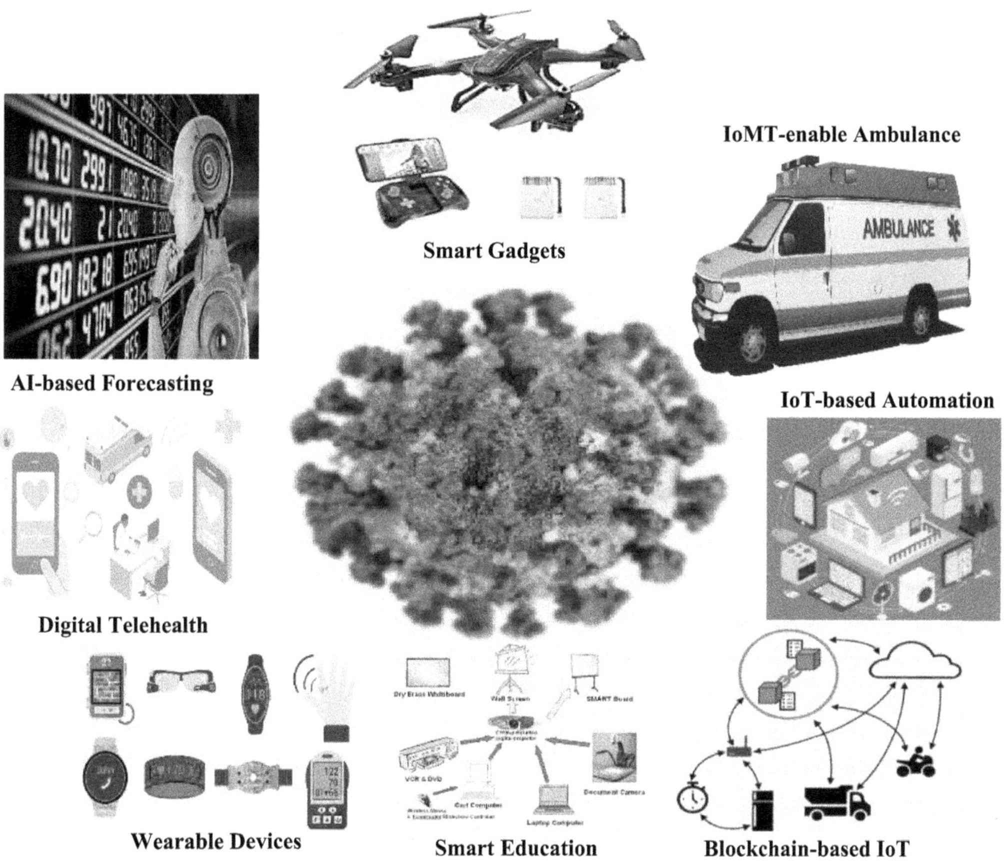

FIGURE 8.1 The major application areas of IoT technology.

technology used today may provide difficulties for this technique. The most significant of them is probably AI-based models, along with allied fields like ML-based algorithms and neural networks. These technologies enable programmers to design apps that can autonomously gather, sort, and utilize the information that is readily available to the public, and compile and arrange knowledge bibliographic databases [171]. Numerous interactive educational platforms, including Quizlet, already employ AI-based models, and AI is used by Querium, Duolingo, and other platforms to offer learners personalized content with or without a human. To assist with questions and maintain student attention, several of these platforms also included decentralized repetition teaching methods and AI-based chatbots, increasing the effectiveness of the process. The use of VR-based technology has increased dramatically since the Oculus Rift, the first widely available commercial VR headset, was released in 2016, which has a significant negative influence on the educational industry in general. VR is able to deliver educational experiences that will not be frequently overlooked, making the entire formative assessment successful by giving educators and administrators a platform for a more fascinating and dynamic encounter. In the case of experiential work of courses, virtual and AR-based educational technologies will be particularly important, where standard laboratories come at a hefty price. In logistics, for instance, these could include automobile simulators, 3D virtualization of warehouse layout, unconventional unit load warehouse design (transverse cross aisles) [172], experimentation and testing, creative combination flow shop settings [173], programming of robots in manufacturing techniques, or modeling of unforeseen events in logistics of redistribution [174].

8.3.2 The Influencing Factors and Trends

The influencing factors of Industry 4.0 and Industry 5.0 are discussed in the section; the factors that influence distance education are the technologies employed in driving the new distance learning paradigm. The factors are the recent technological innovations and weapons used to improve distance learning, especially during the COVID-19 outbreak globally. These influencing factors and trends are the modern technologies used in Industry 4.0 and Industry 5.0, and the most 12 trending-spotting activity that has a great impact on the future of distance and online education and research are presented [16]. These can have an influence on the future of distance learning education and research in the nearest future. The 12 trends are the influencing factors in this modern education system in distance and online teaching are presented in Table 8.1 and the table also discusses the influencing factors.

8.3.3 The Characteristics of Distance Teaching and Learning in the Era of Industry 4.0 and Industry 5.0

Global innovations and dynamism in the context of Industry 4.0 and Society 5.0 had a widespread change in the educational system globally, notably in terms of shifting learning characteristics. The advent of Industry 4.0 is a new age defined by automation, accessibility of information, interconnectivity, and automation. This was accompanied by the concept of Society 5.0 as the current societal perspective, but educational institutions face unique problems.

Industry 4.0 is a significant shift that will impact how employees perform and do business. The incorporation of industrial facilities is part of the transformation, of material supply chains, and service delivery systems that add value to the products or services rendered to customers [201]. The most significant development in Industry 4.0 has been achievements such as AI-based systems, IoT-based systems, energy storage, nanotechnology, biotechnology, quantum computing, robots, automated cars, 3D printing, and so on [202]. Alternatively, Society 5.0 was characterized by a smart environment in which the virtual and real worlds are highly interconnected [203]. Society 5.0 emphasizes humanitarianism by emphasizing the advancement of scientific and technological advances with the primary goal of economic prosperity. According to the author in [204], the characteristics of Industry 5.0 are as follows: (i) it utilizes all available ICT innovations; (ii) it is focused on the community; (iii) it has the involvement of people; (iv) it has the following widely spread values: sustainability, inclusiveness, effectiveness, and intelligence strength; and (v) economic repercussions are developing.

Industry 4.0 and Industry 5.0 will also have a change in the educational profession. Therefore, tremendous technological developments have an impact on the learning process. Educators in the Industry 4.0 and Society 5.0 eras are technologically sophisticated,

TABLE 8.1 The Influencing Factors and Developments of Modern Technologies of Industry 4.0 and Industry 5.0

Relevant Authors	Influencing Factors	Detailed Explanations
[16,26,28,175,176]	Connectivity	Cyber-physical systems integrate the outputs of Industry 4.0 with the machinery that will provide tangible advantages to Society 5.0. The networking capabilities, in conjunction with cloud-based solutions, 5 G cellular networks, wireless energy harvesting, Big Data, and smart network slicing with similar technologies are all examples of emerging technologies that guarantee smooth connectivity in which different systems connect with one another
[16,177–180]	Electrification	Many people have stated that electricity has the ability to provide instantaneous power service because it avoids the transformation inefficiencies involved with the use of fossil resources. This electrification potential is described as the ability to alter several industries. Electric automobiles, for example, seem to become widespread products with revenue growth. Similarly, novel strategies for rechargeable batteries, power-plant architectures, and native electrical vehicle platforms are being developed, and other electrification facilitators are becoming more important.
[177,181,182]	Automation	For a great many years, one of the objectives of every industry has been to automate processes, operations, and instruments. Automation is concerned primarily with the technological innovations that allow equipment, a procedure, or a system automatically execute. Although some regard automation as a danger to the employment market, many studies contend that automation is really beneficial. The connection between automation and the employment market is more complicated, and automation may enhance productivity in the economy alternatively.
[177,183,184]	Data ethics	Information assurance, confidentiality, fairness, accountability, explainability, and accessibility are becoming major concerns in Industry 5.0 due to the increased use of data across several organizations.
[185–188]	Higher Education (HE) Environment	The complexities of the higher education ecosystem have an influence on research and education. These include rising worldwide university rankings, vows by states that they will have world-class universities, and so on, and the creation of provincial administration and reformation units, the establishment of cross-border quality evaluation processes, and the globalization of universities
[177,189–192]	Big Data	Today, the volume of data collected, saved, and analyzed, as well as the discoveries provided by data, are indeed revolutionizing the world in unanticipated ways. This data-driven world is already on, constantly tracking, monitoring, analyzing, and learning. Numerous businesses want to use data to enhance customer engagement, provided with the opportunity, and enter new industries, improve productivity and find innovative solutions to problems

(Continued)

TABLE 8.1 (*Continued*) The Influencing Factors and Developments of Modern Technologies of Industry 4.0 and Industry 5.0

Relevant Authors	Influencing Factors	Detailed Explanations
[178,193]	Sustainable development goals (SDGs)	In 2015, all United Nations Member States endorsed SDGs with an agenda [75] that presented a shared blueprint for people and the planet's wealth and stability either now or tomorrow. These goals emphasize the importance of combining methods to reduce starvation and other hardships that promote education and healthcare, alleviate poverty, and stimulate economic growth, all of which are at the heart of Industry 5.0.
[26,177,181,194–196]	AI	Utilizing interconnected systems, intelligence provides automated decision-making. Intelligent systems are the outcome of a number of technologies, including AI, Big Data analysis, cloud technology, cyber security, high-performance CPUs, and other technologies as examples.
[177–180,197,198]	Technological development	The emergence of modern technological development will surely influence the learning and teaching processes and environments. The utilization of modern technologies like mobile phones, social media, personal computers, and other related disruptive technologies will surely help to sharpen the distance learning educational systems, especially during the pandemic outbreak. Therefore, the introduction of new devices and approaches may result in a succession of modifications in the teaching and learning environments.
[185,186,199]	Lifelong education	Lifelong education is characterized as the quest for new information that is continual, autonomous, and self-motivated. Self-motivation and facilitating continuous improvement enhance social integration and community engagement, and personal growth, assisting the worker in maintaining a competitive advantage and marketability.
[178,193,199,200]	Have faith in the modern technology	Although technology's significance and impact on society grow, maintaining faith in these technological innovations has now become increasingly crucial. The growing use of automation, data, and AI-based systems, as well as shifting power relationships, may have the opposite effect on trust in technology.
[177,186,193]	Labor market	Numerous studies have identified a transformation in the labor market that is affecting basic skilled employees in particular. Improvements in the job market as a result of technological improvements may have unanticipated consequences in education, the economy, and civilization.

and nearly all of them have access to technology that allows them to easily receive information. Their enthusiasm for learning fluctuates as a result of advances and increases in the sort of work [205] stated that information in the technological era has its own personalization, interactivity, and degree of control over the user. They went on to say that education in the age of technology stresses accessibility, allowing individuals to expand their passions and goals, and provide a dynamic social interaction environment where people can learn everything they want and become the type of individual they desire.

HEIs must respond sensibly to changes in education brought about by recent developments in Industry 4.0 and Industry 5.0. The HEIs must pay special attention to how the teaching methods they provide can respond to new problems brought about by current changes. In the age of Industry 4.0, HEIs must be equipped to accept technological advancements and integrate the new Industry 5.0 technological lifestyle into the HEI procedures. Therefore, HEIs must thoroughly examine the characteristics of education in Industry 4.0 and Industry 5.0 in order to deliver excellent education and educational services that are relevant and well-targeted.

The following are the key characteristics of distance education in Industry 4.0 and Industry 5.0 eras.

The World Wide Web (Internet): In the digital age, the educational system and environment are changing dramatically. The presence of the internet in daily life is among the most significant shifts. Mostly everyone nowadays utilize the internet more frequently because they own their devices, which is also true in education. Practically, all today's students utilize the internet to access knowledge and learning resources. They may utilize YouTube, search engines such as Google, and Google Scholars, and other search engines, or education management systems like Zoom, Google Meet, Classroom, Moodle, and One Note, among others. This promotes incredibly quick interaction among students. Learners can rapidly obtain knowledge and share information using a number of social media and communication platforms. The platforms can be used to collect tasks more quickly without needing to acquire or communicate alongside their instructors. Through virtual banking, you may purchase books or other school materials and equipment, pay school fees effortlessly, quickly, and accurately and have a variety of other features that facilitate connection and educational interactions. All of this is possible because of the accessibility of the Internet by both the students and their teachers.

Another characteristic that drives Industry 4.0 and 5.0 in distance educational systems is the AI-based system. AI is tailored in place to bolster specific teaching and learning purposes. The AI-based system is being used to assist students with special needs in their studying, It can even support learners that do not comprehend the educational materials by recommending information and resources that are appropriate for their circumstances. AI support for instructors can be applied to support them in administrative tasks such as evaluating students' work and preparing teaching materials. The AI-based model is extremely adaptable and well-versed in our routines. When students utilize Wikipedia, YouTube, or Google to find specific topics before they even request them, AI will offer suggestions based on our preferences and habits. This obviously makes it extremely simple for learners and staff to obtain timely and accurate information.

8.4 THE FUTURE STRATEGIES FOR DISTANCE LEARNING IN THE ERA OF INDUSTRY 4.0 AND INDUSTRY 5.0

The chapter carefully studies and considers Industry 4.0 and 5.0 in distance learning and teaching during a pandemic, its important trends and influencing factors that have a significant impact on distance education, as well as an examination of the skills required in the future. The chapter outlined four techniques to assist HEIs in redesigning their program and curriculum. The following are the strategies based on the authors in [16]:

Lifelong distance, online, and transdisciplinary education.

Human-centric design modules, resilience, and sustainability.

Hands-on management courses and data fluency.

Human agent with computer interaction experiences.

8.4.1 Lifelong Distance, Online, and Transdisciplinary Education

Meaningful steps toward permanent and interdisciplinary learning opportunities will be important for the future of distance learning education. Industry 4.0 and Industry 5.0 technical improvements necessitate a shift in the present workforce. New abilities will be required of the next generation of learners and their teachers, and perhaps redefine their job requirements in response to technology advances more frequently than present teachers. Teachers must not just be more adaptable in order to meet these demands and adaptive, but they ought to have access to lifelong learning situations.

Industry 4.0 and Industry 5.0 revolutions' results, outcomes, and products cannot be designed and developed without transdisciplinary environments. These are contexts in which societal and environmental considerations are considered jointly instead of independently in various divisions or groups integrated alongside working frameworks, approaches, and processes; for example, combining a holistic approach, participatory design, and future thinking [206,207] should be familiarized and reinvigorated by HEIs.

8.4.2 Human-Centric Design Modules, Resilience, and Sustainability

Many times, throughout history, technological innovation and systems have resulted in adverse reactions and performance has improved. These unforeseen repercussions should be understood by the upcoming generations of learners and instructors, and their consequences for and their consequences for sustainability assessment as part of the Industry 4.0 and Industry 5.0 revolutions and sustainability assessment as part of the Industry 4.0 and Industry 5.0 revolutions

According to [208], "although AI is a software application that we people build, comprehension and controlling the influence of AI in humanity necessitates more than a technical knowledge of its creation". Data and information privacy, technological trust, automation, and data itself are all interconnected ideas, and this interdependence, together with other technical breakthroughs, may result in greater societal transformations. As a result, cultural and ethical concerns concerning these technologies should be incorporated into present educational curricula, and the required procedures and methodologies for development must indeed be explained to the learners as well as create human-centered technological distance learning platforms.

8.4.3 Hands-on Management Courses and Data Fluency

The capacity to use new digital technology is known as digital proficiency, and it necessitates competencies that go beyond what HEIs now provide. Future teachers and instructors need to be proficient not only in new technologies but also in the data and materials gathered. For example, the author in [209] discovered in previous research that the sector does not currently possess systems and procedures to cope with integrated data difficulties like the huge amount of data generated, their quality, accessibility, variety, and availability. The HEI's level of data proficiency indicates the need for a fundamental shift in present education techniques. To remedy the contemporary deficiencies in privacy protection, security, maintenance, and reliability, society requires professionals who really know these deficiencies and therefore will take the appropriate measures to eradicate them. Moreover, data administration, statistics, visualization techniques, ML- and Al-based systems, data integrity, the social ramifications of prospective autonomous vehicles, and automated systems, in addition, should be incorporated into present distance education programs and curricula to deal with the skyrocketing complexities and critical sustainability among these ecosystems.

8.4.4 Human Agent with Computer Interaction Experiences

Distance education's performance in the age of Industry 4.0 and Industry 5.0 will rely on "collaborative understanding" involving humans and smart technologies. This according to authors in [210] will require mutual learning centered around a bidirectional process that includes mutual trade, reliance, activity, or persuasion through human engagement, resulting in the creation of new meanings and introducing new ideas to supplement existing ones, or developing skills and capacities in collaboration with each set of participants. This process of education should be available to the future workforce so that they can differentiate themselves and comprehend how various systems demand primarily human, only machine initiatives or collaborations on task interdependence are permitted. Development of new materials by instructors will not just be the designers of these interactive effort methods or complementary learnings but will also participate in such duties. As a result, distance education programs and curricula should stretch students' perspectives on computer interactions, robots, human agent, and machines and expose them to various modes of engagement and teamwork with the next generation of Industry 4.0 and Industry 5.0 revolutions.

8.5 CONCLUSION AND FUTURE SCOPE

Distance education has evolved in recent years, becoming increasingly prevalent and yet a developing sector. Its ideals have evolved quickly, but technological advancement still necessitates modification and remodeling for today's solutions as well as foreseeable prospects. The pandemic is gradually affecting the public, particularly organizations', and this brings about a change in the perception of distance learning education. However, some people believe that a study performed remotely is less valuable than a face-to-face classroom lecture or full-time education. Distance learning has distinct advantages over conventional in-person classroom teaching and learning. The most important is location and time flexibility and independence. Both allow individuals to learn at any time of day and from any location, allowing knowledge to penetrate isolated locations and communities with limited periods of time for traditional education. One thing that is sure is that the COVID-19 outbreak has caused many institutions to move to the direction of online and distance learning. We can be sure of one thing, however: the COVID-19 pandemic offered an ultimate test for the distance and online learning teaching model, with various countries coping with various levels of the pandemic. Hence, recently as the pandemic continues worldwide, smarter means of communication can be developed by HEIs globally and will continue offering a variety of ways to deliver their operations, and, as COVID-19 fades into antiquity due to advances in medical sectors, we may better what was formerly conventional learning with the distance and online teaching capability from this period. As the COVID-19 issue rages on over the world, HEIs can establish better means of communication and ways of service delivery, and as COVID-19 falls into obscurity, thus improving the conventional ways of teaching and learning transform into distance and online teaching over time. Therefore, this chapter covers the societal changes in distance education in the face of an outbreak, particularly the significance of distance education during the worldwide advent of any pandemic, and the problems given by Industry 4.0 and 5.0 trends. The chapter covers the tactics that demand HEIs all over the world to re-evaluate their position, as well as the role of Industry 4.0 and Industry 5.0 in this current and future pandemic. The value and significance of the art of the longer perspective may be the most essential contributor to this technology, examining the major developments that are influencing the environment around us, comprehending their consequences, and devising methods to assist us to transform our educational system. Future work will further check the practical applications of various Industry 4.0 and Industry 5.0 technologies in distance learning education HEIs.

REFERENCES

[1] Pregowska, A., Masztalerz, K., Garlińska, M., & Osial, M. (2021). A worldwide journey through distance education—from the post office to virtual, augmented and mixed realities, and education during the COVID-19 pandemic. *Education Sciences*, 11(3), 118.

[2] Awotunde, J. B., Jimoh, R. G., Oladipo, I. D., Abdulraheem, M., Jimoh, T. B., & Ajamu, G. J. (2021). Big data and data analytics for an enhanced COVID-19 epidemic management. In: Oliva, D., Hassan, S.A., & Mohamed, A. (eds), *Artificial Intelligence for COVID-19. Studies in Systems, Decision and Control* (Vol. 358, pp. 11–29). Cham: Springer.

[3] Ilmi, Z., Darma, D. C., & Azis, M. (2020). Independence in learning, education management, and industry 4.0: Habitat Indonesia during COVID-19. *Journal of Anthropology of Sport and Physical Education*, 4(4), 63–66.

[4] Alam, G. M., Forhad, A. R., & Ismail, I. A. (2020). Can education as an 'International Commodity' be the backbone or cane of a nation in the era of fourth industrial revolution?- A Comparative study. *Technological Forecasting and Social Change*, 159, 120184.

[5] Awotunde, J. B., Ogundokun, R. O., Ayo, F. E., Ajamu, G. J., & Ogundokun, O. E. (2021). UTAUT model: Integrating social media for learning purposes among university students in Nigeria. *SN Social Sciences*, 1(9), 1–27.

[6] Awad, N., & Barak, M. (2018). Pre-service science teachers learn a science, technology, engineering and mathematics (STEM)-oriented program: The case of sound, waves and communication systems. *Eurasia Journal of Mathematics, Science and Technology Education*, 14(4), 1431–1451.

[7] Awotunde, J. B., Oluwabukonla, S., Chakraborty, C., Bhoi, A. K., & Ajamu, G. J. (2022). Application of artificial intelligence and big data for fighting COVID-19 pandemic. *International Series in Operations Research and Management Science*, 320, 3–26.

[8] Folorunso, S. O., Ogundepo, E. A., Awotunde, J. B., Ayo, F. E., Banjo, O. O., & Taiwo, A. I. (2022). A multi-step predictive model for covid-19 cases in Nigeria using machine learning. In Hassan, S.A., Mohamed, A.W., & Alnowibet, K.A. (eds), *Decision Sciences for COVID-19. International Series in Operations Research & Management Science* (Vol. 320, pp. 107–136). Cham: Springer.
[9] Simamora, R. M. (2020). The challenges of online learning during the COVID-19 pandemic: An essay analysis of performing arts education students. *Studies in Learning and Teaching*, 1(2), 86–103.
[10] Ogundokun, R. O., Daniyal, M., Misra, S., & Awotunde, J. B. (2021). Students' perspective on online teaching in higher institutions during COVID-19 pandemic. *International Journal of Networking and Virtual Organisations*, 25(3–4), 308–332.
[11] Leo, S., Alsharari, N. M., Abbas, J., & Alshurideh, M. T. (2021). From offline to online learning: A qualitative study of challenges and opportunities as a response to the COVID-19 pandemic in the UAE higher education context. In Alshurideh, M., Hassanien, A.E., & Masa'deh, R. (eds), *The Effect of Coronavirus Disease (COVID-19) on Business Intelligence*. Studies in Systems, Decision and Control (Vol. 334, pp. 203–217). Cham: Springer.
[12] Hebebci, M. T., Bertiz, Y., & Alan, S. (2020). Investigation of views of students and teachers on distance education practices during the coronavirus (COVID-19) Pandemic. *International Journal of Technology in Education and Science*, 4(4), 267–282.
[13] Awotunde, J. B., Jimoh, R. G., AbdulRaheem, M., Oladipo, I. D., Folorunso, S. O., & Ajamu, G. J. (2022). IoT-based wearable body sensor network for COVID-19 pandemic. *Studies in Systems, Decision and Control*, 378, 253–275.
[14] Awotunde, J. B., Adeniyi, E. A., Kolawole, P. O., & Ogundokun, R. O. (2022). Application of big data in COVID-19 epidemic. *Data Science for COVID-19: Volume 2: Societal and Medical Perspectives*, 2021, 141–165.
[15] Oz, O., & Ozdamar, N. (2020). Academic's views on industry 4.0 within the scope of open and distance education. *Asian Journal of Distance Education*, 15(2), 58–85.
[16] Broo, D. G., Kaynak, O., & Sait, S. M. (2022). Rethinking engineering education at the age of industry 5.0. *Journal of Industrial Information Integration*, 25, 100311.
[17] Criollo-C, S., Guerrero-Arias, A., Jaramillo-Alcázar, Á., & Luján-Mora, S. (2021). Mobile learning technologies for education: Benefits and pending issues. *Applied Sciences*, 11(9), 4111.
[18] Peters, O. (2010). *Distance Education in Transition: Developments and Issues*. BIS-Verlag der Carl-von-Ossietzky-Univ.
[19] Kurt, R. (2019). Industry 4.0 in terms of industrial relations and its impacts on labour life. *Procedia Computer Science*, 158, 590–601.
[20] Ferri, F., Grifoni, P., & Guzzo, T. (2020). Online learning and emergency remote teaching: Opportunities and challenges in emergency situations. *Societies*, 10(4), 86.
[21] Tomasik, M. J., Helbling, L. A., & Moser, U. (2021). Educational gains of in-person vs. distance learning in primary and secondary schools: A natural experiment during the COVID-19 pandemic school closures in Switzerland. *International Journal of Psychology*, 56(4), 566–576.
[22] Jung, J. (2020). The fourth industrial revolution, knowledge production and higher education in South Korea. *Journal of Higher Education Policy and Management*, 42(2), 134–156.
[23] Penprase, B. E. (2018). The fourth industrial revolution and higher education. *Higher Education in the Era of the Fourth Industrial Revolution*, 10, 978–981.
[24] Simpson, M., & Anderson, B. (2012). History and heritage in open, flexible and distance education. *Journal of Open, Flexible, and Distance Learning*, 16(2), 1–10.
[25] Mohamed Hashim, M. A., Tlemsani, I., & Matthews, R. (2022). Higher education strategy in digital transformation. *Education and Information Technologies*, 27(3), 3171–3195.

[26] Xu, L. D., Xu, E. L., & Li, L. (2018). Industry 4.0: State of the art and future trends. *International Journal of Production Research*, 56(8), 2941–2962.

[27] Abikoye, O. C., Bajeh, A. O., Awotunde, J. B., Ameen, A. O., Mojeed, H. A., Abdulraheem, M., ... & Salihu, S. A. (2021). Application of internet of thing and cyber physical system in industry 4.0 smart manufacturing. *Advances in Science, Technology and Innovation*, 2021, 203–217.

[28] Lasi, H., Fettke, P., Kemper, H. G., Feld, T., & Hoffmann, M. (2014). Industry 4.0. *Business & Information Systems Engineering*, 6(4), 239–242.

[29] Romero, D., Stahre, J., Wuest, T., Noran, O., Bernus, P., Fast-Berglund, Å., & Gorecky, D. (2016). Towards an operator 4.0 typology: A human-centric perspective on the fourth industrial revolution technologies. In *Proceedings of the International Conference on Computers and Industrial Engineering (CIE46)* (pp. 29–31). Tianjin, China.

[30] Xu, X., Lu, Y., Vogel-Heuser, B., & Wang, L. (2021). Industry 4.0 and Industry 5.0—Inception, conception and perception. *Journal of Manufacturing Systems*, 61, 530–535.

[31] Prisecaru, P. (2016). Challenges of the fourth industrial revolution. Knowledge horizons. *Economics*, 8(1), 57.

[32] Javaid, M., Haleem, A., Singh, R. P., Rab, S., Suman, R., & Khan, S. (2022). Exploring relationships between Lean 4.0 and manufacturing industry. *Industrial Robot*, 49, 402–414.

[33] Lu, Y. (2017). Industry 4.0: A survey on technologies, applications and open research issues. *Journal of Industrial Information Integration*, 6, 1–10.

[34] van Someren, T. C., & van Someren-Wang, S. (2017). The instrument: Strategic innovation as a new foundation for Russian innovation system. In *Strategic Innovation in Russia* (pp. 81–161). Cham: Springer.

[35] Kennedy, S. (2020). *China's Uneven High-Tech Drive*. Center for Strategic and International Studies (CSIS).

[36] Bajeh, A. O., Mojeed, H. A., Ameen, A. O., Abikoye, O. C., Salihu, S. A., Abdulraheem, M., ... & Awotunde, J. B. (2021). Internet of robotic things: Its domain, methodologies, and applications. *Advances in Science, Technology and Innovation*, 2021, 135–146.

[37] Zhou, K., Liu, T., & Zhou, L. (2015). Industry 4.0: Towards future industrial opportunities and challenges. In *2015 12th International Conference on Fuzzy Systems and Knowledge Discovery (FSKD)* (pp. 2147–2152). IEEE.

[38] Sartal, A., Bellas, R., Mejías, A. M., & García-Collado, A. (2020). The sustainable manufacturing concept, evolution and opportunities within Industry 4.0: A literature review. *Advances in Mechanical Engineering*, 12(5), 1687814020925232.

[39] Piccarozzi, M., Aquilani, B., & Gatti, C. (2018). Industry 4.0 in management studies: A systematic literature review. *Sustainability*, 10(10), 3821.

[40] Benitez, G. B., Lima, M. J. D. R. F., Lerman, L. V., & Frank, A. G. (2019). Understanding Industry 4.0: Definitions and insights from a cognitive map analysis. *Brazilian Journal of Operations & Production Management [recurso eletrônico]. Rio de Janeiro, RJ*, 16(2), 192–200.

[41] Zeng, J., Huang, J., & Pan, L. (2020). How to balance acute myocardial infarction and COVID-19: The protocols from Sichuan Provincial people's hospital. *Intensive Care Medicine*, 46(6), 1111–1113.

[42] Manogaran, G., Thota, C., Lopez, D., & Sundarasekar, R. (2017). Big data security intelligence for healthcare industry 4.0. In Thames, L. & Schaefer, D. (eds), *Cybersecurity for Industry 4.0*. Springer Series in Advanced Manufacturing (pp. 103–126). Cham: Springer.

[43] Gamage, K. A., Wijesuriya, D. I., Ekanayake, S. Y., Rennie, A. E., Lambert, C. G., & Gunawardhana, N. (2020). Online delivery of teaching and laboratory practices: Continuity of university programmes during COVID-19 pandemic. *Education Sciences*, 10(10), 291.

[44] Sumtsova, O., Aikina, T., Bolsunovskaya, L., Phillips, C., Zubkova, O., & Mitchell, P. (2018). Collaborative learning at engineering universities: Benefits and challenges. *International Journal of Emerging Technologies in Learning (iJET)*, 13(1), 160–177.

[45] Marone, E. M., & Rinaldi, L. F. (2018). Educational value of virtual reality for medical students: An interactive lecture on carotid stenting. *The Journal of Cardiovascular Surgery*, 59(4), 650–651.

[46] Gopal, M., Skobodzinski, A. A., Sterbling, H. M., Rao, S. R., LaChapelle, C., Suzuki, K., & Litle, V. R. (2018). Bronchoscopy simulation training as a tool in medical school education. *The Annals of Thoracic Surgery*, 106(1), 280–286.

[47] Stepan, K., Zeiger, J., Hanchuk, S., Del Signore, A., Shrivastava, R., Govindaraj, S., & Iloreta, A. (2017). Immersive virtual reality as a teaching tool for neuroanatomy. *International Forum of Allergy & Rhinology*, 7(10), 1006–1013.

[48] Maresky, H. S., Oikonomou, A., Ali, I., Ditkofsky, N., Pakkal, M., & Ballyk, B. (2019). Virtual reality and cardiac anatomy: Exploring immersive three-dimensional cardiac imaging, a pilot study in undergraduate medical anatomy education. *Clinical Anatomy*, 32(2), 238–243.

[49] Taubert, M., Webber, L., Hamilton, T., Carr, M., & Harvey, M. (2019). Virtual reality videos used in undergraduate palliative and oncology medical teaching: Results of a pilot study. *BMJ Supportive & Palliative Care*, 9(3), 281–285.

[50] Mayrose, J., & Myers, J. W. (2007). Endotracheal intubation: Application of virtual reality to emergency medical services education. *Simulation in Healthcare*, 2(4), 231–234.

[51] Lorenzo-Alvarez, R., Pavia-Molina, J., & Sendra-Portero, F. (2018). Exploring the potential of undergraduate radiology education in the virtual world second life with first-cycle and second-cycle medical students. *Academic Radiology*, 25(8), 1087–1096.

[52] Maicher, K. R., Zimmerman, L., Wilcox, B., Liston, B., Cronau, H., Macerollo, A.,... & Danforth, D. R. (2019). Using virtual standardized patients to accurately assess information gathering skills in medical students. *Medical Teacher*, 41(9), 1053–1059.

[53] Foo, C. C., Cheung, B., & Chu, K. M. (2021). A comparative study regarding distance learning and the conventional face-to-face approach conducted problem-based learning tutorial during the COVID-19 pandemic. *BMC Medical Education*, 21(1), 1–6.

[54] Awotunde, J. B., Adeniyi, A. E., Abiodun, K. M., Ajamu, G. J., & Matiluko, O. E. (2022). Application of cloud and IoT technologies in battling the COVID-19 pandemic. In Al-Turjman, F. & Nayyar, A. (eds), *Machine Learning for Critical Internet of Medical Things* (pp. 1–29). Cham: Springer.

[55] Chick, R. C., Clifton, G. T., Peace, K. M., Propper, B. W., Hale, D. F., Alseidi, A. A., & Vreeland, T. J. (2020). Using technology to maintain the education of residents during the COVID-19 pandemic. *Journal of Surgical Education*, 77(4), 729–732.

[56] Sleiwah, A., Mughal, M., Hachach-Haram, N., & Roblin, P. (2020). COVID-19 lockdown learning: The uprising of virtual teaching. *Journal of Plastic, Reconstructive & Aesthetic Surgery*, 73(8), 1575–1592.

[57] Arslan, R., Kofoğlu, M., & Dargut, C. (2020). Development of augmented reality application for biology education. *Journal of Turkish Science Education*, 17(1), 62–72.

[58] Labovitz, J., & Hubbard, C. (2020). The use of virtual reality in podiatric medical education. *Clinics in Podiatric Medicine and Surgery*, 37(2), 409–420.

[59] Chang, X. Q., Zhang, D. H., & Jin, X. X. (2016). Application of virtual reality technology in distance learning. *International Journal of Emerging Technologies in Learning*, 11(11).

[60] Yepez, J., Guevara, L., & Guerrero, G. (2020). AulaVR: Virtual Reality, a telepresence technique applied to distance education. In *2020 15th Iberian Conference on Information Systems and Technologies (CISTI)* (pp. 1–5). IEEE.

[61] Ott, M., & Pozzi, F. (2008). ICT and cultural heritage education: Which added value? In *World Summit on Knowledge Society* (pp. 131–138). Berlin, Heidelberg: Springer.

[62] Liu, Y., Fan, X., Zhou, X., Liu, M., Wang, J., & Liu, T. (2019). Application of virtual reality technology in distance higher education. In *ICDEL 2019: Proceedings of the 2019 4th International Conference on Distance Education and Learning, Shanghai, China, May 24–27, 2019* (pp. 35–39).

[63] Shtaleva, N. R., Derkho, M. A., Pribytova, O. S., & Shamina, S. V. (2021). Distant learning: Challenges and risks of 2020. In *IOP Conference Series: Earth and Environmental Science* (Vol. 699, No. 1, p. 012026). IOP Publishing.

[64] Aristovnik, A., Keržič, D., Ravšelj, D., Tomaževič, N., & Umek, L. (2020). Impacts of the COVID-19 pandemic on life of higher education students: A global perspective. *Sustainability*, 12(20), 8438.

[65] Ajagbe, S. A., Awotunde, J. B., Oladipupo, M. A., & Oye, O. E. (2022). Prediction and forecasting of coronavirus cases using artificial intelligence algorithm. In Al-Turjman, F. & Nayyar, A. (eds), *Machine Learning for Critical Internet of Medical Things* (pp. 31–54). Cham: Springer.

[66] Dua, A. B., Kilian, A., Grainger, R., Fantus, S. A., Wallace, Z. S., Buttgereit, F., & Jonas, B. L. (2020). Challenges, collaboration, and innovation in rheumatology education during the COVID-19 pandemic: Leveraging new ways to teach. *Clinical Rheumatology*, 39(12), 3535–3541.

[67] Mishra, L., Gupta, T., & Shree, A. (2020). Online teaching-learning in higher education during lockdown period of COVID-19 pandemic. *International Journal of Educational Research Open*, 1, 100012.

[68] Pennington, Z., Lubelski, D., Khalafallah, A. M., Ehresman, J., Sciubba, D. M., Witham, T. F., & Huang, J. (2020). Letter to the editor "Changes to neurosurgery resident education since onset of the COVID-19 pandemic". *World Neurosurgery*, 139, 734.

[69] İbili, E., Çat, M., Resnyansky, D., Şahin, S., & Billinghurst, M. (2020). An assessment of geometry teaching supported with augmented reality teaching materials to enhance students' 3D geometry thinking skills. *International Journal of Mathematical Education in Science and Technology*, 51(2), 224–246.

[70] Huang, T. L., & Liao, S. (2015). A model of acceptance of augmented-reality interactive technology: The moderating role of cognitive innovativeness. *Electronic Commerce Research*, 15(2), 269–295.

[71] Kečkeš, A. L., & Tomičić, I. (2017). Augmented reality in tourism–research and applications overview. *Interdisciplinary Description of Complex Systems: INDECS*, 15(2), 157–167.

[72] Karagozlu, D., Kosarenko, N., Efimova, O., & Zubov, V. (2019). Identifying students' attitudes regarding augmented reality applications in science classes. *International Journal of Emerging Technologies in Learning (iJET)*, 14(22), 45–55.

[73] Fidan, M., & Tuncel, M. (2019). Integrating augmented reality into problem based learning: The effects on learning achievement and attitude in physics education. *Computers & Education*, 142, 103635.

[74] Hew, K. F., & Cheung, W. S. (2014). Students' and instructors' use of massive open online courses (MOOCs): Motivations and challenges. *Educational Research Review*, 12, 45–58.

[75] Kaščak, J., Telišková, M., Török, J., Baron, P., Zajac, J., & Husár, J. (2019). Implementation of augmented reality into the training and educational process in order to support spatial perception in technical documentation. In *2019 IEEE 6th International Conference on Industrial Engineering and Applications (ICIEA)* (pp. 583–587). IEEE.

[76] Dünser, A., Steinbügl, K., Kaufmann, H., & Glück, J. (2006). Virtual and augmented reality as spatial ability training tools. In *Proceedings of the 7th ACM SIGCHI New Zealand Chapter's International Conference on Computer-Human Interaction: Design Centered HCI*, Christchurch, New Zealand, July 6–7, 2006 (pp. 125–132).

[77] Papanastasiou, G., Drigas, A., Skianis, C., Lytras, M., & Papanastasiou, E. (2019). Virtual and augmented reality effects on K-12, higher and tertiary education students' twenty-first century skills. *Virtual Reality*, 23(4), 425–436.

[78] Javornik, A., Duffy, K., Rokka, J., Scholz, J., Nobbs, K., Motala, A., & Goldenberg, A. (2021). Strategic approaches to augmented reality deployment by luxury brands. *Journal of Business Research*, 136, 284–292.

[79] Hilken, T., de Ruyter, K., Chylinski, M., Mahr, D., & Keeling, D. I. (2017). Augmenting the eye of the beholder: Exploring the strategic potential of augmented reality to enhance online service experiences. *Journal of the Academy of Marketing Science*, 45(6), 884–905.
[80] Picot-Coupey, K., Krey, N., Huré, E., & Ackermann, C. L. (2021). Still work and/or fun? Corroboration of the hedonic and utilitarian shopping value scale. *Journal of Business Research*, 126, 578–590.
[81] Mathwick, C., Malhotra, N., & Rigdon, E. (2001). Experiential value: Conceptualization, measurement and application in the catalog and Internet shopping environment☆. *Journal of Retailing*, 77(1), 39–56.
[82] Kuehn, B. M. (2018). Virtual and augmented reality put a twist on medical education. *Jama*, 319(8), 756–758.
[83] Arizzi, G., Breitenreiter, J., Khalsa, R., Iyer, R., Babin, L. A., & Griffin, M. (2020). Modeling business student satisfaction: Utilitarian value and hedonic value as drivers of satisfaction. *Marketing Education Review*, 30(4), 196–207.
[84] Hsiao, K. L., Huang, T. C., Chen, M. Y., & Chiang, N. T. (2018). Understanding the behavioral intention to play Austronesian learning games: From the perspectives of learning outcome, service quality, and hedonic value. *Interactive Learning Environments*, 26(3), 372–385.
[85] Azizi, S. M., Roozbahani, N., & Khatony, A. (2020). Factors affecting the acceptance of blended learning in medical education: Application of UTAUT2 model. *BMC Medical Education*, 20(1), 1–9.
[86] Mittal, A., Aggarwal, A., & Mittal, R. (2020). Predicting university students' adoption of mobile news applications: The role of perceived hedonic value and news motivation. *International Journal of E-Services and Mobile Applications (IJESMA)*, 12(4), 42–59.
[87] Hazaymeh, W. A. (2021). EFL students' perceptions of online distance learning for enhancing English language learning during Covid-19 pandemic. International Journal of Instruction, 14(3), 501–518.
[88] Boytchev, P., & Boytcheva, S. (2019). Innovative elearning technologies in the open education era. In *CompSysTech '19: Proceedings of the 20th International Conference on Computer Systems and Technologies, Ruse Bulgaria, June 21–22, 2019* (pp. 324–331).
[89] Chen, Y. C., Lu, Y. L., & Lien, C. J. (2021). Learning environments with different levels of technological engagement: A comparison of game-based, video-based, and traditional instruction on students' learning. *Interactive Learning Environments*, 29(8), 1363–1379.
[90] Morgan, C., & Townsend, C. (2022). Why the drive: The utilitarian and hedonic benefits of self-expression through consumption. *Current Opinion in Psychology*, 101320.
[91] Perannagari, K. T., & Chakrabarti, S. (2020). Factors influencing acceptance of augmented reality in retail: Insights from thematic analysis. *International Journal of Retail & Distribution Management*, 48(1), 18–34.
[92] Hsu, S. H. Y., Tsou, H. T., & Chen, J. S. (2021). "Yes, we do. Why not use augmented reality?" customer responses to experiential presentations of AR-based applications. *Journal of Retailing and Consumer Services*, 62, 102649.
[93] Chen, Y., & Lin, C. A. (2022). Consumer decision-making in an augmented reality environment: Exploring the effects of flow via augmented realism and technology fluidity. *Telematics and Informatics*, 75, 101833.
[94] Khan, M. A., Nabi, M. K., Khojah, M., & Tahir, M. (2020). Students' perception towards e-learning during COVID-19 pandemic in India: An empirical study. *Sustainability*, 13(1), 57.
[95] Hoq, M. Z. (2020). E-Learning during the period of pandemic (COVID-19) in the kingdom of Saudi Arabia: An empirical study. *American Journal of Educational Research*, 8(7), 457–464.
[96] Baabdullah, A. M., Alsulaimani, A. A., Allamnakhrah, A., Alalwan, A. A., Dwivedi, Y. K., & Rana, N. P. (2022). Usage of augmented reality (AR) and development of e-learning outcomes: An empirical evaluation of students'e-learning experience. *Computers & Education*, 177, 104383.

[97] Boboc, R. G., Chiriac, R. L., & Antonya, C. (2021). How augmented reality could improve the student's attraction to learn mechanisms. *Electronics*, 10(2), 175.
[98] Kiryakova, G., Angelova, N., & Yordanova, L. (2018). The potential of augmented reality to transform education into smart education. *TEM Journal*, 7(3), 556.
[99] Saleem, M., Kamarudin, S., Shoaib, H. M., & Nasar, A. (2021). Influence of augmented reality app on intention towards e-learning amidst COVID-19 pandemic. *Interactive Learning Environments*, 1–15.
[100] Pal, D., & Vanijja, V. (2020). Perceived usability evaluation of Microsoft Teams as an online learning platform during COVID-19 using system usability scale and technology acceptance model in India. *Children and Youth Services Review*, 119, 105535.
[101] Chytas, D., Johnson, E. O., Piagkou, M., Mazarakis, A., Babis, G. C., Chronopoulos, E.,... & Natsis, K. (2020). The role of augmented reality in anatomical education: An overview. *Annals of Anatomy-Anatomischer Anzeiger*, 229, 151463.
[102] Weeks, J. K., & Amiel, J. M. (2019). Enhancing neuroanatomy education with augmented reality. *Medical Education*, 53(5), 516–517.
[103] San Martin-Rodriguez, L., Soto-Ruiz, M. N., Echeverria-Ganuza, G., & Escalada-Hernandez, P. (2019). Augmented reality for training operating room scrub nurses. *Medical Education*, 53(5), 514–515.
[104] Mahmood, F., Mahmood, E., Dorfman, R. G., Mitchell, J., Mahmood, F. U., Jones, S. B., & Matyal, R. (2018). Augmented reality and ultrasound education: Initial experience. *Journal of Cardiothoracic and Vascular Anesthesia*, 32(3), 1363–1367.
[105] Gómez-Galán, J., Vázquez-Cano, E., Luque de la Rosa, A., & López-Meneses, E. (2020). Socio-educational impact of augmented reality (AR) in sustainable learning ecologies: A semantic modeling approach. *Sustainability*, 12(21), 9116.
[106] Digital Trends. Available online: https://www.digitaltrends.com/virtual-reality/hololens-holoanatomy-award-jackson-hole-science-media-awards/ (accessed on 18 May 2022).
[107] Ruthberg, J. S., Quereshy, H. A., Ahmadmehrabi, S., Trudeau, S., Chaudry, E., Hair, B., ... & Mowry, S. E. (2020). A multimodal multi-institutional solution to remote medical student education for otolaryngology during COVID-19. *Otolaryngology–Head and Neck Surgery*, 163(4), 707–709.
[108] Bin, S., Masood, S., & Jung, Y. (2020). Virtual and augmented reality in medicine. In *Biomedical Information Technology* (pp. 673–686). Elsevier, Academic Press.
[109] Birt, J., Stromberga, Z., Cowling, M., & Moro, C. (2018). Mobile mixed reality for experiential learning and simulation in medical and health sciences education. *Information*, 9(2), 31.
[110] Folorunso, S. O., Awotunde, J. B., Banjo, O. O., Ogundepo, E. A., & Adeboye, N. O. (2021). Comparison of active COVID-19 cases per population using time-series models. *International Journal of E-Health and Medical Communications (IJEHMC)*, 13(2), 1–21.
[111] Ogundokun, R. O., Daniyal, M., Misra, S., & Awotunde, J. B. (2021). Students' perspective on online teaching in higher institutions during COVID-19 pandemic. *International Journal of Networking and Virtual Organisations*, 25(3–4), 308–332.
[112] Awotunde, J. B., Jimoh, R. G., Folorunso, S. O., Adeniyi, E. A., Abiodun, K. M., & Banjo, O. O. (2021). Privacy and security concerns in IoT-based healthcare systems. In Siarry, P., Jabbar, M., Aluvalu, R., Abraham, A., & Madureira, A. (eds), *The Fusion of Internet of Things, Artificial Intelligence, and Cloud Computing in Health Care. Internet of Things* (pp. 105–134). Cham: Springer.
[113] Gardas, B. B., & Navimipour, N. J. (2022). Performance evaluation of higher education system amid COVID-19: A threat or an opportunity? *Kybernetes*, 51(8), 2508–2528.
[114] Basilaia, G., & Kvavadze, D. (2020). Transition to online education in schools during a SARS-CoV-2 coronavirus (COVID-19) pandemic in Georgia. *Pedagogical Research*, 5(4), em0060.

[115] Parra-González, M. E., López Belmonte, J., Segura-Robles, A., & Fuentes Cabrera, A. (2020). Active and emerging methodologies for ubiquitous education: Potentials of flipped learning and gamification. *Sustainability*, 12(2), 602.
[116] Hoq, M. Z. (2020). E-Learning during the period of pandemic (COVID-19) in the kingdom of Saudi Arabia: An empirical study. *American Journal of Educational Research*, 8(7), 457–464.
[117] Asad, M. M., Hussain, N., Wadho, M., Khand, Z. H., & Churi, P. P. (2021). Integration of e-learning technologies for interactive teaching and learning process: An empirical study on higher education institutes of Pakistan. *Journal of Applied Research in Higher Education*, 13(3), 649–663.
[118] Sadrishojaei, M., Navimipour, N. J., Reshadi, M., & Hosseinzadeh, M. (2021). A new preventive routing method based on clustering and location prediction in the mobile internet of things. *IEEE Internet of Things Journal*, 8(13), 10652–10664.
[119] Awotunde, J. B., Folorunso, S. O., Bhoi, A. K., Adebayo, P. O., & Ijaz, M. F. (2021). Disease diagnosis system for IoT-based wearable body sensors with machine learning algorithm. In Kumar Bhoi, A., Mallick, P.K., Narayana Mohanty, M., & Albuquerque, V.H.C.d. (eds), *Hybrid Artificial Intelligence and IoT in Healthcare. Intelligent Systems Reference Library* (Vol. 209, pp. 201–222). Singapore: Springer.
[120] Abdel-Basset, M., Manogaran, G., Mohamed, M., & Rushdy, E. (2019). Internet of things in smart education environment: Supportive framework in the decision-making process. *Concurrency and Computation: Practice and Experience*, 31(10), e4515.
[121] Awotunde, J. B., Ogundokun, R. O., Adeniyi, A. E., Ayo, F. E., Ajamu, G. J., Abiodun, M. K., & Ogundokun, O. E. (2022). Cloud-IoMT-based wearable body sensors network for monitoring elderly patients during the COVID-19 pandemic. In *Biomedical Engineering Applications for People with Disabilities and the Elderly in the COVID-19 Pandemic and Beyond* (pp. 33–48). Elsevier, Academic Press.
[122] Amasha, M. A., Areed, M. F., Alkhalaf, S., Abougalala, R. A., Elatawy, S. M., & Khairy, D. (2020). The future of using Internet of things (loTs) and context-aware technology in E-learning. In *ICEIT 2020: Proceedings of the 2020 9th International Conference on Educational and Information Technology*, Oxford, United Kingdom, February 11–13, 2020 (pp. 114–123).
[123] Zaharov, A. A., Nissenbaum, O. V., Ponomarov, K. Y., & Shirokih, A. V. (2018). Use of open-source internet of things platform in education projects. In *2018 Global Smart Industry Conference (GloSIC)* (pp. 1–6). IEEE.
[124] Awotunde, J. B., Chakraborty, C., & Adeniyi, A. E. (2021). Intrusion detection in industrial internet of things network-based on deep learning model with rule-based feature selection. *Wireless Communications and Mobile Computing*, 2021(2021), 7154587.
[125] Yang, Y., & Yu, K. (2016). Construction of distance education classroom in architecture specialty based on internet of things technology. *International Journal of Emerging Technologies in Learning*, 11(5), 56.
[126] Vharkute, M., & Wagh, S. (2015). An architectural approach of internet of things in E-Learning. In *2015 International Conference on Communications and Signal Processing (ICCSP)* (pp. 1773–1776). IEEE.
[127] Cornel, C. E. (2015). The role of internet of things for a continuous improvement in education. *Hyperion Economic Journal*, 2(3), 24–31.
[128] Abbasy, M. B., & Quesada, E. V. (2017). Predictable influence of IoT (Internet of Things) in the higher education. *International Journal of Information and Education Technology*, 7(12), 914–920.
[129] Zahedi, M. H., & Dehghan, Z. (2019). Effective e-learning utilizing internet of things. In *2019 13th Iranian and 7th National Conference on e-Learning and e-Teaching (ICeLeT)* (pp. 1–6). IEEE.

[130] Ying, L., Jiong, Z., Wei, S., Jingchun, W., & Xiaopeng, G. (2017). VREX: Virtual reality education expansion could help to improve the class experience (VREX platform and community for VR based education). In *2017 IEEE Frontiers in Education Conference (FIE)* (pp. 1–5). IEEE.

[131] Lin, H. C. S., Yu, S. J., Sun, J. C. Y., & Jong, M. S. Y. (2021). Engaging university students in a library guide through wearable spherical video-based virtual reality: Effects on situational interest and cognitive load. *Interactive Learning Environments*, 29(8), 1272–1287.

[132] Saniuk, S., Grabowska, S., & Straka, M. (2022). Identification of social and economic expectations: Contextual reasons for the transformation process of industry 4.0 into the industry 5.0 concept. *Sustainability*, 14(3), 1391.

[133] Skobelev, P. O., & Borovik, S. Y. (2017). On the way from Industry 4.0 to Industry 5.0: From digital manufacturing to digital society. *Industry 4.0*, 2(6), 307–311.

[134] Alvarez-Cedillo, J., Aguilar-Fernandez, M., Sandoval-Gomez Jr, R., & Alvarez-Sanchez, T. (2019). Actions to be taken in mexico towards education 4.0 and society 5.0. *International Journal of Evaluation and Research in Education*, 8(4), 693–698.

[135] Mansor, N. A., Abdullah, N., & Rahman, H. A. (2020). Towards electronic learning features in education 4.0 environment: Literature study. *Indonesian Journal of Electrical Engineering and Computer Science*, 19(1), 442–450.

[136] Choo, O. Z., & Prihadi, K. (2019). Academic resilience as mediator of multidimensional perfectionism and academic performance among gen-z undergraduate students. *International Journal of Evaluation and Research in Education*, 8(4), 637–646.

[137] de la Fuente, J., Urien, B., Luis, E. O., González-Torres, M. C., Artuch-Garde, R., & Balaguer, A. (2022). The proactive-reactive resilience as a mediational variable between the character strength and the flourishing in undergraduate students. *Frontiers in Psychology*, 13, 1–14.

[138] Reinhardt, I. C., Oliveira, J. C., & Ring, D. T. (2020). Current perspectives on the development of industry 4.0 in the pharmaceutical sector. *Journal of Industrial Information Integration*, 18, 100131.

[139] Yli-Ojanperä, M., Sierla, S., Papakonstantinou, N., & Vyatkin, V. (2019). Adapting an agile manufacturing concept to the reference architecture model industry 4.0: A survey and case study. *Journal of Industrial Information Integration*, 15, 147–160.

[140] Javaid, M., & Haleem, A. (2020). Critical components of Industry 5.0 towards a successful adoption in the field of manufacturing. *Journal of Industrial Integration and Management*, 5(3), 327–348.

[141] Rastogi, R., Tandon, N., Rajeshwari, T., Moorjani, P., & Malvi, S. (2022). Yagyopathy holistic science for various solutions: A scientific phenomenon with modern healthcare, QoL and society 5.0. In Srinivasa, K.G., Siddesh, G.M., & Manisekhar, S.R. (eds), *Society 5.0: Smart Future Towards Enhancing the Quality of Society.* Advances in Sustainability Science and Technology (pp. 229–274). Singapore: Springer.

[142] Tanwar, S., Parekh, K., & Evans, R. (2020). Blockchain-based electronic healthcare record system for healthcare 4.0 applications. *Journal of Information Security and Applications*, 50, 102407.

[143] Jovanovic, J., & Devedzic, V. (2015). Open badges: Novel means to motivate, scaffold and recognize learning. *Technology, Knowledge and Learning*, 20(1), 115–122.

[144] McCaig, C., Bowers-Brown, T., & Drew, S. (2007). *Accelerated Learning Programmes: A Review of Quality, Extent and Demand.* HEA: York.

[145] Awotunde, J. B., Ogundokun, R. O., Ayo, F. E., Ajamu, G. J., Adeniyi, E. A., & Ogundokun, E. O. (2019). Social media acceptance and use among university students for learning purpose using UTAUT model. In Borzemski, L., Świątek, J., & Wilimowska, Z. (eds), *Information Systems Architecture and Technology: Proceedings of 40th Anniversary International Conference on Information Systems Architecture and Technology – ISAT 2019. ISAT 2019. Advances in Intelligent Systems and Computing* (Vol. 1050, pp. 91–102). Cham: Springer.

[146] Awotunde, J. B., Ogundokun, R. O., Ayo, F. E., Ajamu, G. J., & Ogundokun, O. E. (2021). UTAUT model: Integrating social media for learning purposes among university students in Nigeria. *SN Social Sciences*, 1(9), 1–27.
[147] Klötzer, C., Weißenborn, J., & Pflaum, A. (2017). The evolution of cyber-physical systems as a driving force behind digital transformation. In *2017 IEEE 19th Conference on Business Informatics (CBI)* (Vol. 2, pp. 5–14). IEEE.
[148] Ferreira, C. M., & Serpa, S. (2018). Society 5.0 and social development. *Management and Organizational Studies*, 5(4), 26–31.
[149] Awotunde, J. B., Misra, S., Ayeni, F., Maskeliunas, R., & Damasevicius, R. (2022). Artificial intelligence based system for bank loan fraud prediction. In Ajith et al., (eds.), *Hybrid Intelligent Systems. HIS 2021. Lecture Notes in Networks and Systems* (Vol. 420, pp. 463–472). Cham: Springer.
[150] Theorin, A., Bengtsson, K., Provost, J., Lieder, M., Johnsson, C., Lundholm, T., & Lennartson, B. (2017). An event-driven manufacturing information system architecture for Industry 4.0. *International Journal of Production Research*, 55(5), 1297–1311.
[151] Stankovic, J. A., Sturges, J. W., & Eisenberg, J. (2017). A 21st century cyber-physical systems education. *Computer*, 50(12), 82–85.
[152] Doherty, C., Kiley, J., & O'Hea, O. (2018). *The Generation Gap in American politics*. Pew Research Center.
[153] Nahavandi, S. (2019). Industry 5.0—A human-centric solution. *Sustainability*, 11(16), 4371.
[154] Chen, S. H., Jakeman, A. J., & Norton, J. P. (2008). Artificial intelligence techniques: An introduction to their use for modelling environmental systems. *Mathematics and Computers in Simulation*, 78(2–3), 379–400.
[155] Rêgo, B. S., Jayantilal, S., Ferreira, J. J., & Carayannis, E. G. (2021). Digital transformation and strategic management: A systematic review of the literature. *Journal of the Knowledge Economy*, 46, 1–28.
[156] Rodríguez-Abitia, G., & Bribiesca-Correa, G. (2021). Assessing digital transformation in universities. *Future Internet*, 13(2), 52.
[157] Abad-Segura, E., González-Zamar, M. D., Infante-Moro, J. C., & Ruipérez García, G. (2020). Sustainable management of digital transformation in higher education: Global research trends. *Sustainability*, 12(5), 2107.
[158] Breque, M., De Nul, L., & Petridis, A. (2021). *Industry 5.0: Towards a Sustainable, Human-Centric and Resilient European Industry*. Luxembourg, LU: European Commission, Directorate-General for Research and Innovation.
[159] Carayannis, E. G., & Morawska-Jancelewicz, J. (2022). The futures of Europe: Society 5.0 and Industry 5.0 as driving forces of future universities. *Journal of the Knowledge Economy*, 1–27.
[160] Giang, N. T. H., Hai, P. T. T., Tu, N. T. T., & Tan, P. X. (2021). Exploring the readiness for digital transformation in a higher education institution towards industrial revolution 4.0. *International Journal of Engineering Pedagogy*, 11(2), 4–24.
[161] Sułkowski, Ł., Kolasińska-Morawska, K., Seliga, R., & Morawski, P. (2021). Smart learning technologization in the Economy 5.0—The Polish perspective. *Applied Sciences*, 11(11), 5261.
[162] Vishnevsky, V. P., Harkushenko, O. M., Zanizdra, M. Y., & Kniaziev, S. I. (2021). Digital and green economy: Common grounds and contradictions. *Science Innovations*, 17(3), 14–27.
[163] Mohamed Hashim, M. A., Tlemsani, I., & Matthews, R. (2022). Higher education strategy in digital transformation. *Education and Information Technologies*, 27(3), 3171–3195.
[164] Hashim, S. A., Yusof, B. N. M., Saad, H. A., Ismail, S., Hamdy, O., & Mansour, A. A. (2021). Effectiveness of simplified diabetes nutrition education on glycemic control and other diabetes-related outcomes in patients with type 2 diabetes mellitus. *Clinical Nutrition ESPEN*, 45, 141–149.
[165] Ochoa, S. F., Fortino, G., & Di Fatta, G. (2017). Cyber-physical systems, internet of things and big data. *Future Generation Computer Systems*, 75, 82–84.

[166] Sung, M. (2015). A study of adults' perception and needs for smart learning. *Procedia-Social and Behavioral Sciences*, 191, 115–120.

[167] Trappey, A. J., Trappey, C. V., Govindarajan, U. H., Chuang, A. C., & Sun, J. J. (2017). A review of essential standards and patent landscapes for the Internet of Things: A key enabler for Industry 4.0. *Advanced Engineering Informatics*, 33, 208–229.

[168] O'donovan, P., Gallagher, C., Bruton, K., & O'Sullivan, D. T. (2018). A fog computing industrial cyber-physical system for embedded low-latency machine learning Industry 4.0 applications. *Manufacturing Letters*, 15, 139–142.

[169] Yun, J. J., Lee, D., Ahn, H., Park, K., & Yigitcanlar, T. (2016). Not deep learning but autonomous learning of open innovation for sustainable artificial intelligence. *Sustainability*, 8(8), 797.

[170] Mourtzis, D., & Vlachou, E. (2018). A cloud-based cyber-physical system for adaptive shop-floor scheduling and condition-based maintenance. *Journal of Manufacturing Systems*, 47, 179–198.

[171] Monyela, M. (2022). Knowledge organisation in academic libraries: The linked data approach. In *Innovative Technologies for Enhancing Knowledge Access in Academic Libraries* (pp. 71–88). IGI Global.

[172] Zhang, Z., Wang, X., Wang, X., Cui, F., & Cheng, H. (2019). A simulation-based approach for plant layout design and production planning. *Journal of Ambient Intelligence and Humanized Computing*, 10(3), 1217–1230.

[173] Azzi, A., Faccio, M., Persona, A., & Sgarbossa, F. (2012). Lot splitting scheduling procedure for makespan reduction and machine capacity increase in a hybrid flow shop with batch production. *The International Journal of Advanced Manufacturing Technology*, 59(5), 775–786.

[174] Stan, M., Borangiu, T., & Răileanu, S. (2021). Data-and model-driven digital twins for design and logistics control of product distribution. In *2021 23rd International Conference on Control Systems and Computer Science (CSCS)* (pp. 33–40). IEEE.

[175] Rais, A. (2018). *Next Stop: Industry 4.0–Three Trends for the Digital* Transformation. Available at: https://www.process-worldwide.com/next-stop-industry-40-three-trends-for-the-digital-transformation-a-712526/

[176] Gürdür, D., Feljan, A. V., El-khoury, J., Mohalik, S. K., Badrinath, R., Mujumdar, A. P., & Fersman, E. (2018). Knowledge representation of cyber-physical systems for monitoring purpose. *Procedia Cirp*, 72, 468–473.

[177] Schwab, K., & Davis, N. (2018). *Shaping the Fourth Industrial Revolution*. Geneva: World Economic Forum.

[178] Sanders, A. K., Falcão, T., Haider, A., Jambeck, J., LaPointe, C., Vickers, C., & Ziebarth, N. (2018). World economic and social survey 2018: Frontier technologies for sustainable development.

[179] Burston, J. (2016) (5). Predictions for Energy in 2030| World Economic Forum. Available at: https://www.weforum.org/agenda/2016/11/5-predictions-for-energy-in-2030/

[180] Capros, P., Mantzos, L., Tasios, N., De Vita, A., & Kouvaritakis, N. (2010). *EU Energy Trends to 2030: Update 2009*. Publications Office of the European Union.

[181] Engell, S., Paulen, R., Sonntag, C., Thompson, H., Reniers, M., Klessova, S., & Copigneaux, B. (2016). Proposal of a European research and innovation agenda on cyber-physical systems of systems 2016–2025. CPSoS Project, TU Dortmund.

[182] Cengarle, V., Törngren, M., Bensalem, S., McDermid, J., Sangiovanni-Vincentelli, A., & Passerone, R. (2013). Structuring of CPS domain: Characteristics, trends, challenges and opportunities associated with CPS. D2. 2 of CyPhERS FP7 Project.

[183] Lomas, N. (2018). Gartner picks digital ethics and privacy as a strategic trend for 2019. TC. https://techcrunch.com/2018/10/16/gartner-picks-digitalethics-and-privacy-as-a-strategic-trend-for–2019.

[184] Thompson, H., Reimann, M., Ramos-Hernandez, D., Bageritz, S., Brunet, A., Robinson, C., ... de Sutter, I. (2018). *Platforms4CPS, Key Outcomes and Recommendations.* EU, Steinbeis-Edition.
[185] Higgins, J. M. (2013). The fourth singularity and the future of jobs. *World Future Review,* 5(1), 11–23.
[186] Rainie, L., & Anderson, J. (2017). *The Future of Jobs and Jobs Training.* Pew Research Center.
[187] Musselin, C. (2018). New forms of competition in higher education. *Socio-Economic Review,* 16(3), 657–683.
[188] Rust, V. D., & Kim, S. (2012). The global competition in higher education. *World Studies in Education,* 13(1), 5–20.
[189] Awotunde, J. B., Jimoh, R. G., Oladipo, I. D., & Abdulraheem, M. (2021). Prediction of malaria fever using long-short-term memory and big data. In Misra, S. & Muhammad-Bello, B. (eds), *Information and Communication Technology and Applications. ICTA 2020. Communications in Computer and Information Science* (Vol. 1350, pp. 41–53). Cham: Springer.
[190] Awotunde, J. B., Jimoh, R. G., Ogundokun, R. O., Misra, S., & Abikoye, O. C. (2022). Big data analytics of iot-based cloud system framework: Smart healthcare monitoring systems. In Misra, S., Kumar Tyagi, A., Piuri, V., & Garg, L. (eds), *Artificial Intelligence for Cloud and Edge Computing. Internet of Things* (pp. 181–208). Cham: Springer.
[191] Papadopoulos, T., Singh, S. P., Spanaki, K., Gunasekaran, A., & Dubey, R. (2022). Towards the next generation of manufacturing: Implications of big data and digitalization in the context of industry 4.0. *Production Planning & Control,* 33(2–3), 101–104.
[192] Eriksson, K. M., Chirumalla, K., Myrelid, P., Ericsson, M., Granlund, A., Håkansson, L., & Johansson, D. (2022). Experiences in running a professional course on digitally-enabled production in collaboration between three Swedish Universities. *Advances in Transdisciplinary Engineering,* 21, 653–664.
[193] Harris, K., Kimson, A., & Schwedel, A. (2018). Labor 2030: The collision of demographics, automation and inequality. *Bain & Company,* 7, 63.
[194] Awotunde, J. B., Folorunso, S. O., Jimoh, R. G., Adeniyi, E. A., Abiodun, K. M., & Ajamu, G. J. (2021). Application of artificial intelligence for COVID-19 epidemic: An exploratory study, opportunities, challenges, and future prospects. *Studies in Systems, Decision and Control,* 358, 47–61.
[195] Calderone, L. (2018). AI could beat humans at everything by 2030| RoboticsTomorrow.
[196] Boyd, R., & Holton, R. J. (2018). Technology, innovation, employment and power: Does robotics and artificial intelligence really mean social transformation? *Journal of Sociology,* 54(3), 331–345.
[197] Margolies, J., Ronanki, R., Steier, D., Tuff, G., White, M., Bhattacharya, A., & Gupta, N. (2018). Tech trends 2018: The symphonic enterprise. *Deloitte Insights,* 162.
[198] Ghobakhloo, M., & Ching, N. T. (2019). Adoption of digital technologies of smart manufacturing in SMEs. *Journal of Industrial Information Integration,* 16, 100107.
[199] Törngren, M., Bensalem, S., McDermid, J., Passerone, R., Sangiovanni-Vincentelli, A., & Schätz, B. (2015). Education and training challenges in the era of cyber-physical systems: Beyond traditional engineering. In *Proceedings of the WESE'15: Workshop on Embedded and Cyber-Physical Systems Education* (pp. 1–5). Taylor & Francis.
[200] Komiyama, H., & Yamada, K. (2018). *New Vision 2050: A Platinum Society.* Springer Nature.
[201] Salkin, C., Oner, M., Ustundag, A., & Cevikcan, E. (2018). A conceptual framework for Industry 4.0. In *Industry 4.0: Managing the Digital Transformation* (pp. 3–23). Cham: Springer.
[202] Schwab, K. (2016). The Fourth Industrial Revolution, Word Economic Forum. Secretaría de Economía (SE) (2012), Plan Nacional Estratégico de la Industria Aeroespacial.
[203] Chin, J., Callaghan, V., & Allouch, S. B. (2019). The Internet-of-Things: Reflections on the past, present and future from a user-centered and smart environment perspective. *Journal of Ambient Intelligence and Smart Environments,* 11(1), 45–69.

[204] Salgues, B. (2018). *Society 5.0: Industry of the Future, Technologies, Methods and Tools*. John Wiley & Sons.

[205] Collins, A., & Halverson, R. (2018). *Rethinking Education in the Age of Technology: The Digital Revolution and Schooling in America*. Teachers College Press.

[206] Gürdür, D., & Törngren, M. (2018). Visual analytics for cyber-physical systems development: Blending design thinking and systems thinking. In *15th Annual NordDesign Conference*. NordDesign.

[207] Broo, D. G. (2022). Transdisciplinarity and three mindsets for sustainability in the age of cyber-physical systems. *Journal of Industrial Information Integration*, 27, 100290.

[208] Dignum, V. (2020). AI is multidisciplinary. *AI Matters*, 5(4), 18–21.

[209] Gürdür, D. (2018). Data and visual analytics for cyber-physical systems: Current situation and strategies for action (Doctoral dissertation, KTH Royal Institute of Technology).

[210] Ansari, F., Erol, S., & Sihn, W. (2018). Rethinking human-machine learning in industry 4.0: How does the paradigm shift treat the role of human learning? *Procedia Manufacturing*, 23, 117–122.

CHAPTER 9

Innovations of Teaching and Testing English to Students of other Majors Online During COVID-19 Pandemic at Mila University Centre, Algeria

Abderrahim Bouderbane

University Centre of Mila

CONTENTS

DOI: 10.1201/9781003322252-9

9.1 INTRODUCTION

In the past 2 years, the world has experienced a drastic change in all walks of life due to the spread of the COVID-19 pandemic. In March 2020, the Algerian authorities called for total confinement after the announcement of confirmed cases in different Algerian cities. This precautious measurement was applied in all sectors; however, minimum service is guaranteed to contain the disease. In the sector of higher education, the Ministry of Higher Education launched distance learning via learning platforms like Moodle and Google Meet to save the rest of the academic year. Distance learning was challenging for both teachers and learners since the university was unprepared for this wholly new situation in terms of training (for teachers and learners), materials, and the shift in teaching techniques and methods. The teaching of English to students of other majors was based primarily on a genre-based approach to teaching English as a foreign language (EFL). Likewise, this approach is based on analyzing typologies of texts in different fields. The aim behind the genre-based approach is to increase knowledge of EAP and ESP to obtain minimum competence in grammar, morphology, vocabulary, and syntax. The teaching of text typologies relying on platforms is problematic because it is not appropriate for the explanation, discussion, evaluation, and consolidation of such types of texts. The Ministry of Higher Education emphasized the use of 'Moodle' to upload lectures, tasks, and even tests for students. However, Algerian universities adopted online teaching to cover primary subjects via Google Meet, while teachers uploaded lectures of all the other subjects on the platform 'Moodle'. It is not sufficient for beginners to submit different types of texts without any explanation. Teachers in this case improvised and sent students 'YouTube' or 'Facebook' page links with extra files to explain the notions and contextualize vocabulary with simplified English.

Objectives of the Chapter: The chapters makes a distinction a distinction between the theoretical adaptations made when teaching ESP and EAP before the COVID-19 pandemic and how teachers design courses for students of other majors. Then, during the COVID-19 pandemic, the adaptations are applied in distance education on the basis of two teaching platforms which are 'Google Meet' and 'Moodle'.

This chapter also compares testing techniques used before the COVID-19 pandemic and during the COVID-19 pandemic to elaborate on the use of the barcode scanning correction systems and some specific tasks both in summative and formative assessments in distance education. The chapter aims at investigating the difference between the theoretical adaptations made for teaching EAP and ESP both in face-to-face teaching and online teaching, specifically during confinement to see how teachers maintain the process online.

In addition, this chapter aims also at discovering the benefits of using barcode scanning readers to correct exam papers in online testing. Correcting papers using barcode readers was used in correspondence to different factors including exam time limits, ESP and EAP course content, and the criteria of assessment and tasks in exams.

Organization of the Chapter

The chapter is organized as follows: Section 9.2 is organized coherently, and it includes the theoretical foundations of ESP and EAP with the approaches and theories taken from famous scholars like Swales (1990), Trimble (1985), and Bhatia (2002). An overview of teaching ESP and EAP online is scrutinized in Section 9.3 to see how courses are generally delivered in distance learning to students in all the departments. In the same vein, Section 9.4 elaborates in detail on the two teaching methods employed online which are 'Google Meet' and 'Moodle'. In Section 9.5, the general testing methods of EAP and ESP are highlighted with emphasis on the types of formal assessment tasks. Sections 9.6 and 9.7 introduce, respectively, the functionality of using a barcode scanning correction system, its utility in correcting ESP and EAP exam papers, and the extent to which online testing tasks are suitable in ESP and EAP exams. Section 9.8 concludes the chapter with improvised ways in which barcode scanning correction can be used in distance learning to improve teaching methods and testing tasks.

9.2 THEORETICAL FOUNDATIONS OF ESP AND EAP

The teaching of English as a foreign language has changed recently in terms of the techniques and methods employed in the classroom. This is due to the openness of Algeria to the external world in different domains, including finance, business, trade, commerce, law, medicine, science, and technology. English is now the international language, and it is the means of communication between scientists from all over the world. Hence, it has become necessary to track down everything new in the academic world. Students and researchers are obliged to acquire English in their subject-specific domains to enhance the quality of teaching and learning and cope with everything new in the academic world.

The teaching of EAP and ESP has become a prominent issue at the university. Generally speaking, the methodology of teaching EAP and ESP is all about text analysis or text typologies. Swales (1990) introduced the genre-specific approach for teaching EAP in different academic contexts. A genre is a specific type of discourse that a specific speech community uses to achieve a message depending on conventions, shared background knowledge, and context. Hence, the genre introduced by Swales categorizes different typologies as they are used in different discourse communities. Genres in general do not refer to 'a pattern of forms or even a method of achieving our own ends' only but also 'what ends we may have' (Miller, 1984; p.160). The discourse community in this context differs from the notion of the speech community which refers to a group of people with the same beliefs, culture, traditions, rituals, and religion (Hymes, 1972).

Hymes' notion denotes specifically communicative competence or competencies. It is divided into grammatical or linguistic, pragmatic and discourse competence, socio-cultural, and strategic competence for keeping communication going. However, Swales' notion of the discourse community entails subject specialists, and he referred to it as 'nomenclature' in academic writing. The nomenclature shares the same norms, background knowledge, conventions, categorizations, and text realization. Meanwhile, Hayes and Flower (1980) distinguished between the knowledge-telling and knowledge-transforming aspects of academic writing. Knowledge-telling is a disorganized way of writing and the basis of which is to transform written messages without any consideration for planning or organization. The speech community incorporates different messages in everyday life communication almost always in the same way. As a result, the patterning of the language is built around sending a message regardless of the way in which people send this message. However, the knowledge-transforming nomenclature in academic writing focuses on content and organization because it is the transformation of knowledge in subject-specific areas. The content is important, but the message is made clear with other surface structure features, like cohesive devices, introducing phrases, and hedges. The features ensure the flow of ideas and the smoothness of language between sentences and paragraphs.

The content is understood when it is organized coherently between the readers, the writers, and the subject matter. In this context, genre-specific approaches to teaching EAP and ESP focus on developing features integrated with academic writing. The features are divided into textual and meta-textual features. Textual features are taught using register analysis because they occur between words, phrases, and sentences (Dudley-Evans & St. John, 1998). The meta-textual features are referred to as meta-discourse, and they are advocated in texts to ensure clarity and directness, intended emphasis, cohesion, and coherence. Meta-textual features are textual clues and cohesive devices the function of which is to link words, phrases, and clauses (Halliday & Hassan, 1985). Academic writers employ other discourse markers like ellipses, omissions, and substitutions to incorporate smoothness, clarity of language, and ideas at the same time. A recent study conducted by Hyland (2005) classified meta-discourse markers into interactive and interactional. Interactive markers are used to organize the content coherently and they are transition markers, frame markers, discourse makers, and code glosses. Writers generally use such types of markers in academic English to prepare readers for important information. Hyland also indicated other equally important markers including hedges, boosters, attitude markers, and engagement markers. They reflect the involvement of writers in the text itself.

In light of all theories introduced in ESP and EAP domains, teachers at the University Centre of Mila adopted teaching methods for students of other specialties including business, economics, law, science and technology, computer science, engineering, physics, and chemistry. The exploration of teaching scientific texts, or typologies, according to the notions and rhetorical functions introduced by Wilkins (1976) and Munby (1978) demonstrates the proposition of information in different texts. The analysis is also based on identifying specific lexicon or vocabulary, and cohesive markers (Bhatia, 2002). The following Table 9.1 summarizes the main aspects of the analysis of text types in EAP and ESP texts:

The teachers and students analyze vulgarized texts in different domains, and the Algerian Ministry of Higher Education selected them specifically to meet the needs of the students. The purpose of this is to build minimum competence in scientific English.

TABLE 9.1 Analysis of Text Features for Subject-Specific Students

Notions	Rhetorical Functions	Cohesive Devices	Specific Lexicon	Communicative Purposes
- Time and place	- Identifying or defining - Explaining - Exemplifying - Argumenting - Illustrating	- They are used to link sentences, paragraphs, and ideas.	- Words phrases and expressions which are peculiar in this subject area	- The objectives of the texts entail specifically the characterization organization, and meanings of the texts

Accordingly, the texts are written in simplified English with a clear communicative purpose which is exhibited in the rhetorical functions of the text. The texts are also organized in terms of language and ideas coherently using different cohesive markers. For that, the students ought to analyze the content and be familiar with the language and text features.

9.3 TEACHING ESP AND EAP DURING THE COVID-19 PANDEMIC

During the COVID-19 pandemic, teaching has witnessed a turning point in the methods and techniques applied to maintain the process of learning. However, the universities in Algeria were not fully prepared for this wholly new situation as distance learning requires appropriate software and applications. Most third-world countries are staggering to provide minimum digital innovations for their institutions under difficult conditions like internet disruptions, technical problems, and lack of expert assistance. Therefore, the Ministry of Higher Education in Algeria divided courses into compulsory and secondary in the process of teaching. Compulsory courses belong to the fundamental unit one of the subjects, they are important and should be taught by all means possible. Henceforth, they are taught once a week on the Google Meet platform, and students are encouraged to log on and attend lectures online. Besides, teachers are obliged to provide lectures in the form of PDFs on the platform Moodle. The problematic issue in this new teaching context occurs with secondary subjects of unit two which are less important including methodology and English.

All students, at the beginning of confinement, were asked to use their identification card numbers to log on to the platform and download lectures. The platform Moodle recorded about 3000 log on attempts, i.e., ¼ of the total number of students throughout confinement. This is due to mainly two reasons: the inaccessibility of some students in the platform, as in some lectures, teachers should invite students even after logging on, and the invisibility of some lectures since teachers blocked the option of guest-viewing to protect the lectures and distance learning from scammers and unauthorized access.

The university asked engineers to control the log-on processes by omitting the option of inviting, and classifying students to groups and levels of both bachelor's and master's degrees. The process resulted in about 500 groups and 49 cohorts of levels for the master's and bachelor's degrees. This classification is based on the manual importation of students' names from the university's database which is called 'Progres' (it is used for enrollment purposes). It assisted students with low computer skills to log on and interact easily on the platform.

Table 9.2 indicates the number of groups in each level with the total number of students. This varies depending on the level in every department at the university.

TABLE 9.2 Total Numbers of Groups and Students in all the University

Level	Cohorts	Students
First year	170	4204
Second year	130	3517
Third year	100	2029
First-year master	60	1514
Second-year master	35	900
Total	495	12,164

Considering the teaching of English in other departments, the teaching of EAP and ESP is a compulsory course in 49 levels in the 3 faculties of science and technology, management, economics, commercial science, and letters and languages. However, it is excluded from the department of Arabic (students study French), and it is a subject in the English department in the 3rd-level labeled English for science and technology. Most of the teaching–learning interaction process is embodied in sharing documents and tests on the platform. However, a lot of other available options are customized to perform other substantial activities like dragging and dropping, adding URLs of videos, and attaching lecture annexes for comments, questions, and additional information.

The sharing of lectures is based on information transformation between students in different kinds of social media communities. That is, the transformation of information about any new lectures, tasks, activities, and even assessments is necessary to reach a wider audience. Therefore, the teachers responsible for subjects at the university created new groups (Facebook, and Whats App) to ensure the spread of updates among students. Hence, updates are always announced on Facebook about new lectures, tasks, and even activities.

9.4 TEACHING METHODS

9.4.1 Platform Moodle

Generally speaking, the teaching of EAP and ESP is based on register analysis where students are exposed to different texts to increase knowledge of subject-specific areas (Trimble, 1985). Trimble detailed explicit steps of teaching ESP rhetorically by introducing texts up to levels. The first level identifies the objectives of the discourse (text), like presenting new information, experimenting, or giving details. The second level identifies the general rhetorical functions employed to achieve the objectives, like stating the problem or introducing theoretical frameworks. The third level builds up texts with rhetorical functions, such as defining, illustrating, exemplifying, and explaining. The last level wraps up the content by determining relationships between elements of the texts like cause and effect, time, and space.

The adaptation of the Swales model of teaching EAP and Trimble's model of teaching ESP is very common among English as foreign language teachers. On the platform, teachers follow almost the same steps, but electronically and in parts. As most ESP and EAP specialists assume, the teaching of ESP and EAP is based on three phases, mainly, the warm-up, presentation, and consolidation. The warm-up prepares the students psychologically to take in knowledge. In the presentation phase, the teacher explains the lecture by means of examples, explanations, and illustrations. At the end of the lecture, students are consolidated to evaluate their strengths and weaknesses. The phases are generally consecutive because they complete one another.

Generally speaking, the same steps are kept in online classes, and precipitation occurs only in consolidation face. It can be used sometimes at the beginning of the lecture instead of the warm-up phase to increase interactivity and raise students' awareness of the important sections of the lecture. The teacher uploads the lecture on the platform 'Moodle' as the first phase in online teaching. Teachers of EAP and ESP in the faculties of economics, and science and technology uploaded five lectures each semester for every level in the bachelor's and master's degree. Teachers are obliged to upload a lecture every 15 days and over a period of 3 months. If the lecture has annexes, the teacher should provide them in the same week, and inform the students about them on social networking. In the preceding week, the students expect other kinds of joint files on Moodle. Since classes are online, teachers support the texts with videos, PowerPoint presentations, and links in each appropriate section on the platform. These additional data are added to substitute the missing part in teaching (face-to-face) and pave the way for discussions in Google Meet classes.

Here, teachers add links, videos, and links to web pages or websites which are related directly to the content of the text in each lecture. They are considered as additional materials which are designed specifically to explain difficult words or vocabulary, or define the text and summarize its main rhetorical functions with visual–verbal demonstrations. These teaching materials are used to explain, give examples, and provide additional information for the students. They are generally in simplified English and directed to lay people. Some teachers upload their own PowerPoint presentations as they adjust the content to meet the needs of the students and achieve the required objectives of the texts or lectures. All these additional materials belong to the presentation phase and teachers responsible for the subject emphasized their use.

The second week is for consolidating students' knowledge with tasks and activities. The activities reflect the link between the theory and practice of the texts. The consolidation of the texts is divided into two types: grammatical exercises and language quizzes to test the student's language knowledge of grammatical correctness, discourse markers and cohesive devices, adjectives, and adverbs. The second type of consolidation is used to test language production through composition and writing. Hence, students here are engaged in two main activities, mainly paraphrasing and summarizing. This is a type of synthesizing activity where students build up new knowledge based on previous knowledge or background knowledge. Luoma (2004) categorized these types of processes as bottom-up and top-down processes. In the top-down process, learners deal with the analysis of text features from general to specific (essays, paragraphs, and sentences). However, the bottom-up process builds up new knowledge from previous knowledge through composition and writing. These techniques embody analysis and synthesis, as the former refers to breaking the language down into parts, and the latter refers to reconstructing new knowledge from the parts found in the analysis.

The whole semester is about 3 months in total, and every 15 days, teachers upload lectures on the platform together with links to relevant teaching materials like websites, videos, and PowerPoint presentations. The additional files are uploaded in the platform 'Moodle' in the first week, and in the second week, at least two evaluation tasks are uploaded to consolidate learning. Most EAP and ESP teachers at the Mila University Centre are subject specialists and not EFL teachers. The faculties selected teachers with advanced English proficiency skills or teachers who studied abroad to give courses to students in their specific fields. However, EFL specialists monitor EAP and ESP teachers to guide the learning process during the COVID-19 pandemic.

Table 9.3 has some significant knowledge concerning both face-to-face and online teaching. The scientific majors are more oriented toward distance learning, and they are

TABLE 9.3 Number of Uploaded Lectures, Websites, Videos, PowerPoint Presentations, Tasks, and Activities on the Platform

Department	Uploaded Lectures (Total)	Websites (Total)	Videos (Total)	PowerPoint Presentations (Total)	Tasks and Activities (Total)
Economics	5	17	4	0	13
Management	5	22	7	4	14
Commerce	5	18	4	3	11
Computer science	5	33	18	5	18
Science and technology	5	14	8	5	16
French	5	9	6	1	10

more prone to using videos and websites during learning. This characteristic is an application of the subject matter in the field of study. Additionally, it is noticed that teachers in the department of data processing or computer science interact positively in distance learning. They use almost all types of online options available in the platform even when they teach their students English for science and technology.

The teachers in the science majors are assimilating the learning context where English is just a means of communication, but the use of technological devices is a necessity for a good learning environment. As a matter of fact, it is noticed that the teachers of EAP focus on written materials in general. Visual–verbal interaction is not always applicable as learners are required to read and write to understand notions and functions on the basis of definitions, explanations, and examples. Also, the tasks and evaluation activities are incorporated on the basis of paraphrasing and summarizing because it is a question of memorizing definitions, understanding explanations, and using their own words, phrases, and expressions to develop ideas in different ways. Oxford (1990) referred to this learning strategy as 'linguistic' since learners are more interested in reading and writing exercises. Linguistic learners like words and dictionaries, and are interested in the definition of words in different contexts. Unlike EAP learners, English for science and technology learners are called visual–verbal, and they engage in a different set of learning activities like listening and watching videos. This type of learning is 'learning by doing' where learners are also involved in bodily movements using hands, and things in a given learning environment.

Hence, the new learning environment is applicable to mainly all science majors. The teachers find it appropriate to apply different techniques and methods and vary their means of education to contextualize learning. For EAP learners, the context determines the extent to which options and utilities in the platform are used. The organization and characterization of information in EAP discourse involve the identification of steps in academic writing. Academic texts are generally organized coherently like establishing background knowledge, discussing theories and concepts, giving examples, and then establishing arguments or counter-arguments (Bahatia, 2002). According to Trimble (1985), the rhetorical functions of EST texts require the procedural development of information. That is, in order to meet the objectives of text analysis their rhetorical functions, illustrations of experiments, or the functionality of a new apparatus must be demonstrated. Hence, videos and PowerPoint presentations are the best tools for the job.

9.4.2 Google Meet Sessions

The teaching of EAP and ESP has received little attention during the COVID-19 pandemic. Consequently, EAP and ESP, methodology, and information technology were classified as second-class subjects. The administration of the University Centre of Mila obliged teachers to schedule online sessions for all second-class subjects to wrap up lessons and prepare for examinations. Teachers were expected to answer students' questions, explain difficult things, and give instructions to students about the examination. In fact, students and teachers interact on the Moodle platform through the chat option or on Facebook. The chat option on Moodle is rarely used among students, and they preferred to interact directly with teachers on Facebook. The chat option is added to leave questions and inquiries about tasks and evaluation activities from both teachers and learners in continuous assessment.

The online sessions scheduled for ESP and EAP courses are generally in the afternoon. The reason is that teachers are kept busy with subject-specific lessons in their fields. The departments called for respecting the time patterning of the same face-to-face classes. In face-to-face teaching, the classes last for an hour and a half, and the first lecture of the day starts at 8 o'clock and finishes at 9.30. The weekend is Friday and Saturday, but in online teaching, it is only Friday. The most problematic issue for learners is scheduling classes at the end of the day. Most students are preoccupied with different activities like working. The university's database recorded the lowest log-on records in secondary subjects specifically 'English and Methodology'. Therefore, students interacted most on the platform than in Google Meet.

There are plenty of technical problems in online teaching in third-world countries for it deserves practice. Researchers now, in online teaching, are tackling more complicated and developed hi-tech means including the assimilation of virtual reality (VA) and augmented reality (AR) to increase the adaptation of real-life learning environments (Bonner & Reinders, 2018). It is also used in cloud computing technologies to improve the interaction between students and teachers at all levels (Kakoulli-Constentinou, 2018). Besides, the uses of computer games are designed to increase ESP knowledge of some specific areas like shipping, surfing, medicine, and biology. Recently, the teaching of ESP is applied in different social networking like Facebook, and Twitter since they attract a lot of users around the world (Plutino, 2017; Rosell-Aguilar, 2018). The teaching of ESP and EAP using hi-tech devices requires proficiency in both English and computer skills to achieve the task appropriately.

9.5 TESTING METHODS

The process of evaluating EAP and ESP students at the Mila University Centre is achieved on the basis of two different tasks namely: continuous assessment and final exams.

9.5.1 The Continuous Assessment Activities

The evaluation in continuous assessment is made on the platform Moodle with tasks and activities uploaded after the lectures. The purpose of this evaluation is to consolidate students' knowledge and test their interaction and participation on the platform Moodle.

Hence, most of the activities subject to students in EAP and ESP are grammatical correctness, vocabulary building, and syntax. Douglas (2013: p.368) claimed that when he referred to 'precision, context and interaction between specific purpose language, and specific purpose background knowledge'. In fact, the background knowledge in this evaluation is related to what has already been taught because every subject matter in EAP and ESP has a wide choice of subject-specific words, phrases, and expressions. If teaching and testing objectives are different, this results in a 'backwash' effect, and it may affect the outcome of learners negatively (Fulcher & Davidson, 2007). Thus, assessment in ESP is communicative in terms of subject-specific language items (Douglas, 2000). Researchers in second language testing (Fulcher & Davidson, 2007; Luoma, 2004) considered the formative assessment of language for subject-specific purposes as it is used to build learners' knowledge on the basis of real contexts taken from different areas. However, Hutchinson and Waters (1987) presented the testing of ESP from a genre-specific approach. This type of testing includes the analysis of language-specific items of every discourse community, i.e., in terms of language features, tenses, specific lexicon, and vocabulary.

Other types of activities, which are practiced specifically among EAP students, are summarizing and paraphrasing. The teachers indulged the learners with memorizing strategies in an attempt to be able to distinguish between definitions, explanations, and examples for the application of some theories in subject-specific domains, like economics, management, and commerce. The teaching of EAP focuses mainly on clear rhetorical functions to maintain the communicative purposes of the texts. Therefore, the consolidation phase of teaching focuses on rhetorical functions and cohesive devices to strengthen ideas and notions which are used to negotiate the meaning in EAP discourse.

9.5.2 Final Exam Evaluation

All EAP and ESP final exams are summative achievement tests that are scheduled at the end of each semester or at the end of the year to test the learners' overall general language ability. The testing criterion of the final exams seeks to evaluate language knowledge and topic knowledge. It is the assessment of the actual use of language. This kind of assessment can be implemented in speaking in particular. The activities and tasks included are performance-based and completely integrated with subject-specific content. The point is that performance-based assessment is built around a social learning environment that encourages learning, communication, achieving shared goals, and achieving feedback between the learners and the teacher.

Concerning the examination, 49 cohorts of levels took the examination for both master's and bachelor's degrees. Before the examination, the students received detailed instructions from their teachers about the procedures of the examination and how to answer the questions. The instructions are highly recommended to avoid all types of intricacies, such as managing time, forgetting to write their name or forgetting to answer any part in the examination. Table 9.4 summarizes the data about EAP and ESP exams:

Henceforth, EAP and ESP exams are divided into two parts. In the first part, students received the examination texts in their subject-specific fields a night before the examination. The students read the text and prepare themselves by analyzing the text and

TABLE 9.4 Data of EAP and ESP Exam Organizations

Department	Examined Students (Total)	EAP/ESP Teachers (Total)	Teachers as Moderators (Total)	Teachers as Invigilators (Total)
Economics	2350	10	2	18
Management	1700	8	2	15
Commerce	2500	11	3	18
Data processing	1560	8	2	15
Science and technology	2815	13	4	21
French	1200	5	2	12

understanding its language features. On D-day, the students bring the text and take the examination at the university and not online. The point behind this procedure is to save the allotted time (1 hour) of the examination, and give the students a chance to concentrate on the questions. In each amphitheater, three teachers invigilate the examination, two of them supervise it, and the third one works on providing the students with barcodes (QR) in their exam papers. The EAP and ESP exams are multiple-choice and students tick the right answers appropriately.

The second part of the examination is correcting papers using the barcode (QR) automatic correction system. This system was used for the first time during the coronavirus pandemic. It allows teachers to correct papers easily, effectively, and efficiently through a model answer inserted in the system. This system works by assimilating the correct answers of the students in their exam papers. The student chooses only one correct answer because the system does not accept two correct answers. The examination is composed of various questions to test the students in a genre-specific approach. For example, in notions and functions, students select among a number of notions in the text, and students divide them into notions of time and place. Concerning rhetorical functions, students select the type of rhetorical functions from the texts of the examination. Students are also given some definitions, and they have to select the exact words which correspond with the definitions. There is also an exercise about collocations and the meaning of scientific expressions taken from the texts, and how they are used in specific contexts for different communicative purposes.

9.6 BARCODE SCANNING CORRECTION

The use of barcode reading is not new in educational contexts. In fact, the use of QR or barcodes in education facilitates the learning process. QR is capable of connecting mobile devices and computer technology with written texts (De Pietro & Frontera, 2012). Barcodes are squares that can carry information, texts, images, and even videos. This means that barcodes have the possibility to enrich the process of paper-based learning, and they can also be used in fieldwork, or for outdoor activities to support learning with materials and contents (Law & So, 2010). Other researchers investigated different uses of barcodes in learner-generated content to generate different content and demonstrate their effects on paper-based tasks (Mikulski, 2011). Depending on the nature and usage of barcodes, we can say that they fit many purposes in the teaching of EAP and ESP contexts. Lee et al.

(2011) incorporated QR and mobile phones into field trips, which are designed specifically for biology students, to obtain different information like texts and images about different elements in nature. Even in paper-based tasks, additional information about the content of the material can be added ubiquitously in the form of links, website images, and even short videos. Such technology is used to save time and space. Teachers use it to avoid the details and examples in a given subject matter and develop the content from different perspectives. It is also very useful in communicative tasks due to its information, application, and functionality.

Figure 9.1 represents the digits that students should blacken▪ vertically from left to write according to the students' card number identification; for example, if the identification number is 367455890901, the student has to blacken 3, 6, and then 7, and so forth. In fact, no two words are the same among all students. However, Figure 9.2 is an example of answering options, in correspondence with exam tasks, and in the same way, students blacken the correct answers in the options provided. A barcode scanner correction system is used to identify the students' personal information and scan the correct options. As a result, students are always instructed to pay careful attention when blackening the digits and options.

0	0	0	0	0	0	0	0	0	0	0	0
1	1	1	1	1	1	1	1	1	1	1	1
2	2	2	2	2	2	2	2	2	2	2	2
3	3	3	3	3	3	3	3	3	3	3	3
4	4	4	4	4	4	4	4	4	4	4	4
5	5	5	5	5	5	5	5	5	5	5	5
6	6	6	6	6	6	6	6	6	6	6	6
7	7	7	7	7	7	7	7	7	7	7	7
8	8	8	8	8	8	8	8	8	8	8	8
9	9	9	9	9	9	9	9	9	9	9	9

FIGURE 9.1 Students' identification number digits.

Question 1:	A	B	C	D
Question 2:	A	B	C	D
Question 3:	A	B	C	D
Question 4:	A	B	C	D
Question 5:	A	B	C	D
Question 6:	A	B	C	D
Question 7:	A	B	C	D
Question 8:	A	B	C	D
Question 9:	A	B	C	D

FIGURE 9.2 Answering options for tasks.

It is not very common among researchers and educationalists to use QR or barcodes in testing and evaluation. The evaluation system at Mila University Centre is based on a three-step barcode decoding read system. First, the barcodes are printed and fixed in the students' exam papers during the examination. Second, a model answer is saved on the computer with correct answers for all EAP and ESP exams. The teachers made the exercises particularly suitable for ticking so that the computer reads the data easily. On each barcode's label, there is a specific number that distinguishes the students in every department. The labels are scanned using a barcode scanner which deciphers students' answers according to the model answer saved in the computer. Third, the marks of the students are imported into the university's database to calculate averages between continuous assessment and the final exams. Final averages are obtained, while the final marks of EAP and ESP model answers are displayed to students 3 days after the final examination. In order to calculate the time of correcting the papers, we used the following formula:

$$\text{Total Time} = \text{number of copies}\left(\text{single time} + \text{set up time}\right)$$

In this formula, single time refers to the time that the software takes to process the information in the computer, while setup time refers to the time of scanning the paper and decoding the answers. Statistically, the single time equals ≤0.45 ms, and the setup time equals≤2seconds respectively. The total number of exam papers is 12,125 in all departments, and the total time of correction is calculated as follows:

$$\text{Time} = 12125(0.45 + 2)$$

$$= 12125(2.45)$$

$$= 29706.25 \text{ Seconds}$$

The total time of the correction of papers in seconds is 25,884.25. In order to measure the time in minutes, we divide the total time by3600, which is the total number of seconds for 1 hour.

$$\frac{29706.25}{3600}$$

$$= 8.251$$

The total time then is 8 hours and 41 minutes. Hence, the following graph summarizes the results of the total time of correcting papers in each department (Figure 9.3).

The number of papers corrected in each department plays an important role in determining the amount of time for barcode scanner exam paper correction. The total amount of correction time is 8 hours and 41 minutes, and we divided it between a single time and

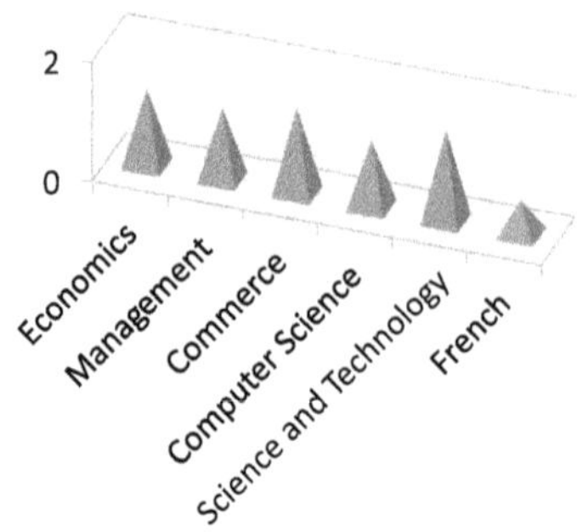

FIGURE 9.3 Barcode correction time in each department.

set up time. Hence, the formula of calculation is very simple because the higher the number of students in the department, the longer period the barcode scanner takes to correct the papers. Consequently, the Department of Science and Technology recorded the longest period of barcode scanning time of 1 hour and 51 minutes, as well as the Department of Commerce recorded also a long time of 1 hour and 41 minutes. The other departments recorded less time, specifically the Department of Economics with 1 hour and 6 minutes, the Department of Management with 1 hour and 18 minutes, and the Department of Computer Science with 1 hour and 10 minutes. The Department of French language recorded the lowest time of 57 minutes only.

EAP and ESP final exams revealed transparency and easiness in handling and correcting the papers. Besides, students agreed that the contexts of the exercises are easy and functional while answering correctly relies on comprehending parts of the text and understanding the smoothness of ideas and language. Hence, the students in EAP and ESP final exams employed skimming and scanning techniques to answer all the questions. At first, they skim the text at home to understand the general idea and supporting details. During the examination, students read the questions of the exercises and skim the text again quickly (make a quick selective reading) to locate important information, like notions and functions, and identify the meaning of words in specific contexts. In order to analyze the texts rhetorically and understand the meaning, students should locate different clues such as cohesive devices and discourse markers. These clues help students determine what academic writers want to say next, and how to say it using specific vocabulary, grammar, and text features as well.

The process of evaluation using barcodes in EAP and ESP exercises is easy and efficient. Teachers appreciated the operation because the evaluation was easy, and so many papers with different exercises were corrected in a very short period of time. Even the tasks can be handled easily for students, and they achieve different communicative purposes; i.e., they summarize exactly the content of the lectures uploaded on Moodle. Therefore, there is no backwash or other unwanted variables like time management or tricky questions.

The texts of the exams are composed of three parts: an introduction, a body paragraph, and a conclusion. For example, the text in the examination of the Department of Commerce was about the nature of commerce. In the beginning, the text defines commerce. In the body paragraphs, the text develops types of commerce which are wholesale and retail trade, the means of transportation of merchandise, and insurance of commercial activities. Therefore, the exercises vary from identifying the notions of place and time. Then, identify the rhetorical functions employed in each part of the text. Finally, they ask questions about specific vocabulary, grammar, and syntax.

Besides saving time and effort, barcodes defined boundaries between online tests and in-class tests. In online tests, the sharing of the exams online with students opens up possible ways to cheat in these exams. Academic honesty is not a big deal for students; they have to get good marks to pass the exams, and they cannot do it with without dragging and dropping correct answers. It is possible for students to receive assistance or help whenever they need it, but it is unethical and dishonest when they copy information from other students or rely on other people to answer instead of them. In addition, third-world countries are still unable to provide the appropriate means to take exams online. The slow speed of the internet, internet disruptions, and technical problems hinder the process of online evaluation. As a result, students complained all the time when they took online exams until some of them are repeated in classrooms because of technical problems in terms of accessing, receiving, and sending back exams.

Online teaching takes time and deserves a lot of practice. Thanks to online teaching, both asynchronous and synchronous learning is activated. Students are kept busy sometimes waiting for the timing of lectures to be announced and some other times for lectures and tasks to be uploaded on the platform. However, online exams have changed the criterion of examination, while its possible means and procedures are in constant change. Hence, barcodes seem to be the appropriate solution to avoid technical problems, keep the originality of exams, and test the students' proficiency.

Before the examination, students received extensive instruction about the procedures of the examination and how to answer the questions. Most of the instructions focused on demonstrating the difference between the automatic correction of the computer and how to answer. A computer is a machine and it does not think, reflect, analyze or blink an eye when it does the correction of papers. Therefore, respecting the instruction is very necessary to accomplish the examination successfully. However, the problematic issue faced in this examination is that some students did not attend Google Meet lectures, and they were not aware of the instructions of the examination. As it is known, the instructions given to students focused mainly on writing ticks correctly and answering the questions appropriately. However, some students did not use the tick correctly and did not answer the questions appropriately. In this case, the barcode scanner does not read the answers appropriately, and they were considered wrong or canceled answers. The teachers deliberated the complaints of the students and corrected the papers manually. The following Table 9.5 summarizes the results of the complaints of students in every department:

The students of computer science committed fewer mistakes than all other students in the other branches. They are mostly aware of these procedures with high-tech devices, and they recorded the lowest rate of (12) mistakes committed when answering questions in the examination. High rates are recorded in the other majors, specifically in the departments of economics (44), commerce (48), and management (35). Hence, instructional

TABLE 9.5 Complaints of Students in EAP and ESP Examinations

Department	Total Number of Problems	Problems of Correct Answers	Problems of Missing Names	Problems of Missing Marks
Economics	44	35	3	9
Management	35	28	4	3
Commerce	48	40	2	5
Data processing	18	12	4	2
Science and technology	29	21	4	4
French	23	18	2	3

discourse for students is beneficial for guidance as it gives an idea about the regulations of the examination.

In order to calculate the accuracy of the barcode scanner correction of exam papers, we need to calculate first the rate of problems that were made in every department. Hence, the following formula calculates the rate of problems:

$$\frac{\text{Number of mistakes}}{\text{total number of papers}} \times 100$$

$$\frac{197}{12125} \times 100$$

$$0.016 \times 100 = 1.62$$

Statistically, the calculation of the number of mistakes by the barcode scanner is as follows:

$$\frac{144}{12125} \times 100$$

$$0.011 \times 100 = 1.18$$

Consequently, the exact rate of the barcode scanner equals the rate of the total number of mistakes subtracted from the rate of the barcode scanner rate: 1.62–1.18 =0.44.

The following Figure 9.4 illustrates the rates of mistakes:

The statistical measures indicated that the accuracy rate is 0.44, and the scanner is expected to achieve a value as high as (+1) when accuracy is strong and (–1) when it is weak. To sum up, the difference between the rate of the barcode scanner equals the difference between (–1) and (+1) in0.44which is smaller> than 0.56.

9.7 ONLINE TESTING TASKS

Concerning the questions in EAP and ESP exams, students complained verbally about the similarity of options to select from in the answers. In the distinction between the rhetorical

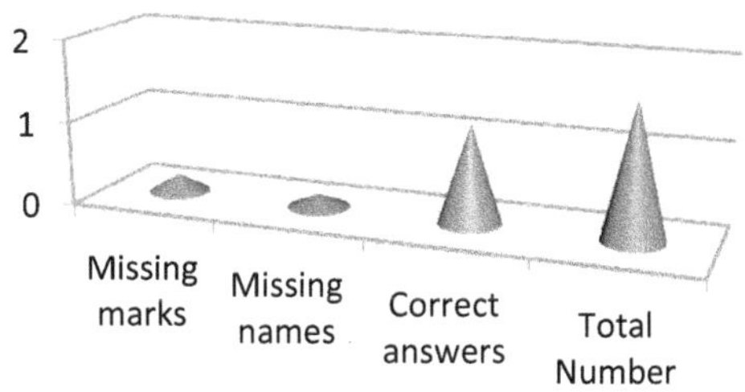

FIGURE 9.4 Rates of barcode correction problems.

functions of defining and identifying, students find it difficult to answer correctly and identify them. Other students complained about the meaning of some difficult scientific words, and the possible combinations of collocations in scientific and academic texts. Such exercises in ESP and EAP require familiarity with possible combinations of words, phrases, and expressions in the expressions. Meanwhile, students need schema knowledge or topic familiarity to understand the meaning of the words and their usage in different contexts. The following exercise is an excerpt from the examination of the Department of Science and Technology:

Task One

Replace the word written in **bold** by word or expression in the list (A, B, C, or D)

1. The university is supported by private funds as well as tuition income and grants.
   A. Generated ☐, B: funded ☐, C:backed up ☐, D:preserved ☐
2. You must wash fruits before you eat them since a lot of bacterial residues are present on unwashed fruits.
   A. Viruses ☐, B: germs ☐, C: insects ☐, D: pesticides ☐.
3. When the explorer set sail, he had clear guidance from the government regarding what route to take.
   A. Instruction ☐, B:image ☐, C: view ☐, D:navigation ☐
4. If you don't feel good you should go to a hospital. The general practitioner will prescribe some medications.
   A. Specialist ☐, B: doctor ☐, C:nurse ☐, D:medical stuff ☐

This exercise requires background information about the context and usage of words. The definitions are provided and the intended words are written in bold. The words belong to different subjects in ESP including bacteria, the general practitioner, supported, and guidance. The context of these words determines the extent to which language functions are used. Cognitive strategies like guessing, evaluating, finding clues, and keywords are important to obtain the exact meaning. The testing is based on both language proficiency as in the words 'guidance' and 'supported', and schema knowledge as in the words 'general practitioner' and 'bacteria'. In addition, some other areas are difficult for being subject-specific with cognitive and strategic implications. The following exercise is an excerpt from the examination of the Department of Management:

Task Two

Match a word from column A with its synonym from column B

1. To catalogue	☐ A:to spot
2. To annotate	☐ B: to organize alphabetically or otherwise
3. To flag	☐ C: To add comments to a document
4. To cite	☐ D: To mark for someone's attention
5. To network	☐ E: a disadvantage
6. A downside	☐ F:to quote or reference another work
7. To download	☐ G:to make contact socially or professionally
8. To notice	☐ H: to get from the internet

The exercise in the examination of the Department of Management was designed specifically to test students' proficiency in the domain of management, with much emphasis on managerial and administrative words and phrases. Students use the words to express the exact meaning, and other words cannot replace them. All the students managed to find correct definitions of the words: to download, to cite, and to notice. However, some students complained about the difficulty of other words like to annotate, to flag, and to catalog as they could not lament the correct answers. Teachers believe that the lack of background knowledge is due to the coronavirus pandemic. They tend to focus on face-to-face teaching in comparison to online teaching, where there is less interaction like explanations, examples, and reading comprehension tasks.

At the level of the department, the administration found difficulties when importing the marks from the software responsible for correcting the exam papers to the so-called 'Progres'. The latter is the software responsible for calculating the total averages of students. During the importation of marks, the department divided again the cohorts of levels into groups as they study the other subjects because the software 'Progres' imports students' marks in groups. Henceforth, the administrators copied the marks first into groups using Excel files, and then the Excel files are imported again into 'Progres'. The teachers displayed the marks for the students and gave them a period of 3 days to check the marks (of all subjects) and make a complaint if they find out any intricacies. The teachers then deliberated again and corrected the missing marks manually until they fixed all the problems.

Many studies were conducted to develop the process by which teachers take students automatically. As an example, Gomez (2015) used barcode technology to evaluate the usefulness of taking absences through a system known as the system development life cycle (SDLC). It allows the dean and other staff members to monitor absences on the basis of the relation between the teacher's load report, attendance monitoring log, and a summary of the attendance report. The tracking system of barcode reading was mainly implemented during the COVID-19 pandemic as an act of social distancing, and it is used to reduce human contact. Elaskeri et al. (2021) integrated a barcode tracking system with a university database called Edugate. The latter is a database in which information about students is entered manually to help track the student's profile (personal information, area of study, and marks).The aim is to make it easy to track down the students' attendance without any bodily interaction by focusing on the student's personal information. Attendance in these studies was part of the evaluation system.

9.8 CONCLUSION AND FUTURE SCOPE

The teaching of EAP and ESP at the Mila University Centre had experienced a drastic change specifically after the spread of the COVID-19 pandemic. At the Mila University Centre, two online platforms were used to accomplish the task of delivering EAP and ESP courses to students. The Moodle platform was used mainly to upload lectures, tasks, and activities for students throughout the whole year. In addition to uploading lectures for students, teachers can upload other files, like videos, audio files, PowerPoint presentations, and web links. These features are very effective in the teaching of EAP and ESP courses because they enhance learners' background knowledge in different situations. The teachers relied on genre approaches to analyze features of the texts, such as identifying the notions and rhetorical functions, cohesive devices, vocabulary, and lexicon. In these secondary subjects, online classes took place only at the end of the semester, where teachers scheduled only two lessons to explain difficult things, answer questions, and give instructions for the examination.

The barcode scanning correction system is precise and concise. It allows learners to retrieve information easily because the tasks are designed to test specific communicative purposes. Henceforth, this technology maintains the adaptation of specific tasks to improve subject-specific knowledge according to the students' needs. It helps in determining specific genres in ESP and EAP contexts. The fact is that tasks of exams (designed for barcodes correction) limit the scope of teaching for specific communicative purposes. Tasks can be divided into four categories: text organization tasks, subject-specific words and expressions tasks, identifying communicative purposes tasks (mainly rhetorical functions), and cohesion and coherence tasks (linking words). Teachers use these tasks in distance education to test the creativity and skills of students.

The evaluation of EAP and ESP lectures took place at the university to avoid a lot of unwanted variables, like slow-speed internet, technical problems, and honesty disintegration. Meanwhile, the evaluation was based on sending texts to students before the exam and answering questions about the texts in the exam. Barcodes were used to correct exam papers and evaluate the answers automatically. Accordingly, the questions are multiple-choice (in the form of quizzes), and students just tick the right answers. However, students' bad handwriting and carelessness resulted in misreading the data on the computer. For that, the handwriting should be clear and direct to avoid all types of intricacies. The evaluation of students' exam papers was a very successful operation, and it looks like Google

Form exams but taken in class to ensure the authenticity of the examination. Therefore, barcode scanning is a prerequisite system in online assessment, and universities can elaborate on barcode correction systems as a testing method to enhance testing criteria. This barcode scanning is useful in online education, specifically when we know that it is not used for assessment only but also for other learning purposes like providing additional information, summarizing, and paraphrasing specific words, phrases, and expressions in different contexts. It can also raise the bar of online assessment in different contexts where technological devices (like AR, games, and virtual reality) are pedagogically implemented. The use of barcode scanning technology will revolutionize MALL with its multi-tasking and easy-to-use functionality; teaching and learning can be adapted to assimilate real-life contexts. It is a useful tool in times of unexpected crises like COVID-19 as it is used in asynchronous learning with various technologies like e-books, e-lessons, and learning platforms. Therefore, the barcode correction system is very useful in CALL, MALL, and autonomous-asynchronous learning.

REFERENCES

Bahatia, V. K. (2002). Applied genre analysis: Analytical advances and pedagogical procedures. In A. M. Johns (Eds). *Genre in the Classroom: Multiple Perspectives*. Mahwah, NJ: Erlbaum Associates, pp. 279–283.

Bonner, E., & Reinders, H. (2018). Augmented and virtual reality in the language classroom: Practical ideas. *Teaching English with Technology*, 18(3), 3–53.

De Pietro, O., & Fronter, G. (2012). Mobile tutoring for situated learning and collaborative learning in AIML application using QR-code. *Sixth International Conference on Complex, Intelligent, and Software Intensive Systems*, pp. 799–805. doi: 10.1109/CISIS.2012.154.

Douglas, D. (2000). *Assessing Language for Specific Purposes*. Cambridge: CUP.

Douglas, D. (2013). ESP and assessment. In B. Paltridge & S. Starfield (Eds), *The Handbook for English for Specific Purposes*, pp. 367–383. Malden, MA, Wiley-Blackwell.

Dudley-Evans, T., & St. John, M.J. (1998). *Developments in English for Specific Purposes: A Multi-Disciplinary Approach*. Cambridge: Cambridge University Press.

Elaskeri, S., et al. (2021). Using barcode to track student attendance and assets in higher education institutions. *The 12th International Conference on Ambient Systems, Networks and Technologies*, pp. 226–233. Warsaw: Poland, 23–26 March.

Fulcher, G., & Davidson, F. (2007). *Language Testing and Assessment: An Advanced Resource Book*. New York: Routledge.

Gomez, J. A. (2015). Faculty attendance monitoring system: An improved feature with barcode scanner. *Journal on Human Development*, 9(1), 50–61. doi:10.31219/osf.io/smz6f.

Halliday, M. A. K., & Hassan, R. (1985). *Language, Context and Text Aspects of Language in a Social Semiotic Perspective*. Victoria, Deakin University Press.

Hayes, J.R., &Flower, L. (1980). Identifying the organization of writing processes. In L. W. Gregg, & E. R. Steinberg (Eds). *Cognitive Processes in Writing: And Interdisciplinary Approach Hillsdale*, pp. 3–30.NJ: Lawrence Erlbaum.

Hutchinson, T., &Waters, A. (1987). *English for Specific Purposes: A Learning-Centered Approach*. Cambridge: Cambridge University Press.

Hyland, K. (2005). *Metadiscourse: Exploring Interaction in Writing*, London: Continuum.

Hymes, D. (1972). On communicative competence. In J. P. Pride and Holmes (Eds.). *Sociolinguistics*. London: Penguin Books, pp.269–293.

Kakoulli-Constantinou, E. (2018). Teaching in clouds: Using the G-Suite for education for the delivery of two English for specific and academic purposes courses. *The International Journal of Teaching English for Specific and Academic Purposes*, 6(2), 305–317. doi: 1022190/jtesap1802305c.

Law, C., & So, S. (2010). QR codes in education. *Journal of Educational Technology Development and Exchange*, 3(1), 85–100. Retrieved from http://www.sicet.org/journals/jetde/jetde10/7-So.pdf.

Lee, J.-K., Lee, I.-S., &Kwon, Y.-J. (2011). Scan &learn! Use of quick response codes & smartphones in a biology field study. *The American Biology Teacher*, 73(8), 485–492. doi: 10.1525/abt.2011.73.8.11.

Luoma, S. (2004). *Assessing Speaking*. Cambridge: Cambridge University Press.

Mikulski, J. (2011). 10 ways to use QR codes in the classroom. *Classroom in the Cloud*, 2011. Retrieved from http://www.classroominthecloud.net/2011/06/10-ways-to-use-qr-codes-in-classroom.html.

Miller, C. R. (1984). Genre as social action. *Quarterly Journal of Speech*, 1984(70), 151–67.

Munby, J. (1978). *Communicative Syllabus* Design. Cambridge: Cambridge University Press.

Oxford, R. (1990). *Language Learning Strategies: What Every Teacher Should Know*. Cambridge: Cambridge University Press.

Plutino, A. (2017). Teachers as awakeners: A collaborative approach in language learning and social media. In Carman Alvarez-Mayo, Angela Gallagher-Brett, &Franck Michel (Eds.). *Innovative Language Teaching and Learning at University: Enhancing Employability*, pp. 115–125. Research-Publishing. Net. doi: 10.14705/rpnet.2017.innoconf2016.661.

Rosell-Aguilar, Fernando. (2018). Twitter as a formal and informal language learning tool: From potential to evidence. In Fernando Rosell-Aguilar, Tita Beaven, and MaraFuertas Gutierrez (Eds.). *Innovative Language Teaching and Learning at University: Integrating Informal Learning into Formal Language Education*, pp. 99–106. Research-publishing.net. http://researchpublishing.net/manuscript?10.14075/rpnet.2018.22.780.

Swales, J. (1990). *Genre Analysis: English in Academic and Research Settings*. Cambridge: Cambridge University Press.

Trimble, L. (1985). *English for science and Technology: A Discourse Approach*. Cambridge: Cambridge University Press.

Wilkins, D. (1976). *Notional Syllabuses*. Oxford: Oxford University Press.

CHAPTER 10

Distance Learning and Higher Education Hybridization

Opportunities and Challenges for Students with Disabilities

Cristina Dumitru

University of Piteşti

CONTENTS

DOI: 10.1201/9781003322252-10

10.1 INTRODUCTION

The world has changed significantly, and the higher education landscape has evolved dynamically worldwide since COVID-19. Universities need to keep pace with the dynamics of the rapidly evolving environment by assessing the unexpected challenges and opportunities that may occur. The medical crisis, the economic and social insecurity, and the fast-paced changing labor work demands forced higher education to rethink its purposes and roles, to transform, and shift to a totally digital learning environment. Not only universities survived COVID-19 but they also managed to reinvent themselves, as 'drivers, mediators, and critical observers of the change' itself (van't Land, Corcoran, & Iancu, 2021, p.2). The massive use of innovative technology facilitated access to education and enabled students and teachers to continue the learning process remotely. Universities entered a new digital era that generated new opportunities for both students and teachers, leading to innovative directions for research initiatives. Once entering this 21st-century digital race, several transformations were triggered not only in the educational area but in other various aspects of life (we went working remotely, attending online concerts, visiting virtual museums, creating Zoom birthday parties, etc.). To provide ongoing learning, universities are in the process of digitalizing their services and designing mainly postgraduate programs for the online environment. Moreover, digitalization reshaped us irreversibly, with little probability of going back to the pre-COVID-19 era, a word 'that does not use technology in teaching and learning' (Thormann & Kaftal Zimmerman, 2012, p. 12).

Significant steps toward higher education digitalization were made even before the COVID-19 pandemic. Additionally, digitalization is an important factor in the increase of students' enrollments as university programs were able to respond better to students' demands. Compared to other public sectors, innovative technology-enabled higher education to rapidly adapt to pandemic crises by shifting learning and teaching to virtual classrooms and tackling scientific knowledge, accessibility, use, openness, and transparency issues. The transition to distance learning was developed based on traditional ways of knowledge transfer. Nonetheless, it seems that it is not enough as learning is strongly impacted by the environment in which it develops (onsite and/or digital). Furthermore, it is crucial more than ever to understand how students learn in an online environment and how the teaching process should be adapted according to the challenges and opportunities of distance learning. Universities are aware that connecting with their students is an important step toward raising educational quality. Therefore, several studies and research were undertaken to explore the ways universities are incorporating new and sophisticated possibilities into their teaching, including artificial intelligence, machine learning, chatbots, and virtual assistants, augmented and virtual realities, new ways of connecting, engaging, interacting, networking, and exchanging (online academic mobilities) (Bygstad et al., 2022; Pu et al., 2022; Tripathi & Jauhari, 2020). In addition, the process of digitalization itself introduced to the educational landscape different gaming technologies. COVID-19 is considered a trigger and a driver for universities to reimagine, transform, and re-enact physical and online learning environments 'to facilitate new ways of on-campus connection, collaboration, and pedagogical practice' (Macken et al., 2021, p.7).

Objectives of the Chapter

In this context of the shifting to a hybrid space, the present chapter aims to contribute to a better understanding of learning dynamics (students' learning behaviors) and engagement of students with disabilities in distance education settings. This kind of understanding can help teachers to explore the learning behaviors of students with disabilities. Moreover, it can help them to prioritize and provide effective support to all students. Throughout the chapter, the following aspects are covered:

1. To understand the background and related aspects of distance education and to analyze the impact of digitalization on higher education and on students with disabilities.
2. To explore the challenges of online education on students with disabilities and to understand the complexity of factors that have an influence on students' performance in online education.
3. To describe some pedagogical strategies and the main principles of distance education and provide some recommendations for higher education institutions for better accommodation of their services for students with disabilities.

Chapter Organization

The chapter is organized into seven units following the objectives launched in the introduction part, as follows: Section 10.2 highlights *distance education and online learning* covering some differences between distance education and online learning, and some historical points of online and remote learning in higher education. Section 10.3 enlightens *distance learning in higher education* with a focus on the impact of distance learning on higher education. Section 10.4 gives a detailed overview of *digital inclusion* mentioning the impact of digitalization on reducing barriers to accessing knowledge for students with specific vulnerabilities. Section 10.5 highlights *disability in higher education* mentioning all the challenges that students with disabilities still face in online learning environments. Section 10.6 elaborates on *how to ensure a successful higher education program* with a focus on identifying the main factors that have an impact on students' performance and academic learning in online settings, what pedagogical strategies can be implemented for successful academic learning, how should be a course designed to accommodate specific educational needs of students with disabilities, what are the main principles of online education. Section 10.7 elaborates *recommendations for higher education institutions and faculty* to ensure an inclusive and diverse learning environment that will be sensitive to their students' needs and will explore various ways to accommodate the learning environment to students learning needs. And finally, section 10.8 concludes the chapter with future scope.

10.2 DISTANCE EDUCATION AND ONLINE LEARNING

Distance education and digital transformation in higher education is not a new topic since it has been on the policy agenda for at least 30 years. Technology can affect society in all its layers. Education adopts new and innovative technologies, mainly as an exploratory tool to increase students' motivation to acquire knowledge through active learning technologies.

What do distance education and online education mean? *Distance learning* is often used in synonymy with distance education, online teaching and learning, e-learning, virtual classrooms, web-based training, computer-assisted learning, technology-enhanced instruction, technically mediated learning, and teaching and learning with technology. Nonetheless, variation in the style of delivery differs depending upon several characteristics such as venue, the role of the student and the teacher, participants, methods, tools utilized, etc. The use of new technology in education can take various forms and serve various purposes, ranging from improving with sophisticated facilities the traditional learning environment to transitioning to a fully online learning environment. *E-learning* is a large concept and can refer to the use of solely online or both online and onsite formats of learning, also known as blended learning. *Open learning* is another term used in connection with distance learning, mostly because an open learning environment will imply certain flexibility in providing various options for the learner to engage in the learning process, time and place included; thus, the term distance learning can be included in the term of open learning (UNESCO, 2020). There is an extensive and still growing interest in understanding and researching the effects of open and distance learning (Caliskan, 2012) on academic performance, learning satisfaction, motivation, and mental health. *Distance education* represents a well-designed set of instructional opportunities, technologically mediated, where knowledge transfer and learning experience are aimed similarly to onsite learning. In order to facilitate the knowledge and competence transfer to occur, being a two-sided process, both teachers and learners need to meet some conditions. The learning experience is rigorously designed based on a qualitative curriculum proposal aiming to generate specific learning outcomes, defined by competencies. Contrary to onsite learning where the educational process is based on direct interaction, in online settings, the interaction between teachers–students and students–students is technologically mediated. Both in online and in onsite learning, the educational process consists of two important phases:

1. **Input**: The process of transferring the educational content to the learners, planned and guided by the instructor.
2. **Output**: The process of acquisition of the educational content in the form of constructing knowledge, performed actively by the students and facilitated by the teacher.

A systematic literature review (Bernard et al., 2004), after analyzing a great number of articles on distance learning, identifies three elements that define distance learning: (i) geographic distance and time separation between instructors and learners; (ii) educational content to be delivered, that is planned, well-thought and organized; (iii) provision of the transactional environment for the educational exchange to occur.

Peters (2017) explains the difference between *distance learning*, which is linked mostly to the digital tools utilized for the connection and interaction between the learners and their teachers or instructors, and *distance education*, which refers primarily to the methodological and instructional aspects of delivering knowledge and providing learning opportunities for both students and teachers. Wheeler (2012, p. 1018) considers 'distance learning to be an outcome of distance education.' All the above-mentioned types of learning are based on an interactive knowledge-acquisition process, technologically mediated. For a learning process to occur it is important for the learner to be in its zone of proximal development (Vygotsky, 1930, 1977), which is the potential zone where the learner needs a more knowledgeable other to acquire knowledge. However, in online learning, the teacher still has the role of the knowledgeable other. Moreover, all the tools and digital instruments available can work as learning facilitators (broadcast, podcast, web-based instruction, video-conferencing, wikis, blogs, and social networking, just to note some). Higher education requires more autonomous and independent learning, and the incorporation of new technologies can assist and support students, especially students with disabilities.

The history of distance learning, a form of the asynchronous study program, available in the early 1800s, starts with the correspondence courses, at the University of London in 1858 (UK), McGill University in 1889 (Canada), the University of Chicago, and Penn State University in 1892 (USA), and Universities of Alberta and Saskatchewan in 1912 (Canada). Distance learning was pushed by the increased demand of learners to overcome time and space constraints and join an educational program regardless of their geographic location (Mehlenbacher & Mehlenbacher, 2019). An example of the incipient form of distance education is learning by correspondence, which can be tracked 'as far as the instructional writings (epistles) of St Paul to the early Christian church' (Wheeler, 2012, p. 1018). Initially, distance education comprised printed and post-mailed courses; afterward, radio broadcasts in the 1940s improved the educational input for home-based study groups of St. Francis Xavier University (Canada) (Haughey, 2013). Later both printing and audio- and video-conferencing were combined to reach isolated learners. Since the development of the world wide web, cloud-based learning management systems (Sheikh Ibrahim et al., 2015) simulated traditional teaching and proposed several sophisticated features to design educational material, provided support and immediate feedback, assessed and helped self-assessment of the learning progress, shared learning resources, facilitated co-creation of learning materials, etc.

10.3 DISTANCE LEARNING IN HIGHER EDUCATION

COVID-19 required higher education institutions to rapidly shift to remote course delivery. The digital era revolutionized distance learning through new technological tools enabling learners to connect and communicate and interact simultaneously and synchronously, moving it from primarily solitary activity to a more collaborative and interactive learning experience. Moreover, the role of learners has also changed. For example, they can actively interact with the educational content, and even engage in the co-creation of the educational content itself by giving input into the development of learning content and materials, utilizing ICT tools based on collaborative use of information, open educational resources (OERs), interaction and cooperation. All actors participating in the process become co-creators of the educational content. Technologically enhanced learning provides opportunities for learners to develop independent study skills, and metacognitive skills through self-directed learning and simultaneously engage in collective, collaborative, and cooperative learning experiences (Wheeler, 2012).

In distance learning, education is provided regardless of the space and time conditions. Distance learning, also referred to as distance education, e-learning, or online learning, represents an educational approach that occurs when the professors and students are separated by space and time (Todri et al., 2021). The opportunities and threats of integrating technology use in higher education were extensively studied (O'Byrne et al., 2021), even before going exclusively online. Universities are trying to effectively embed technology in their programs and services to better cover the needs of students to act upon an increasingly digital future (Wang & Hannafin, 2005). Teaching and learning during remote learning imply the use of online tools, flexible hours in implementing academic activities, and free and unlimited access to online educational content made available by the universities themselves (Buzzai et al., 2021). With the global pandemic COVID-19, universities are primarily using the hardware and software necessary to transform their on-campus courses for online delivery (Buzzai et al., 2021). They are also increasing the use of specific instructions or media, such as videos or multiformat materials (e.g., text, audio, quizzes, hands-on exercises (Lee et al., 2021), micro-learning, gamification, etc.). Moreover, higher education institutions propose a wide range of unbundling degree programs based on micro-credentialing, modular learning, horizontal and vertical stackable courses, and others (Macken et al., 2021).

Even without the pandemic challenges and changes, students are comfortable with and prefer to utilize technology in interacting with their colleagues, teachers, and university staff. Students recognized that the technology utilized for remote courses was primarily laptops and smartphones (Prokes & Housel, 2021). The advantage identified was the ability to accomplish many transactions whenever they need to and from wherever they were located, increasing chances to engage and to continue learning. Some students mostly have utilized online libraries, and the number of students who request onsite library services has decreased significantly. The library concept changed a lot and had to reinvent itself to survive by offering online and digital services. Another asset of incorporating online tools in delivering university programs is the option to track and store the student's performance which enhances immediate and continuous feedback and allows students to be more aware of their academic performances (Curum & Khedo, 2021). Often, for smaller universities, ensuring valid and qualitative online learning requires some serious investments in the infrastructure. Nonetheless, the increased demands from the workplace, academic assignments, and stressful everyday-life events push university students to be more likely to misuse the internet to find comfort and to escape from reality.

According to Aagaard and Lund (2019), it is getting harder and harder to predict how the future workplace will evolve and how universities should prepare future professionals. Nevertheless, experts envisage the need for even more developed interdisciplinary skills, as well as transformative competence which is the ability of teachers and students to continuously reform and update their teaching practice, which is a sine qua noncondition in order to find ways out of non-standard situations. Moreover, the main mission of education has not dramatically changed and the link between higher education and the labor market can be considered crucial. It is expected that the process of developing online services for pursuing a university program will continue evolving. Even though learning activities in the physical environment will be encouraged, hybrid formats and the use of digital components will increase significantly, even in smaller universities. Students will choose the higher education institution based on the availability of hybrid formats, which will force universities to adapt and (re)consider the use of a wide range, of mixed and online formats for academic studies and mobility programs.

Higher education understands that new technology changed the educational and research landscapes and the new context of digital and digital education has to be embraced by being open to providing flexible solutions and formats for its students to access study programs and educational, research, and professional activities. Flexibility is expected regarding the time and place of learning context, various modalities in engaging and participating in the learning process, a diverse range of instructional methodologies, and other variables linked to the learning environment. From this perspective and from the new pedagogical paradigms, valuing the potential and inner resources of every individual, there is a growing interest of universities in ensuring an open and flexible learning environment, accessible to everyone, mostly in post-COVID-19 era, any place and anytime, which refers interchangeably to open learning, e-learning, flexible learning, and distance learning.

The role of the teacher in future higher education is presumed to be replaced by artificial intelligence (AI), which will provide more customizable learning paths to the individual learning needs of each student. It is expected that the main roles of the teacher will change while dealing more with planning, researching, decision-making, and ethical and moral values shaping (Folostina & Dumitru Tabacaru, 2021).

10.4 DIGITAL INCLUSION AND DISABILITY: OPPORTUNITIES AND CHALLENGES

10.4.1 The Promise of Digitalization

There are many advantages to incorporating distance learning into higher education learning just to mention some, increasing accessibility to educational content, providing various opportunities for co-working and co-creating qualitative educational resources, encouraging vulnerable students (students with disabilities, or those already working and older students) to engage and continue their studies. Another important aspect is cost-effective learning in online settings (Chirikov et al., 2020), a very good teacher can reach thousands of students, via online teaching platforms. Distance learning can be a valid solution for updating and upgrading the knowledge and skills of professionals already working through providing university programs while applying new knowledge and skills directly to their practice. Higher education institutions that offer distance learning programs can be very tempting for lower tuition fees, a more flexible study pace, and an intercultural richer learning experience. Simultaneously, universities can improve their services by incorporating technology from the online university admission process to course registration and scheduling, reporting, seminars, and conference organizations. Clear instructions and an easily accessible schedule for completing each educational and research activity are needed to better navigate a distance learning program, which would counterbalance the lack of physical encounters. Additionally, to communicate collaboratively the interactions between faculty and students can be ensured through alternative forms.

Distance learning brought a significant change in understanding mobility programs, staff and students' mobility programs, and ways to integrate the benefits of engaging in mobilities even if traveling is not possible by designing online exchange and learning mobility programs. During COVID-19 confinement, some aspects were taken into consideration:

a. The growth of willingness of universities to integrate digital tools in their programs and consider a hybrid format.

b. Readiness of students for digital mobility formats.

c. Mixed and online formats of academic mobility should be considered as new tools for its further development.

d. 'Virtual mobility' as an alternative and additional form of international student mobility.

e. Academic mobility, including incoming and outgoing international academic student mobility, and teacher and research mobility.

At the same time not just the concept of mobility was challenged but also the concept of the university campus is now discussed, and serious steps are being taken for heavily incorporating new digital technologies into education.

a. A campus is not only a physical building but also a place where communities are formed, a place of memories, and connections. Social interaction between students and direct personal communication cannot be transferred to digital formats.

b. Most students feel safe when they are on campus but must limit their contact with family and friends. Together with social restrictions and general pressure during a pandemic, this has a negative impact on mental health.

c. The spatial design of campuses should help to improve their efficiency in terms of education; for this it is necessary to ensure safe social interaction in public spaces, disease prevention, and health promotion measures, creating an environment conducive to professional and social interaction, as well as strengthening functional and social ties with the outside world.

d. Among other things, an online university program should cover the implementation of a system of psychological safety and emotional comfort on campus, as well as the preservation of a culture of research and laboratory practice.

10.4.2 Impact of Digitalization on Students with Disabilities

Digitalization can have great potential to improve the learning achievements of students with disabilities, their engagement, and active participation in the learning process by reducing and even removing barriers to accessing educational content, encouraging autonomy and independence, and facilitating learning. For students with disabilities searching for new educational materials, books, recent articles, good practices, and upcoming conferences are nowadays possible with the opportunities provided by virtual encounters. Distance learning significantly expanded the horizons of opportunities for instructors and learners to broaden their knowledge and skills with fewer costs and access portals of big universities, otherwise impossible to access. Definitely, the best learning experience remains to actually join and physically attend the desired university, but less is better than nothing. Accessing free online university courses has introduced learners and those with disabilities to a world of non-traditional ways of learning opportunities where they can share information and create knowledge. Additionally, online learning improved the collaboration between students and teachers due to reducing physical barriers worldwide. Higher education policies are determined by the continuous increase in requirements for accessibility and quality of education, by the demand of the economy and society for more personalized, flexible, and adaptive learning pathways, and by the need to prepare future society members ready to manage a high degree of future uncertainty. There are broad trends in the development of universal and inclusive universities: (i) development and incorporation of various forms of digital education, including hybrid learning, e-learning, we-learning, on-the-job training, and learning enterprises and (ii) updating the educational content in the light of new requirements toward its practical orientation.

Today, digital society trends trigger the need to invest and build a strong digitalized higher education based on (i) the digital economy and the new requirements of the labor market (the main source of educational goal setting for professional development and training); (ii) the new SMART digital technologies (Liu & Zheng, 2021) that shape the digital environment and constantly develop and (iii) the digital generation needs (a new generation of students with specific socio-psychological characteristics). Currently, the process of digitalization is evolving dynamically and actively, converging digital technologies with practices, including educational ones. The purpose of higher education policies, in the context of technologically and digitally enhanced learning environments, is determined by the possibility of using them as a basis for:

i. Setting scientific research priorities in the field of education digitalization.

ii. Developing a conceptual, strategic educational framework that clearly defines the tasks, organization, and content of the process of digitalization of higher education at the regional and international levels, in higher education networks and clusters.

iii. Proposing methodological recommendations on the implementation of the digitalization process and on the organization of the educational process in the digital environment for academia.

iv. Setting criteria, mechanisms, and tools for the examination of the quality of educational programs, learning outputs, research products, as well as the educational process in the context of the digitalization of education.

Higher education has been embracing distance education for some years and yet is struggling to identify ways to provide distance learning and rethink teaching to accommodate the educational needs of all students. Universities took learning for granted, as university teachers were expected to have well-equipped students with study skills and efficient learning techniques. Going online, expanding the range of students in virtual classrooms (re)questioned the way learning happens. University teachers have to learn to address the diversity of the new generation of learners. Universities should adapt to their students' educational needs and define the best-suited learning options for all learners. Universities should become more flexible to align with the different educational goals, to meet students' special educational needs, to respond to labor market trends, especially as the procedural knowledge tends to be left to training, traineeships, or work placements, while higher education is constructed more on epistemic matters (Mehlenbacher & Mehlenbacher, 2019). In view of current demands, higher education has to explore innovative ways to transform teaching practices to better respond to the unpredictability and complexity of nowadays world and to be relevant for all its students. Students benefit from receiving teacher support and they value collaborative activities, especially in online settings (Dumitru Tăbăcaru et al., 2022).

Some students who are unable to attend onsite classes due to mobility or behavioral issues and who find the school environment difficult to manage to prefer online learning that has no physical boundaries. Nonetheless, besides physical access, there is much work to be done with facilitating access to the knowledge building process. Students with disabilities need to have individualized learning paths, based on their strengths. Teachers have to be receptive to students' needs and resources and be able to individualize the approach and the student's ability to manage time. Distance education flexibility provides students with disabilities with more opportunities for learning and engaging in learning than onsite learning. However, distance learning cannot by itself provide personalized educational paths by itself as teachers still have a significant role in designing learning experiences, directing and guiding students learning behaviors, finding the best and viable resources, and exploring the optimal structure that supports each student's individual needs.

10.5 DISABILITY IN HIGHER EDUCATION

10.5.1 Institutional Answers to Inclusion

Since the ratification of the *UN Convention on the Rights of Persons with Disabilities* (United Nations, 2020) and the extensive promotion of inclusion *The Americans with Disabilities Act* (The Americans with Disabilities Act of 1990 and Amendments Act of 2008, 2009), *The Salamanca Statement and Framework for Action on Special Needs Education* (UNESCO, 1994), *The Dakar Framework for Action* (Education Forum, 2000), the number of students with disabilities enrolled in higher education institutions has increased significantly. Nonetheless, disability has long been a source of concern for universities, academia, and researchers. The COVID-19 pandemic has exacerbated student inequalities, considering the lack of social support, the lack of means to get online, low digital skills, and not being adequately equipped or supported to continue higher education even in online settings

(Chadwick & Wesson, 2016). Given the increasing number of students with disabilities in higher education and the desire of universities to keep the enrollment rate high, it is crucial to understand how learning happens, how learning should be facilitated in online settings, and what are the facilitators and the barriers to the academic success of students with disabilities in an online environment (Fichten et al., 2020). Students with disabilities, or with insufficiently developed life management and self-control skills, face great difficulties in online settings. In the long run, society will benefit from independent and active members; hence, the need for higher education policies to promote and foster equality and equal opportunities for all its students here included equal access, optimal internet infrastructure, and future employment prospects. Widening access to higher education for students with disabilities has been a priority for the educational political agenda since the ratification of the *UN Convention on the Rights of Persons with Disabilities* (United Nations, 2020). Since the COVID-19 pandemic, the inclusion of disabilities in higher education revitalized the interest of higher education institutions for inclusion. The principal elements of higher education policy for the post- COVID-19 era should be flexibility, accessibility, adaptability of students, and the well-being of all actors involved in the educational process, teachers included. The philosophy of the educational process is evolving in strong connection with the changing role of the student in the digital age. International, national, and regional education policies should reflect this trend through the adoption of relevant regulations, new incentive systems, and performance indicators, as well as efficient allocation of funds. Comparative international studies on the transformation of higher education in the pandemic and post-pandemic period should also be supported.

Online learning develops new possibilities for further continuing higher education for students with disabilities. Nevertheless, steps need to be taken to respond to demands such as affordable tuition costs, immediate employment value to prospective students, and better consideration of academic needs. Higher education programs should provide valuable opportunities for learning and career advancement through a structured and comprehensive approach to train knowledgeable and professional workers for the unpredictable labor market. Higher education institutions need to equip students with relevant competencies to act actively in a global knowledge market force. The dynamics of the current labor market urge universities to propose and implement better services for students' needs and labor market trends. In order to respond effectively to these challenges, universities should put in place effective strategic planning processes. According to Curum and Khedo (2021), this new learning paradigm delivered through online platforms encourages education across different contexts and triggers the obligation to adapt the learning contents with appropriate instructional design principles to identify key elements which can better prepare professionals for the always-changing labor market.

The adaptation to the online environment would imply instant sharing of learning content, and opportunities to engage and participate in chat forums or webinars for knowledge exchange outside classroom space and time (Curum & Khedo, 2021). As students tend to be regularly connected to social media platforms and prefer to navigate online, learning is conducted in a self-paced mode with advanced interactions through online communities. Studies (Creemers et al., 2013) revealed the need to develop reflection and critical thinking as important elements in all aspects of learning and performance, so much needed in a technologically enhanced and globalized world that is constantly changing. The purpose of higher education remains to just educate (Macken et al., 2021), whatever education means now. Universities should develop and maintain a lifelong relationship with learners (Macken et al., 2021) as learning is seen as lifelong, contextual, relevant, meaningful, and engaging (Nugent et al., 2020). Other key elements that have to be tackled in (re)designing distance study programs and their adaptation to the needs of students with disabilities are motivation, support in learning, positive attitude, and group learning techniques with opportunities to introduce students to real-world experiences before graduation. Research findings on teacher effectiveness, how learning happens, and on how the process of learning is determined and influenced in the digital environment, and 'how online instruction influences and undermines traditional models of instruction' (Mehlenbacher & Mehlenbacher, 2019, p. 7) should be taken into account while designing online study programs. However, there is a lack of existing analysis on the influence factor of technology acceptance of mobile-based teacher professional training (Peng, 2019). Poorly designed courses increase confusion, and frustration and discourage learners (Curum & Khedo, 2021). Perceptions of students on instruction and instructional practices are especially important in higher education (Pekrun et al., 2002), and the data from their learning needs assessment to provide important implications for study program design due to the identification of effective pedagogical techniques that ensure students' content learning, cognition, psychosocial development, and success in college (Mayhew et al., 2016; Seifert et al., 2017).

10.5.2 Higher Education Perspectives on Diversity

In the last decades, significant steps were taken in working toward inclusivity in higher education for students with disabilities, such as specific learning disorders, attention-deficit and hyperactivity disorder (ADHD), autism specific disorders, and physical, visual, or auditory disabilities. The development of higher education programs is considering the need to facilitate access to a more diverse range of students, including students with disabilities, fostered by the rapid digitalization process. Universities are becoming inclusive and distance education proved to be an opportunity for students with disabilities to access higher education. Digital education is presumed to soon become the new formal way of higher education (Baten et al., 2022), at least for some programs. The inclusive approach to education and better support services during pre-university training increased the chances of students with disabilities to pursue higher education (Dumitru Tabacaru, 2019). Studies (Cumming & Rose, 2021; Santos et al., 2018; Schwartz et al., 2021) show that the number of students with disabilities enrolled in higher education programs has been increasing over recent years. Nonetheless, there is a lot of exclusion, much exclusion also in online learning even though a great number of tools facilitate students learning and help them overcome barriers to learning caused by societal attitudes, organization, and formatting of the educational material (Chadwick & Wesson, 2016). The incorporation of online learning in higher education facilitates access to a more diverse type of students, coming from diverse socio-economic backgrounds with a broad range of experiences (Coman et al.,

2020). Therefore, it is imperative to prepare teachers with knowledge and competencies to address the needs of all those diverse students (Gleason & Jaramillo Cherrez, 2021). Higher education institutions seek to provide equity, translated into equal opportunities to all their students to achieve learning outcomes. How can universities achieve that?

1. By adhering to an inclusive philosophy in providing education in a diverse and rich learning environment.
2. By adapting the learning environment to the diversity of students.
3. By providing multiple means to engage with educational content, materials, and resources (Meyer et al., 2014).
4. By hearing the voice of teachers and students in the process of curriculum elaboration, providing ways for both actors to be actively engaged in designing and elaborating the curriculum.

Moreover, distance education highlighted the learners and their needs also in higher education, by giving access to educational content at flexible times and places, for as long and as often as needed for the student. The indicator that best explains the growing interest of students in open and flexible formats of program studies is the attendance rate (Lu & Cutumisu, 2022), and studies report high attendance at remote university programs, as well as at conference events (Lu & Cutumisu, 2022; Pérez-López & Ibarrondo-Dávila, 2020; Wheeler, 2012).

10.5.3 Disability and Challenges for Higher Education

Distance learning has indeed made the learning process more flexible and enriched it with several opportunities, as mentioned above. However, it has serious disadvantages and faces some challenges mostly caused by the rapid adoption of online education without having enough time for adaptation: just to mention some – difficulties in adjusting to the new learning environment, lack of clear systems to assist learners to adapt, lack of teachers' skills and competencies to transfer the educational content to the remote learning environment, and ensure the quality of online teaching. In the online learning context, there are limited monitoring and control mechanisms to ensure that the experience is of good quality for every learner. Another limitation is the rigid mindset of thinking about education and the manner in which learning should occur. Changing teachers' mindsets so that they can be more flexible to try to innovate, think creatively, and be open to adapting to new ways of learning might be a reasonable aim for future teacher preparation and development (Cullen, 2020).

Considering the growing number of students with disabilities in higher education, universities explore strategies to accommodate and support the learning of all their students, including students with disabilities. Even though the number of students with disabilities is growing, they remain underrepresented in higher education. Moreover, they are among the most marginalized, vulnerable, and excluded groups on campus, in academic and research activities. The main barriers identified are restrained access to educational infrastructure, experiencing various forms of stigma, and discrimination (Fuller et al., 2004). Therefore, students with disabilities face difficulties related to their deficiencies that often interfere with learning (Longobardi et al., 2019; Zingoni et al., 2021). The lack of support mechanisms and services during their studies. Additionally, this may have a negative impact on their academic achievement (del Bianco & Grigg Mason, 2021). As stated before, the issues faced by students with disabilities can be divided into two groups:

1. Issues associated with the symptomatology of the deficiency and specific characteristics that affect learning and are reflected in academic results and performance;
2. Issues related to the format, content, and effectiveness of activities organized for students.

Universities should take into account the barriers to learning and accessing educational content, as well as students' difficulties in transitioning to university studies. During university preparation, educational activities usually presume more independent and autonomous learning. Mechanisms and services to facilitate students' participation and engagement in their personal educational and professional development process should also be equally explored. Extensive research is done to understand how learning happens, and how students with specific learning disorders or students with autism learn. Moreover, research aim at identifying what kind of accommodations are needed for physically, visually, or auditory impaired students, to ensure a diverse and inclusive learning environment and to reduce significantly the gaps with students without disabilities (Cinquin et al., 2019; Clouder et al., 2020; Coman et al., 2020). More and more universities are adjusting their study programs, educational infrastructure, and support services to better assist their students (del Bianco & Grigg Mason, 2021; Edwards, 2014). With the need to integrate distance learning in higher education, universities had the opportunity to test a variety of innovative technological solutions that proved to be useful to facilitate access to all educational and methodological materials for students with disabilities. In addition to full access to the educational content, to all educational materials and learning resources available, students can also participate in a wide range of online events (lectures, consultations, webinars, student conferences, etc.) or test several supporting apps that can help them with the learning planning mechanism, monitoring and assessment process, and development of research skills. University teachers can also provide educational and methodological assistance to students, in the form of office hours scheduled per student, in the post-COVID-19 era, mediated by video-conferencing platforms.

A great extent of studies reported that the emergency transition to remote education led to a decrease in learning effectiveness and motivation (Raes et al., 2020; Sharma et al., 2021). During the period of self-isolation and distance learning, the burden on teachers, students, parents, students of colleges and universities, and professors increased dramatically. The amount of teachers' workload escalated, as well as the time spent at the computer, for both teachers and students. There was a need for the urgent development of various platforms to be used in the educational process. Learning in the digital era revealed some challenges that have to be addressed:

a. What does teaching mean and what are the responsibilities and roles of teachers in the new learning environment?
b. What is the status of the academic profession in the digital age?
c. How to address students' problems related to equal opportunities?
d. How to develop support services for students who face inequalities?
e. Online education and graduate qualification: what are the expectations and work placement prospects for students?
f. Should we (re)think about the aim of education?

The most crucial aspect when it comes to providing qualitative university programs is enhancing the teaching skills of university professors to adapt efficiently to the growing diversity of students (students with disabilities, students already engaged in the workforce but willing to (re)professionalize, international students, students with a diverse range of learning needs, cultural and linguistic backgrounds). At the same time, students are rapidly becoming co-creators of their own learning process, and key participants in their educational process: digitalization and AI increase the opportunities for students to receive a more personalized learning pathway. Students' and teachers' experiences show that interaction and communication did not decrease during online learning, but, on the contrary, increased, taking new forms.

Simultaneously, some students, especially those with disabilities need to gain competencies in digital literacy to fully participate and benefit from a learning environment that is extensively mediated by technologies. Relevant learning experiences in the online environment are possible based on a mastery of specific digital competencies, such as planning skills, self-regulatory and self-monitoring skills, time-management skills, and sound knowledge about technologies. For both teachers and students, the online learning environment might be challenging, therefore they may need practical training (Mahlangu, 2018). Studies report higher levels of academic stress which has led to higher levels of academic and online procrastination for students with specific learning disorders (Niazov et al., 2021).

Certainly, the online learning environment is losing the incentives offered by the physical space to learning, such as physical proximity, real-time authentic feedback of learners, the ability to share some unique and unrepeatable learning experiences, effects, and values that build up a learning community, which in online settings are more difficult to transmit and to create. Moreover, studies show that reduced social contact with instructors and other learners can result in motivation loss (Badali et al., 2022; Poellhuber & Roy, 2019; Sinclair & Kalvala, 2016). While online learning lacks informal learning contexts, making the online learning community less engaging than the onsite learning fosters genuine connections. Consequently, peer feedback tends to be offered more superficially in online discussions (e.g., 'good work!') than in face-to-face encounters when the feedback tends to be more elaborated and personalized. Students might see online learning as a one-way activity without a face-to-face audience with whom they can engage. Moreover, their engagement level is dropping without a well-designed instructional flow of academic support and coaching, such as facilitation and support groups, through onsite activities and virtual support, virtual activities, and face-to-face support, through fully digital display, discussions, and peer tutoring or peer interaction. The same goes for teachers, having to overcome the challenge of missing the regulatory feedback from their audience.

Students with disabilities are more vulnerable to becoming confused and dropping out of universities, mostly if they do not receive immediate and regular feedback; furthermore, students during distance education tend to skip readings and if they engage in reading, the reading process is less active than with the readings in the printed format. Experience has shown that students with disabilities get tired quickly enough during online classes and lose engagement and, over time, motivation to learn. Teachers experience more often a lack of regulatory feedback during lecturing, impeding knowledge transfer and maintaining an optimal motivational level. Teachers need to get skilled in facilitating both oral and written discourse also in online settings, mostly because it is most likely to be skipped or presumed to be done independently by students, while in face-to-face interaction, it is done more naturally during discussion.

Students with disabilities tend to feel isolated during their university studies. Moreover, isolation is increasing in online settings as well. Universities should design meaningful learning experiences based on interaction to increase their sense of belonging and foster the development of an academic community of learners. A wide range of studies (Bolatov et al., 2021; Fatimah & Mahmudah, 2020; Jha et al., 2021; Moy & Ng, 2021) suggests that the COVID-19 pandemic pushed us to remote learning, and it was an important factor in triggering various mental health difficulties, more particularly for students with disabilities, that are more likely to feel isolated, disengaged and disoriented.

10.6 CAN DISTANCE LEARNING BE SUCCESSFUL?

10.6.1 Factors Influencing the Performance in Distance Education

Studies regarding the impact of online or hybrid university program designs yielded mixed results. An OECD report (OECD, 2020) expresses concern for students' welfare and their academic achievement due to the excessive use of the internet (social and digital technologies), for example, for preparing their assignments, thesis, and other study activities. Long hours spent on the internet distract students from their studies, resulting in them neglecting their responsibilities (Buzzai et al., 2021). Improper design of learning content in distance education was identified as one main reason for the cognitive burden (Chu, 2014). Word digitalization, new routines, procedures, and different approaches to knowledge construction changed significantly the structure of the study program. It should be designed to make learners aware of their potential for further development

(Azmuddin et al., 2017; Binali et al., 2021). Awareness might bring insecurities because of a better understanding of one's own deficits (Boshuizen et al., 2004). Therefore, it is crucial to provide support and create meaningful learning opportunities for (re)training and (re)practicing learning strategies (Ericsson, 2004). De Byl and Hooper (2013) identify key attributes to be included: specific focused goals, appropriate challenging tasks, clear instructions, rapid and constant feedback, affirmation of performance, social interaction, safety from failure, curiosity, and novelty. The identified attributes are fundamental aspects of instructional design and are thus relevant for application in the construction of the online platform and learning activities for students with disabilities.

Learning opportunities can be organized in group-work activities; research (Davis, 1984) indicates the importance of group stability to facilitate and enhance effective work on various learning tasks. The types of assignments proposed most frequently during higher education programs could be research projects, class presentations, written reports, and group examinations (Davis, 1984). Course organization is based on neurocognitive research on learning and on the relation between cognitive and metacognition skills, which will imply for the first-study year to build up the theoretical background by providing visualizing lecture information (Garrett et al., 2007) and for the following study years, it will imply them to focus on the theoretical transfer to practice and build up a learning context for interpreting and discuss research results. Chickering and Gamson (1989) explored what makes good instructional practice and they identified the following principles:

1. To encourage student–faculty contact,
2. To encourage cooperation among students,
3. To encourage active learning,
4. To give prompt feedback to students,
5. To emphasize time on task,
6. To communicate high expectations,
7. And to, show respect toward diverse talents and various ways of learning.

Some empirical studies confirmed the validity of the above-mentioned instructional principles for predicting outcomes and a strong correlation was found between good teaching practices and students' cognitive development, and career achievement (Astin, 2005; Loes et al., 2018; Pascarella et al., 2011). Moreover, with the outbreak of COVID-19, online learning is seen to be very popular. In light of new dynamic changes in higher education and the increasing popularity of remote learning a new hybrid learning environment occurs. In the process of transferring the teaching contents from onsite to education at distance, studies are needed to identify appropriate teaching models and techniques, adequate technological tools to cope with remote learning barriers, and support strategies to maintain a sufficient level of engagement, motivation, and participation of students and teachers as well.

10.6.2 Pedagogical Strategies for Distance Learning

Digital learning happens in a digital environment, which is mediated by technological tools. Hence, it is necessary when planning educational activities to consider the two sides of the process of digitalization:

1. Firstly, to build up an efficient digital environment, by developing digital learning tools, online courses, and electronic educational resources.
2. Secondly, (re)thinking about the educational process, how learning happens in the online format, what are the roles and responsibilities of teachers and learners, how to build up learners' community, and how education should be provided, guided, monitored, and assessed.

Thus, the digitalization of the educational process is, on one hand, a deep counter-transformation of the educational process in all its elements, and, on the other hand, the provision of digital technologies and the means utilized in the educational process. The goal of transforming the educational process is to create a flexible and adaptive learning environment to meet the needs of all learners, including students with disabilities, and ensure the maximum beneficial use of digital technologies in pedagogical and instructional design. The aim of this educational transformation is to find the most effective way to acquire the proposed learning outcomes.

What should we expect from a digital university program?

a. Full-fledged personalization of the educational process, based on the unlimited possibilities to gather lots of data on the learning behaviors of each student and propose individual educational pathways that will ensure continuous personalized monitoring of the educational achievements.

b. Expanding opportunities for using various individual and collective forms of organizing educational activities, with time and place flexibility.

c. Providing various ways of engagement and active participation in online learning, increasing the pace of learning, streamlining online time, and maintaining sustainable learning motivation among different groups of students at all stages of the educational process, including through repetitive activities.

d. Developing a sustainable interest in the chosen type of professional activity.

e. Exploring, creating, and testing various opportunities for reducing learning barriers for vulnerable students, including students with disabilities.

f. Providing continuous and immediate feedback, quick and objective assessment of learning outcomes during online educational activities and after completing learning tasks, based on cumulative assessment technologies (rating and portfolio).

g. Ensuring the efficient acquisition of the necessary knowledge, skills, and competencies necessary for obtaining professional qualifications.

h. Freeing paperwork for faculty and other academic and non-academic staff from routine operations, the overall saving of the teacher's working time.

i. Increasing openness and transparency of educational materials, creating collaboratively OERs, developing feedback mechanisms, and developing educational networking.

j. Ensuring the availability of educational programs for people living in remote and hard-to-reach areas or students with physical impairment or other socio-economical disadvantages.

Educational goal setting in the context of digitalization should be based (i) on the maximum use of constantly emerging new opportunities that arise in connection with the use of digital technologies, (ii) on the development of readiness for continuous change (adaptability, tolerance for uncertainty), which requires a certain transformation of the usual system of values, (iii) on promoting social responsibility in the triad human – technology – society, on the understanding of an internal border between the virtual and real worlds, the development of the ability to differentiate these worlds and the corresponding types of responsibility, (iv) on the pedagogical support of the process of network socialization of the student, which is aligned with the process of his traditional socialization in the real world, and the formation of a culture of network communication, and (v) designing the concept of digital hygiene skills (Sklar, 2017), (vi) development of the ability to critically analyse information and filter information noise, advertising, custom-made information stuffing, etc.

Pedagogical strategies for distance learning for students with disabilities are based on the Universal Design for Learning principles (Cumming & Rose, 2021; Meyer et al., 2014), a framework that promotes various ways of presenting and transferring knowledge, various ways of engagement in learning activities and various ways of expressing and assessment of educational outcomes. Universal and inclusive universities will provide personalized educational pathways, digital pedagogical technologies, and various learning experiences. Efficient learning for students with disabilities can be achieved first, through the individualization of education, designing individual educational pathways emerging from common educational goals, utilizing adaptive learning technologies, creating a rich educational environment for independent work, and fostering self-education and self-development of students. Digital pedagogical technologies provide a rich number of individualized learning pathways, including content, the pace of mastering the educational material, the level of complexity, the method of presenting educational material, the form of organization of educational activities, the study group structure, organization, the number of needed exposures to educational content, the level of academic support needed, the degree of openness and transparency for other participants in the educational process, etc. Individualization in higher education is a challenge, but for students with disabilities, there should be the chance to be included and to contribute back to society. Individualization can be implemented for specific students that need a customized educational process to ensure an optimal level of learning motivation and a high rate of competence acquisition.

Building a digital educational environment and distance education process is a complex task that should be evidence-based and developed in a new pedagogical direction, digital instructional design, and methodology. The process of education digitalization is based on the organization of the learning process in a digital educational environment, by using basic concepts and principles of traditional (pre-digital) didactics as a science of learning, supplementing, and transforming them in relation to the conditions of the digital environment.

In the first stage, for educational purposes, it was decided to introduce elements of gamification in the academic discipline. For online courses, video-conferencing cloud platforms can be utilized such as Zoom, BigBlueButton, Google Meet, and many others. The technical characteristics of the video-conferencing platforms made it possible to divide the educational group of students into subgroups and separate them into separate session halls, limiting the time spent in the study groups, and where they, in turn, could work on the assignments. For example, for the Fundamentals of Special Pedagogy course, 2nd-year undergraduates were divided into groups, and they were instructed for the same task (with similar objectives) to identify barriers in learning for students with specific disabilities. Each group received an individual task regarding a specific disability (specific learning disorders, autism, ADHD, and others). The outputs of each group were presented and discussed in the plenary. Each member of the group had a specific role in accordance with the legend and went through all the stages of the task: they chose the disability, discussed the difficulties and barriers faced by a student with that disability, identified specific strategies that can reduce or eliminate those barriers in learning, and explored educational activities suitable for the selected disability. All groups' findings were presented using online platforms.

Distance learning in higher education requires independent study skills, which are challenging for students with disability, particularly for students with specific learning disorders or students with autism specific disorders. Digital technologies make it possible to plan, organize, and monitor the learning process. Moreover, it provides many educational materials, self-tutoring programs, and apps. Distance learning is enriched by OERs, various opportunities for study, self-development, socialization, and learning experiences. In the rich digital learning environment, it is vital for students with disabilities to receive support in setting appropriate educational goals and making specific decisions regarding the educational pathway. In the context of remote learning of online courses, students are required to be able to independently organize their educational activities at all stages of the educational process. The extent to which a student will benefit from the resources proposed by a distance learning study program depends on the student's own subjective activity and educational independence. Students with disabilities might have low motivation drive and reduced instrumental-activity readiness to use the potential of the digital educational environment in the learning process. In this regard, university teachers need to have a more personalized approach to students' learning. They can provide academic support and coaching, to guide students that need direction to use the resources of the digital educational environment actively and effectively. With the low educational independence of students, the creation of a digital learning experience, well-supplied with various opportunities, is important, but not enough for organizing a pedagogically effective digital educational process. Thus, an academic tutor might be an essential element, especially for students with disabilities. Moreover, it is necessary to utilize a set of motivational tools, including those provided by digital technologies, such as continued and immediate feedback, use of gamification, various interaction and networking experiences, collaborative study groups, etc.

The research work of students is a mandatory part of the educational process. Therefore, universities should invest in planning their study programs to provide the chance for students to experience and do science and research. The scientific interest of students should be revitalized. Student involvement in research activities should be an important part of the study program and should not be ignored when remotely. Electronic opportunities can be a solution to the problem. Higher education programs propose and organize several research activities for students to develop and master research skills: publication of articles, writing projects, participation in research competitions, writing scientific reports, participation in scientific conferences, grant applications, etc.

10.6.3 Course Design in the Online Learning Environment

The current situation with a large-scale transition to remote education, and the pandemic crisis is forcing us to fundamentally rethink traditional approaches in the provision of educational services. The COVID-19 pandemic made it possible to see the positive aspects of this form of education and clearly revealed some limitations. In this new context, when planning and designing educational activities for the online environment, clear criteria should be established to meet all the dimensions required for a qualitative educational process. Distance learning transforms and reorganizes the space and time of learning; hence, several aspects should be discussed in the new context. Mehlenbacher and Mehlenbacher (2019) identified several dimensions of all instructional contexts, including the context of distance learning as follows:

i. Learner(s)

ii. Instructor(s)

iii. Instructional tasks

iv. Social dynamics

v. Instructional activities

vi. Learning environment

vii. Artifacts/materiality.

Consequently, while designing courses and other educational activities for online settings, instructors should address all the above-mentioned dimensions. We will discuss below some of them that we found relevant for ensuring an inclusive university education for students with disabilities.

Firstly, in these educational settings, innovation and teaching methods are important to connect students and teachers. Efficiency and efficacy should be the main focus in the

TABLE 10.1 Teacher Role in the Distance Learning Format

Role	Function
Teacher	Knowledge transfer, facilitation, and assimilation of educational material according to the curriculum.
Co-participant	Active interaction with the student, implementation of joint projects, and research.
Coach	Support and assistance in achieving set goals, during the decision on emerging problems, practice self-study.
Monitoring	Continuous evaluation of student performance, need for adjustment.

evaluation and implementation of program studies. Students' learning is strengthened through the practice of focused learning strategies and through valuable directions to procure the information, abilities, and capabilities explicit to a specific occupation. Studies (Ha et al., 2019; Kumar et al., 2021; Sahlaoui et al., 2021) discussed the role of soft skills in software development curricula, the need to provide flexible online information, and communication technologies. An effective distance learning environment will tend to be highly goal structured, with instructors paying careful attention to student learning processes and progress throughout the course. The challenge with online courses is that learners need to be more intrinsically motivated than students in face-to-face settings where set schedules, collegiality, instructor presence, and environmental proximity all contribute as extrinsic motivations. Secondly, the role of students and teachers in distance education is changing. Teachers are not the only holders of the knowledge as information is made available through technology and students can be actively involved in the process of co-creation of the educational content. In Table 10.1 above, the roles of the teacher are presented.

Thus, the teacher's actions in relation to the student are expanding, since simply transferring the Infobase is possible without visiting an educational institution and bypassing communication with the teacher. The central role of the teacher in online settings remains to facilitate learning, design and construct the learning experience, and propose certain instructional tasks and activities to enhance the engagement of the learners. Successful online learning experiences are the result of successful teachers that act more as facilitators for their students (Keegan, 2005), academic coaches that provide guidance in navigating through the immense world of opportunities and challenges of distance learning that provide essential tools for a safe online learning environment, foster cooperation, and collaboration through discussion forums, peer coach groups, project-based learning activities, co-creation of educational content and materials, etc. Distance education does not reduce the significant impact of the teacher, but, on the contrary, strengthens it. There are a great number of studies that found high dropout rates from MOOC courses mainly as a lack of teacher connection, and delayed feedback (Alsolami, 2020; Goopio & Cheung, 2021; Onah et al., 2014).

> It is unfortunate, but true, that some academics teach students without having much formal knowledge of how students learn. Many lecturers know how they learn best, but do not necessarily consider how their students learn and if the way they teach is predicated on enabling learning to happen.
>
> *(Heather Fry et al., 2003, p.9).*

The challenge with online learning is mediating the interaction between the instructors and the students, as well as the interaction between peers, to retrieve their existing understandings, and experiences for a more efficient way to integrate the new material. It is well known that for efficient learning it is crucial to base our instruction design on anticipating students' educational needs, their prior knowledge, readiness for learning, self-regulatory and self-monitoring skills, as well as communication skills (Kris Acheson et al., 2022; Piaget, 1947). Academic success is more likely to be achieved if it is supported by motivational, self-regulatory, and cognitive processes, linked to achievement emotions (Camacho-Morles et al., 2021). Pekrun and Linnenbrink-Garcia (2014) suggested that students develop the necessary skills for academic success in different learning contexts such as attending classes, completing assignments, individual studying, or taking exams. Achievement emotions and their incentive value (positive or negative) have to be taken into account while designing training activities to avoid loss of curiosity, frustration, boredom, and lack of control over the activity because demands exceed individual capabilities or high control/low demands mostly implying no sufficient challenge (Pekrun et al., 2002; Vygotsky, 1977) and reduce the incentive value of a learning activity. Another challenge for course design in online settings is to create authentic tasks and educational activities, connected to real-world solving problems. An effective and motivational distance learning environment will highlight the relevance of the specific instructional tasks and activities, relevant connection with solving real-world and authentic issues, will foster collaboration to get insights from different perspectives, and leverage distance learning opportunities.

Undoubtedly, increasingly globalized learning audiences sustained through distance learning bring a diversity of students together along with a diversity of instructors which can enrich tremendously the learning experience if designed properly. Having students with disabilities in their lectures might pose some challenges for teachers, such as the need to better assess students' educational needs, mostly from the perspective of barriers in accessing educational content, and barriers related to students' technological skills, as some underperformers might be the result of either low technological ability, poor instructional design or some specific characteristics related to students' disability. Students with lower technological skills are more likely to report having experienced lethargy and laziness during online courses, difficulties with focus and being self-reliant when studying (Alomyan, 2021), and information processing challenges that include fatigue, burnout, academic stress, and anxiety. A first-aid helper for students with disabilities with difficulties in maintaining focus is visual aids. They benefit tremendously from concept and mind maps, diagrams, tables, infographics, videos, and any other visual learning aid. Hence, there might arise the need for university teachers to get training on how to integrate these visual powerful learning tools into their educational materials.

Online education pushes toward the reduction of the duration of courses, which is explained by the need to maximize screen exposure and the opportunity to personalize learning pathways. The modularization of the learning process will continue further: from educational (professional) modules to micromodules, which, in the context of digital transformation, become the main didactic units.

A single information space provides extensive opportunities to study a specific subject, or course, search for specific information, and get to know something new. Distance learning opportunities allow students to be interlocutors and co-authors – to comment, indicate their position, argue, and share what they know. Teachers respectively should learn to accept diverse opinions, give them enough academic supervision and coaching, self-assessment activities, and efficient communication tools that allow students to navigate through the incredible amount of information around them and be able to identify the most relevant ones. Therefore, it is expected that higher education institutions dispose of:

a. Modern highly effective pedagogical innovations, implemented in the digital environment,
b. Continuous professional development of teachers,
c. Development of applied and use of digital services,
d. Infrastructural conditions for ongoing changes.

When developing tasks, teachers have to consider using different formats. In the context of the digitalization of higher education, the role of active and interactive forms and methods of teaching is increasing. 'Passive' forms of educational work, such as lectures, are noticeably reduced, as well as homogeneous activities. More and more, interactive pedagogical technologies and methods are used, with a complex structure based on an active internal scenario and the students' own activity. Some examples can be noted, such as project activities of students, in all its variants, game learning technologies, case solving, group discussions and discussions, peer tutoring, study groups, and individual reflection, interactive communication, project based learning, etc. All these technologies allow the student to train, among other things, digital skills required in a digital society: a (i) the ability to use appropriate equipment and programs; (ii) the ability to meaningfully use digital technologies in work, study, and everyday life; (iii) the ability to critically evaluate digital technologies and motivations for inclusion in the digital environment.

To conclude, digitalization is transforming how the learning process in higher education happens: the complexity of methods and techniques of teaching should be adequate to the complexity of the teaching aids used. The variety of forms of educational activities in the online environment increases significantly, gaining a dynamic character (rapid groups shift, spatially distributed training teams, various scenarios for a rapid transition from team to individual activities). There is a risk that new technologies are simplifying our daily tasks, and there is strong contamination of simplifying the educational content. Global digitalization processes lead to the dominance of visual-figurative and visual-logical thinking. The use of predominantly infographic, visual-logical type of thinking makes it possible to provide a quick overview of the educational content, which turns out to be extremely in demand in a dynamic and uncertain world. It might favor surface learning, giving an illusion of knowledge, but in the exponential expansion of knowledge and research, it might be an adaptation process.

10.6.4 Principles of Distance Learning

Instructional design in online settings relies heavily on traditional didactic principles of education, although a number of new principles are introduced (Keegan, 2005; Sidikova, 2021; Soldatov & Soldatova, 2021):

Personalization, a valuable principle for effective online learning, implies the free choice of the student in setting educational goals, designing learning pathways, determining the pace and level of development of certain elements of the educational program, preferred technologies, forms, and methods of learning, teaching and assessment, the composition of the study group, etc. *Digital footprint* technology (Bernardo & Duarte, 2022) helps track the personal development indicators and learning outcomes of each student, recorded in the process of included assessment, and thus generate more customized learning paths. In the history of pedagogy, this principle has a long history, although digital technologies made it possible to advance in practice in building personalized mass education.

Purposefulness requires the use of only digital technologies and teaching aids that ensure the achievement of the goals of the educational process. The expediency principle prohibits the digitalization of inefficient pedagogical technologies, as well as the use of digital tools as a self-sufficient 'pedagogical panacea' (Jain et al., 2021), without a clear understanding of the educational goals that must be achieved using these tools.

Flexibility and adaptability can be achieved through the built-in system for diagnosing individual styles and strategies of learning and other psychological and pedagogical features, as well as the current psychophysiological state of students, automatically performing flexible settings for each specific student (including the order, method, and pace of presentation of the educational material, the level and nature of pedagogical support, including in the form of personalized recommendations, the number of repetitions, the level of complexity of tasks, etc.).

The principle of interactivity and active learning can be, to a certain extent, correlated with the traditional didactic principle of consciousness and activity. Its requirement is to build an educational process based on the process of active multilateral communication carried out in various forms (real, virtual network) between students, teachers, and other subjects involved in the educational process (customers and users of the results of students' project activities, external experts and consultants, employees of employers, etc.). The use of this principle implies the priority of organizing educational work in teams, based on social mechanisms of learning – communication, cooperation, competition, mutual learning, and mutual evaluation.

Practice orientation, successively associated with the traditional didactic principle of linking learning with life experience, requires setting the goals, content, technologies, methods, and techniques to the current and future requirements of the economy and of labor market trends. In the conditions of a practice-oriented educational process, the idea of the fundamental core of education is changing, there is a fundamental and practice-oriented synthesis. In the context of the digitalization of the educational process, another requirement of the principle of practice orientation is the formation of a unified digital environment for a professional educational organization and an employer (in the future, a digital ecosystem of a vocational education cluster).

The principle of multisensorial learning is significantly expanded due to technical developments. In addition, the principle of polymodality requires the involvement in the educational process not only of the visual (visual) and auditory (auditory) but also of the motor (kinaesthetic) channel of perception. In the educational process of online learning, it might involve the use of not only the keyboard and mouse but also numerous manipulators, joysticks, pedals, and other means of manual and foot control of educational and professional equipment (simulators, devices, and machines equipped with sensors and effectors, etc.). The most complete and comprehensive consideration of the principles of polymodality, interactivity, and practice orientation is ensured by the use of augmented reality tools in the educational process.

The use of digital technologies, in combination with the pedagogical technology of criterion-level assessment, ensures the objectivity and transparency of the assessment. Moreover, it creates sustainable learning motivation, thanks to the immediacy of evaluative reinforcement and targeted support. The construction of evaluation as a diagnostic-forming process requires the rejection of the repressive function of evaluation. The student has the right to make a mistake and to correct it, including through any necessary number of reinforcing repetitions.

10.7 RECOMMENDATIONS FOR HIGHER EDUCATION INSTITUTIONS TO ACCOMMODATE EDUCATIONAL NEEDS OF STUDENTS WITH DISABILITIES

Technology transformed higher education by changing not only the course design and delivery but also the way universities interact with students (Hainline et al., 2010). Challenging economic times, the increased demand for international education, and the need to prepare students for a global lifelong learning society has greatly increased pressure from society on higher education. The COVID-19 pandemic has drawn the light on the students as important actors in the educational process, mostly on the need to tackle the difficulties students are facing in these conditions. Universities become aware of the insufficient progress made toward the digitalization of education and aim to provide access to all students. Strategic planning should be discussed towards ensuring the necessary infrastructure to enhance students' access to online learning and research opportunities. At the same time, it is needed to develop teaching methods allowing students to work together and make use of learners and academic communities to enrich their educational and professional development by incorporating digital tools and environments. Many digital technologies have relevant instructional potential, such as:

i. Freedom to search for information in the global information network, access to unlimited educational opportunities for designing personal educational pathways, based on the individual educational needs and characteristics of each student, including the choice of the format and the level of complexity of the educational material and the learning and work pace, the number of reinforcing repetitions, the nature of educational support, partners, game surroundings, etc.

ii. Interactivity, the ability to provide multi-subjectivity in the process of communication and interaction.

iii. Multimedia (polymodality), the ability to comprehensively use various channels of perception (auditory, visual, and motor) in the educational process.

iv. Hypertext, freedom of movement through the text, a concise presentation of information (including in the form of infographics), the modularity of the text and the optionality of its continuous reading, the reference nature of information, the folding-expanding of information, the use of cross-references, etc.

Among the educationally significant digital technologies can be attributed: telecommunication technologies, including those that ensure the convergence of communication networks and the creation of new generation networks, technologies for processing large amounts of data (Big Data), and the 'digital footprint', AI, virtual and augmented reality, electronic identification and authentication technologies, cloud technologies, internet of things, distributed registry technologies (including blockchain), digital technologies for specialized educational purposes - edtech (educational technologies), and others. In addition, a wide range of digital manufacturing technologies is needed to build an effective learning environment. The use of digital technologies creates new opportunities for building the educational process and solving a wide range of educational tasks. SAMR model (Puentedura, n.d.) helps to describe how digital technologies can affect teaching and learning:

1. **Substitution**: Digital technologies replace traditional ones (for example, typing in Word).

2. **Augmentation**: Serve as an optimizing tool in meeting educational goals (for example, assessing learning outcomes by using Google forms, Kahoot! mobile applications; Plikers, etc.)

3. **Modification**: Significant functional changes in the instructional design (for example, the use of blended learning technologies or a flipped classroom).

4. **Redefinition**: Setting and solving new pedagogical tasks posed by the new environment and the digital society.

Another important aspect to tackle is mental health issues, studies (Bolatov et al., 2021; Fatimah & Mahmudah, 2020; Jha et al., 2021; Lischer et al., 2021; Moy & Ng, 2021) suggested that COVID-19 pandemic, pushed to remote learning, was an important factor in triggering various mental health difficulties.

Universities started to include more and more support services to facilitate the inclusion of students with disabilities in higher education, starting with promoting disability education, training, and awareness for academic and general staff (Brett, 2016).

Table 10.2 highlights the type of support necessary for students with disabilities in higher education.

TABLE 10.2 Type of Support Necessary for Students with Disabilities in Higher Education

Type of Support	Strategies for Inclusion
Legal	• Policies on student needs assessment (including documentation of disability and health conditions); • Clear educational policies
Financial	• Financial consultation • Financial aid for students' accommodation on university campuses
Academic	• Providing educational material in various formats • Accommodating curriculum • Academic coaching • Academic supervision • Peer-tutoring • Study groups for developing metacognitive skills, organizational abilities, and time-management skills • Counseling for practicing self-assessment skills • Assessment accommodations • Alternative study forms • Innovative teaching and learning methods
Cultural	• Intercultural learning opportunities and experiences • Leisure activities • Extracurricular options and integration into the university community life • Volunteer programs
Technological	• Inclusive online learning environment • Availability of assistive technology • Special equipment and facilities • Promoting distance education opportunities
Mental health	• Motivational coaching • Preventive intervention and support for mental health issues • Emotional and psychological support • Counseling groups • Independent living skills • Balanced ratio of study and recreational activities
Social	• Advisers/contact teachers or peers to help students with disabilities • Support and mentoring groups • Students' association • Social community programs
Logistic	• Flexible schedules • Accommodating course requirements • Information to students with disabilities about services available

The forecast of future learning in the digital area predicts some steps that have to be taken to avoid negative scenarios:

a. Universities should actively use AI, innovative technologies, and other digital tools to actively engage students.

b. Excessive computer time should be avoided to promote face-to-face socialization and cooperation to maintain the well-being of both teachers and students.

c. Instead of keeping the traditional form of lecturing and just transferring it to an online setting, universities should develop new formats of teaching and knowledge transfer, by using for example micro-learning modules (video clips, project-based learning, self-assessment activities, and peer-tutoring groups).

d. Extensive research is needed to make better decisions to improve higher education programs by enriching the academic experience for all students, regardless of disability, including ethical standards.

e. Infrastructure should be deployed on campus so that all students can participate in virtual collaboration formats and receive support at the national/regional and university levels. Sufficient and sustainable public investment in university infrastructure is needed to support digital and green campus development.

f. Greater institutional autonomy, such as the ability to more flexibly allocate internal funding or acquire/own/sell property, is key to successful campus transformation.

Students should be able to get an initial virtual acquaintance with the educational environment, especially students with disabilities who report having difficulties with the transitional process to university life (Dymond et al., 2017; van Hees et al., 2015) or to employment (Nolan & Gleeson, 2017; Sánchez & Justicia, 2016). Students benefit from opportunities to be acquainted visually with the Educational Portal, to receive information about admission, to learn about distance learning areas, to read answers to frequently asked questions, to be able to contact faculty and other university staff and ask questions, etc.

10.8 CONCLUSION AND FUTURE SCOPE

Various factors are changing the educational landscape. Universities should promote and do learning, and learn themselves, become able to respond and adapt to the changing environment, shape the future. For higher education, the global medical crisis was not only a trial period but also the beginning of a new chapter. In response to the COVID-19 pandemic, higher education institutions innovated by embracing online teaching and (re) questioning effective ways of teaching and knowledge transfer. Obviously, even before the pandemic, distance education was taken into consideration, while blended learning was regarded as the most promising in view of the impending digital transformation. The sudden transition to distance learning showed the readiness and preparedness of higher education institutions for digitalization. Following COVID-19, extensive research was done to improve the process of remote education and make learning accessible to all students. Since the concept of distance learning was introduced, methods for delivering instruction have rapidly changed as the development and use of technological tools have grown intensively. The landscapes for distance learning are changing in ways that are difficult to predict with consequences that are even harder to anticipate. A question arises from the extension of the digital education phenomenon regarding how much autonomy students should be given by universities and should be discussed from the perspective of the educational system as a whole, considering the need to encourage students to acknowledge and assume responsibility for their own learning. Effective learning is ensured in the case of taking responsibility for one's own learning process. Nonetheless, there is room for discussion between the idealistic approach of transferring complete control over one's own study and the practical approach that militates for a 'certain amount of structure designed to maintain the impetus of their studies and to motivate them toward completing their assignments within a set time' (Wheeler, 2012, p. 1019). Wheeler (2012) is raising a reasonable question that higher education and other distance learners' providers should answer, to establish guidelines and a framework based on the optimal balance of freedom and control, to ensure certain quality standards, and at the same time not inhibit creativity and innovation.

REFERENCES

Aagaard, T., & Lund, A. (2019). Digital agency in higher education: Transforming teaching and learning. *Digital Agency in Higher Education*, London: Routledge, pp. 120–132.

Acheson, K., Dirkx, J. M., & Shealy, C. N. (2022). High impact learning in higher education Operationalizing the self-constructive outcomes of transformative learning theory. In Effrosyni Kostara, Andreas Gavrielatos, & Daphne Loads (Eds.), *Transformative Learning Theory and Praxis New Perspectives and Possibilities.* New York: Routledge.

Alomyan, H. (2021). The impact of distance learning on the psychology and learning of university students during the covid-19 pandemic. *International Journal of Instruction, 14*(4). https://doi.org/10.29333/iji.2021.14434a.

Alsolami, F. J. (2020). A hybrid approach for dropout prediction of MOOC students using machine learning. *International Journal of Computer Science and Network Security, 20*(5), 54–63.

Astin, A. W. (2005). How college affects students. Volume 2: A third decade of research (review). *The Review of Higher Education, 29*(1). https://doi.org/10.1353/rhe.2005.0057.

Azmuddin, R. A., Nor, N. F. M., & Hamat, A. (2017). Metacognitive online reading and navigational strategies by science and technology university students. *GEMA Online Journal of Language Studies, 17*(3). https://doi.org/10.17576/gema-2017-1703-02.

Badali, M., Hatami, J., Banihashem, S. K., Rahimi, E., Noroozi, O., & Eslami, Z. (2022). The role of motivation in MOOCs' retention rates: A systematic literature review. *Research and Practice in Technology Enhanced Learning, 17*(1), 5. https://doi.org/10.1186/s41039-022-00181-3.

Baten, E., Vlaeminck, F., Mués, M., Valcke, M., Desoete, A., & Warreyn, P. (2022). The impact of school strategies and the home environment on home learning experiences during the covid-19 pandemic in children with and without developmental disorders. *Journal of Autism and Developmental Disorders.* https://doi.org/10.1007/s10803-021-05383-0.

Bernard, R. M., Abrami, P. C., Lou, Y., & Borokhovski, E. (2004). A methodological morass? How we can improve quantitative research in distance education. *Distance Education, 25*(2). https://doi.org/10.1080/0158791042000262094.

Bernardo, N., & Duarte, E. (2022). Immersive virtual reality in an industrial design education context: What the future looks like according to its educators. *Computer-Aided Design and Applications, 19*(2). https://doi.org/10.14733/CADAPS.2022.238-255.

Binali, T., Tsai, C. C., & Chang, H. Y. (2021). University students' profiles of online learning and their relation to online metacognitive regulation and internet-specific epistemic justification. *Computers and Education, 175*. https://doi.org/10.1016/j.compedu.2021.104315.

Bolatov, A. K., Seisembekov, T. Z., Askarova, A. Z., Baikanova, R. K., Smailova, D. S., & Fabbro, E. (2021). Online-learning due to COVID-19 improved mental health among medical students. *Medical Science Educator, 31*(1). https://doi.org/10.1007/s40670-020-01165-y.

Boshuizen, H. P. A., Bromme, R., & Gruber, H. (Eds.) (2004). *Professional Learning: Gaps and Transitions on the Way from Novice to Expert.* Dordrecht: Kluwer Academic.

Brett, M. (2016). Disability and australian higher education: Policy drivers for increasing participation. In A. Harvey, C. Burnheim, & M. Brett (Eds.), *Student Equity in Australian Higher Education* (pp. 87–108). Singapore: Springer. https://doi.org/10.1007/978-981-10-0315-8_6.

Buzzai, C., Filippello, P., Costa, S., Amato, V., & Sorrenti, L. (2021). Problematic internet use and academic achievement: A focus on interpersonal behaviours and academic engagement. *Social Psychology of Education*, *24*(1). https://doi.org/10.1007/s11218-020-09601-y.

Bygstad, B., Øvrelid, E., Ludvigsen, S., & Dæhlen, M. (2022). From dual digitalization to digital learning space: Exploring the digital transformation of higher education. *Computers and Education*, *182*. https://doi.org/10.1016/j.compedu.2022.104463.

Caliskan, H. (2012). Open and distance learning. In N. M. Seel (Ed.), *Encyclopedia of the Sciences of Learning* (pp. 2513–2513). Boston, US: Springer. https://doi.org/10.1007/978-1-4419-1428-6_5156.

Camacho-Morles, J., Slemp, G. R., Pekrun, R., Loderer, K., Hou, H., & Oades, L. G. (2021). Activity achievement emotions and academic performance: A meta-analysis. *Educational Psychology Review*, *33*(3). https://doi.org/10.1007/s10648-020-09585-3.

Chadwick, D., & Wesson, C. (2016). Digital inclusion and disability. In A. Attrill & C. Fullwood (Eds.), *Applied Cyberpsychology* (pp. 1–23). London: Palgrave Macmillan. https://doi.org/10.1057/9781137517036_1.

Chickering, A. W., & Gamson, Z. F. (1989). Seven principles for good practice in undergraduate education. *Biochemical Education*, *17*(3). https://doi.org/10.1016/0307-4412(89)90094-0.

Chirikov, I., Semenova, T., Maloshonok, N., Bettinger, E., & Kizilcec, R. F. (2020). Online education platforms scale college STEM instruction with equivalent learning outcomes at lower cost. *Science Advances*, *6*(15). https://doi.org/10.1126/sciadv.aay5324.

Chu, M. (2014). Preparing tomorrow's early childhood educators: Observe and reflect about culturally responsive teachers. *YC Young Children*, *69*(2), 82–87.

Cinquin, P. A., Guitton, P., & Sauzéon, H. (2019). Online e-learning and cognitive disabilities: A systematic review. *Computers and Education*, *130*. https://doi.org/10.1016/j.compedu.2018.12.004.

Clouder, L., Karakus, M., Cinotti, A., Ferreyra, M. V., Fierros, G. A., & Rojo, P. (2020). Neurodiversity in higher education: A narrative synthesis. In *Higher Education* (Vol. 80, Issue 4). https://doi.org/10.1007/s10734-020-00513-6.

Coman, C., Ţîru, L. G., Meseşan-Schmitz, L., Stanciu, C., & Bularca, M. C. (2020). Online teaching and learning in higher education during the coronavirus pandemic: Students' perspective. *Sustainability (Switzerland)*, *12*(24), 1–22. https://doi.org/10.3390/su122410367.

Creemers, B., Kyriakides, L., & Antoniou, P. (2013). Going beyond the classical dichotomy related to the content of teacher training and professional development. In *Teacher Professional Development for Improving Quality of Teaching* (pp. 43–61). Dordrecht, Netherlands: Springer. https://doi.org/10.1007/978-94-007-5207-8_4.

Cullen, J. (2020). *Prosumerism in Higher Education—Does It Meet the Disability Test?* (D. Burgos, Ed.; pp. 105–121). Singapore: Springer. https://doi.org/10.1007/978-981-15-4276-3_7.

Cumming, T. M., & Rose, M. C. (2021). Exploring universal design for learning as an accessibility tool in higher education: A review of the current literature. *Australian Educational Researcher*. https://doi.org/10.1007/s13384-021-00471-7.

Curum, B., & Khedo, K. K. (2021). Cognitive load management in mobile learning systems: Principles and theories. *Journal of Computers in Education*, *8*(1). https://doi.org/10.1007/s40692-020-00173-6.

Davis, E. A. (1984). Why some groups fail: A survey of students' experiences with learning groups. *Journal of Management Education*, *9*(4). https://doi.org/10.1177/105256298400900409.

de Byl, P., & Hooper, J. (2013). Key attributes of engagement in a gamified learning environment. *30th Annual Conference on Australian Society for Computers in Learning in Tertiary Education, ASCILITE 2013, Sydney, Australia*.

del Bianco, N., & Grigg Mason, L. (2021). Specific learning disorders in higher education: The University of Arizona case study. *Education Sciences and Society*, *1*, 219–229. https://doi.org/10.3280/ess1-2021oa11930.

Dumitru Tabacaru, C. (2019). The situation of dyslexic children in romanian inclusive schools. *The European Proceedings of Social & Behavioural Sciences*, 948–954. https://doi.org/10.15405/epsbs.2019.04.02.117.

Dumitru Tăbăcaru, C., Dumitru G. & Diaconu, M. B. (2022). Student's voice – What do students appreciate and expect to receive from a master program, a cross-sectional analysis, *Cogent Education*, *9*(1), https://doi.org/10.1080/2331186X.2022.2107298.

Dymond, S. K., Meadan, H., & Pickens, J. L. (2017). Postsecondary education and students with autism spectrum disorders: Experiences of parents and university personnel. *Journal of Developmental and Physical Disabilities*, *29*(5), 809–825. https://doi.org/10.1007/s10882-017-9558-9.

Education Forum. (2000). *The Dakar Framework for Action. UNESCO*, April.

Edwards, A. D. (2014). *Serving the Needs of Students on the Autism Spectrum in Higher Education: A Study of Leadership and Support Services*. Doctoral dissertation. Drexel University, USA.

Ericsson, K. A. (2004). Deliberate practice and the acquisition and maintenance of expert performance in medicine and related domains. *Academic Medicine*, *79*(10 SUPPL.). https://doi.org/10.1097/00001888-200410001-00022.

Fatimah, S., & Mahmudah, U. (2020). How e-learning affects students' mental health during covid-19 pandemic: An empirical study. *DWIJA CENDEKIA: Jurnal Riset Pedagogik*, *4*(1). https://doi.org/10.20961/jdc.v4i1.41991.

Fichten, C., Olenik-Shemesh, D., Asuncion, J., Jorgensen, M., & Colwell, C. (2020). Higher education, information and communication technologies and students with disabilities: An overview of the current situation. In J. Seale (Ed.), *Improving Accessible Digital Practices in Higher Education* (pp. 21–44). Cham: Springer International Publishing. https://doi.org/10.1007/978-3-030-37125-8_2.

Folostina, R. & Dumitru, C. (2021). Digital communication in the inclusive classroom. In V. Trif, (Ed.). *Basic Communication and Assessment Prerequisites for the New Normal of Education*. Hershey: IG Global Publishing. https://doi.org/10.4018/978-1-7998-8247-3.ch014.

Fry, H., Ketteridge, S., & Marshall, S. (2003). *A Handbook for Teaching & Learning in Higher Education (Second)*. London: Kogan Page.

Fuller, M., Healey, M., Bradley, A., & Hall, T. (2004). Barriers to learning: A systematic study of the experience of disabled students in one university. *Studies in Higher Education*, *29*(3). https://doi.org/10.1080/0307507041000l682592.

Garrett, J., Alman, M., Gardner, S., & Born, C. (2007). Assessing students' metacognitive skills. *American Journal of Pharmaceutical Education*, *71*(1). https://doi.org/10.5688/aj710114.

Gleason, B., & Jaramillo Cherrez, N. (2021). Design thinking approach to global collaboration and empowered learning: Virtual exchange as innovation in a teacher education course. *TechTrends*, *65*(3), 348–358. https://doi.org/10.1007/s11528-020-00573-6.

Goopio, J., & Cheung, C. (2021). The MOOC dropout phenomenon and retention strategies. *Journal of Teaching in Travel and Tourism*, *21*(2). https://doi.org/10.1080/15313220.2020.1809050.

Ha, N., Nayyar, A., Nguyen, D., & Liu, C. (2019). Enhancing students' soft skills by implementing CDIO based integration teaching mode. *Proceedings of the 15th International CDIO Conference, Aarhus, Denmark*.

Hainline, L., Gaines, M., Long Feather, C., Padilla, E., & Terry, E. (2010). Changing students, faculty, and institutions in the twenty-first century | Association of American Colleges & Universities. *Peer Review*, *12*(3), 7–10.

Haughey M. (2013). *Distance learning. The Canadian Encyclopedia*. Edmonton: Hurtig Publishing.

Jain, S., Lall, M., & Singh, A. (2021). Teachers' voices on the impact of covid-19 on school education: Are ed-tech companies really the panacea? *Contemporary Education Dialogue*, *18*(1). https://doi.org/10.1177/0973184920976433.

Jha, I. P., Awasthi, R., Kumar, A., Kumar, V., & Sethi, T. (2021). Learning the mental health impact of COVID-19 in the United States with explainable artificial intelligence: Observational study. *JMIR Mental Health*, *8*(4). https://doi.org/10.2196/25097.

Keegan, D. (2005). *Theoretical Principles of Distance Education*. London: Routledge. https://doi.org/10.4324/9780203983065.

Kumar, A., Krishnamurthi, R., Bhatia, S., Kaushik, K., Ahuja, N. J., Nayyar, A., & Masud, M. (2021). Blended learning tools and practices: A comprehensive analysis. *IEEE Access*, *9*. https://doi.org/10.1109/ACCESS.2021.3085844.

Lee, Y. M., Jahnke, I., & Austin, L. (2021). Mobile microlearning design and effects on learning efficacy and learner experience. *Educational Technology Research and Development*, *69*(2). https://doi.org/10.1007/s11423-020-09931-w.

Lischer, S., Safi, N., & Dickson, C. (2021). Remote learning and students' mental health during the Covid-19 pandemic: A mixed-method enquiry. *Prospects*. https://doi.org/10.1007/s11125-020-09530-w.

Liu, T., & Zheng, H. (2021). A study of digital interactive technology and design mode promoting the learners' metacognitive experience in smart education. *International Journal of Information and Education Technology*, *11*(10). https://doi.org/10.18178/ijiet.2021.11.10.1555.

Loes, C. N., Culver, K. C., & Trolian, T. L. (2018). How collaborative learning enhances students' openness to diversity. *Journal of Higher Education*, *89*(6). https://doi.org/10.1080/00221546.2018.1442638.

Longobardi, C., Fabris, M. A., Mendola, M., & Prino, L. E. (2019). Examining the selection of university courses in young adults with learning disabilities. *Dyslexia*, *25*(2). https://doi.org/10.1002/dys.1611.

Lu, C., & Cutumisu, M. (2022). Online engagement and performance on formative assessments mediate the relationship between attendance and course performance. *International Journal of Educational Technology in Higher Education*, *19*(1), 2. https://doi.org/10.1186/s41239-021-00307-5.

Macken, C., Hare, J., & Souter, K. (2021). *Seven Suggestions for the Future of Higher Education*. https://doi.org/10.1007/978-981-16-4428-3_8.

Mahlangu, V. P. (2018). The good, the bad, and the ugly of distance learning in higher education. In *Trends in E-Learning*. https://doi.org/10.5772/intechopen.75702.

Mayhew, M. J., Rockenbach, A. N., Bowman, N. A., Seifert, T. A. D., Wolniak, G. C., Pascarella, E. T., & Terenzini, P. T. (2016). How college affects students: 21st century evidence that higher education works. In *How College Affects Students*. San Francisco: Jossey-Bass.

Mehlenbacher, B., & Mehlenbacher, A. R. (2019). Distance learning. In A. Tatnall (Ed.), *Encyclopedia of Education and Information Technologies* (pp. 1–12). Cham: Springer International Publishing. https://doi.org/10.1007/978-3-319-60013-0_66-1.

Meyer, A., Rose, D. H., & Gordon, D. (2014). *CAST: Universal Design for Learning: Theory & Practice*. Wakefield: CAST Professional Publishing.

Moy, F. M., & Ng, Y. H. (2021). Perception towards e-learning and COVID-19 on the mental health status of university students in Malaysia. *Science Progress*, *104*(3). https://doi.org/10.1177/00368504211029812.

Niazov, Z., Hen, M., & Ferrari, J. R. (2021). Online and academic procrastination in students with learning disabilities: The impact of academic stress and self-efficacy. *Psychological Reports*. https://doi.org/10.1177/0033294120988113.

Nolan, C., & Gleeson, C. I. (2017). The transition to employment: The perspectives of students and graduates with disabilities. *Scandinavian Journal of Disability Research*, *19*(3). https://doi.org/10.1080/15017419.2016.1240102.

Nugent, A., Lodge, J., Carroll, A., Bagraith, R., MacMahon, S., Matthews, K. E., & Sah, P. (2020). Higher education learning framework: Evidence informed model for university learning. In *Higher Education Learning Framework: Evidence Informed Model for University Learning*. Brisbane: The University of Queensland. https://doi.org/10.14264/348c85f.

O'Byrne, L., Gavin, B., Adamis, D., Lim, Y. X., & McNicholas, F. (2021). Levels of stress in medical students due to covid-19. *Journal of Medical Ethics*, *47*(6). https://doi.org/10.1136/medethics-2020-107155.

OECD. (2020). Digital economy 2020. In *Organization for Economic Cooperation and Development*. Paris: OECD Publishing.

Onah, D. F. O., Sinclair, J., & Boyatt. (2014). Dropout rates of massive open online courses: Behavioural patterns MOOC dropout and completion: Existing evaluations. *Proceedings of the 6th International Conference on Education and New Learning Technologies (EDULEARN14), Barcelona, Spain*.

Pascarella, E. T., Salisbury, M. H., & Blaich, C. (2011). Exposure to effective instruction and college student persistence: A multi-institutional replication and extension. *Journal of College Student Development, 52*(1), 4–19. https://doi.org/10.1353/csd.2011.0005.

Pekrun, R., Goetz, T., Titz, W., & Perry, R. P. (2002). Academic emotions in students' self-regulated learning and achievement: A program of qualitative and quantitative research. *Educational Psychologist* (Vol. 37, Issue 2). https://doi.org/10.1207/S15326985EP3702_4.

Pekrun, R., & Linnenbrink-Garcia, L. (2014). Introduction to emotions in education. In *International Handbook of Emotions in Education*. New York: Routledge. https://doi.org/10.4324/9780203148211.

Peng, D. (2019). Mobile-based teacher professional training: Influence factor of technology acceptance. In Maiga Chang, Elvira Popescu, Kinshuk, Nian-Shing Chen, Mohamed Jemni, Ronghuai Huang, J. Michael Spector, & Demetrios G. Sampson (Eds.), *Foundations and Trends in Smart Learning. Lecture Notes in Educational Technology*, pp. 161–170. Singapore: Springer. https://doi.org/10.1007/978-981-13-6908-7_23.

Pérez-López, M. C., & Ibarrondo-Dávila, M. P. (2020). Key variables for academic performance in university accounting studies. A mediation model. *Innovations in Education and Teaching International, 57*(3). https://doi.org/10.1080/14703297.2019.1620624.

Peters, M. A. (2017). Open and distance learning. In M. A. Peters (Ed.), *Encyclopedia of Educational Philosophy and Theory* (pp. 1682–1682). Singapore: Springer. https://doi.org/10.1007/978-981-287-588-4_100711.

Piaget, J. (1947). The psychology of intelligence. In *The Psychology of Intelligence*. Totowa, NJ: Littlefield, Adams. https://doi.org/10.4324/9780203278895.

Poellhuber, B., & Roy, N. (2019). Understanding participant's behaviour in massively open online courses. *International Review of Research in Open and Distance Learning, 20*(1). https://doi.org/10.19173/irrodl.v20i1.3709.

Prokes, C., & Housel, J. (2021). Community college student perceptions of remote learning shifts due to covid-19. *TechTrends, 65*(4). https://doi.org/10.1007/s11528-021-00587-8.

Pu, R., Tanamee, D., & Jiang, S. (2022). Digitalization and higher education for sustainable development in the context of the Covid-19 pandemic: A content analysis approach. *Problems and Perspectives in Management, 20*(1). https://doi.org/10.21511/ppm.20(1).2022.03.

Puentedura, R. R. (n.d.). *The SAMR Model: Background and Exemplars*. http://www.hippasus.com/rrpweblog/archives/2012/08/23/samr_backgroundexemplars.pdf.

Raes, A., Vanneste, P., Pieters, M., Windey, I., van den Noortgate, W., & Depaepe, F. (2020). Learning and instruction in the hybrid virtual classroom: An investigation of students' engagement and the effect of quizzes. *Computers and Education, 143*. https://doi.org/10.1016/j.compedu.2019.103682.

Sahlaoui, H., Alaoui, E. A. A., Nayyar, A., Agoujil, S., & Jaber, M. M. (2021). Predicting and interpreting student performance using ensemble models and shapley additive explanations. *IEEE Access, 9*. https://doi.org/10.1109/ACCESS.2021.3124270.

Sánchez, M. T. P., & Justicia, M. D. L. (2016). Transition to employment of university students with disabilities: Factors for success. *Universitas Psychologica, 15*(2). https://doi.org/10.11144/Javeriana.upsy15-2.teus.

Santos, C. G., Alba, E. F., Cerezo, R., & Núñez, J. C. (2018). Learning disabilities in higher education: A challenge for the university. *Publicaciones de La Facultad de Educacion y Humanidades Del Campus de Melilla, 48*(1). https://doi.org/10.30827/publicaciones.v48i1.7328.

Schwartz, A. E., Hopkins, B. G., & Stiefel, L. (2021). The effects of special education on the academic performance of students with learning disabilities. *Journal of Policy Analysis and Management, 40*(2), 480–520. https://doi.org/10.1002/pam.22282.

Seifert, T. A., Bowman, N. A., Wolniak, G. C., Rockenbach, A. N., & Mayhew, M. J. (2017). Ten challenges and recommendations for advancing research on the effects of college on students. *AERA Open, 3*(2), 233285841770168. https://doi.org/10.1177/2332858417701683.

Sharma, M., Onta, M., Shrestha, S., Raj Sharma, M., & Bhattarai, T. (2021). The pedagogical shift during covid-19 pandemic: Emergency remote learning practices in nursing and its effectiveness. *Asian Journal of Distance Education, 16*(1), 98–110.

Sheikh Ibrahim, M., Salleh, N., & Misra, S. (2015). Empirical studies of cloud computing in education: A systematic literature review. *Lecture Notes in Computer Science (Including Subseries Lecture Notes in Artificial Intelligence and Lecture Notes in Bioinformatics), 9158*. https://doi.org/10.1007/978-3-319-21410-8_55.

Sidikova, S. D. (2021). Theoretical and didactic principles of distance learning. *Academicia: An International Multidisciplinary Research Journal, 11*(1), 541–548. https://doi.org/10.5958/2249-7137.2021.00084.7.

Sinclair, J., & Kalvala, S. (2016). Student engagement in massive open online courses. *International Journal of Learning Technology, 11*(3). https://doi.org/10.1504/IJLT.2016.079035.

Sklar, A. (2017). Sound, smart, and safe: A plea for teaching good digital hygiene. *LEARNing Landscapes, 10*(2). https://doi.org/10.36510/learnland.v10i2.799.

Soldatov, B., & Soldatova, N. (2021). Distance education didactic principles application in teaching. *E3S Web of Conferences, 273*, 12031. https://doi.org/10.1051/e3sconf/202127312031.

The Americans with Disabilities Act of 1990 and Amendments Act of 2008. (2009). *Laws/Regulations*. U.S. Government Publishing Office. https://www.govinfo.gov/app/details/COMPS-803/

Thormann, J., & Kaftal Zimmerman, I. (2012). *The Complete Step-by-Step Guide to Designing and Teaching Online Courses*. New York: Teachers College.

Todri, A., Papajorgji, P., Moskowitz, H., & Scalera, F. (2021). Perceptions regarding distance learning in higher education, smoothing the transition. *Contemporary Educational Technology, 13*(1). https://doi.org/10.30935/cedtech/9274.

Tripathi, M., & Jauhari, S. (2020). Role of EDUTECH in digitalisation of higher education. *BSSS Journal of Management*. https://doi.org/10.51767/jm1109.

UNESCO. (1994). The Salamanca statement and framework for action on special needs education. In *The Salamanca Statement and Framework for Action on Special needs Education* (Issue June). Published at World Conference on Special Needs Education: Access and Quality, Salamanca, Spain, 1994, by España. Ministerio de Educación y Ciencia, pp. 1–47. https://unesdoc.unesco.org/ark:/48223/pf0000098427

UNESCO. (2020). *Open and Distance Learning: Trends, Policy and Strategy Considerations*. E. Khvilon & M. Patru (Eds). Paris: UNESCO.

United Nations. (2020). Convention on the rights of persons with disabilities (CRPD) | United Nations enable. In *Department of Economic and Social Affairs Division for Inclusive Social Development*. Article 24, https://www.un.org/development/desa/disabilities/convention-on-the-rights-of-persons-with-disabilities/convention-on-the-rights-of-persons-with-disabilities-2.html

van Hees, V., Moyson, T., & Roeyers, H. (2015). Higher education experiences of students with autism spectrum disorder: Challenges, benefits and support needs. *Journal of Autism and Developmental Disorders, 45*(6), 1673–1688. https://doi.org/10.1007/s10803-014-2324-2.

van't Land, H., Corcoran, A., & Iancu, D. C. Eds. 2021. *The Promise of Higher Education*. Cham: Springer. https://doi.org/10.1007/978-3-030-67245-4.

Vygotsky, L. S. (1930). *Mind and Society*. Cambridge, MA: Harvard University Press.

Vygotsky, L. S. (1977). The development of higher psychological functions. *Soviet Psychology, 15*(3). https://doi.org/10.2753/rpo1061-0405150360.

Wang, F., & Hannafin, M. (2005). Design-based research and technology-enhanced learning environments. *Educational Technology Research and Development, 53*(4), 5–23.

Wheeler, S. (2012). Distance learning. In N. M. Seel (Ed.), *Encyclopedia of the Sciences of Learning* (pp. 1018–1020). Boston, US: Springer. https://doi.org/10.1007/978-1-4419-1428-6_432.

Zingoni, A., Taborri, J., Panetti, V., Bonechi, S., Aparicio-Martínez, P., Pinzi, S., & Calabrò, G. (2021). Investigating issues and needs of dyslexic students at university: Proof of concept of an artificial intelligence and virtual reality-based supporting platform and preliminary results. *Applied Sciences (Switzerland), 11*(10). https://doi.org/10.3390/app11104624.

CHAPTER 11

E-Learning

A Catalyst to Promote the Digital Citizenship

Pratiksha Dixit
Springfields

Usha Pathak
DAV(PG) College

CONTENTS

DOI: 10.1201/9781003322252-11

11.1 INTRODUCTION

In the fast-changing world, every individual concerned with the academic world, students, teachers, parents, and administrators must change their approach to modern technology and adapt themselves in a better way to the paradigm shift from the traditional to a digital phase of education. Technology is also changing its functional aspect very fast and, therefore, is becoming relevant in today's times adapting to this change is becoming very efficient (Yengin, Karahoca, Karahoca, & Ozcinar, 2010). The chapter aims to investigate how the e-learning method is proving beneficial to the overall education sector and what are the emerging trends and methodologies leading to preparing learners for digital citizenship. The chapter highlights the thoughts and solutions about e-learning to describe its importance and strategies for its implementation. Electronic learning, or e-learning, is a mode of education that comprises contemporary methods of communication including the computer and its networks or any other electronic media inclusive of multimedia, compact discs, satellites, or other new education technologies, audio–visual resources, and interactive programs, search engines, e-libraries, and internet site, whether accomplished in the classroom or at remote places. It is an objective-driven teaching–learning in which the two users at both ends interact with each other through these resources and achieve their specific objectives. As cited by (Sujit Kumar Basak et al., 2018) the fundamental perspectives of e-Learning are the cognitive, emotional, behavioral, and contextual perspectives. E-learning is considered to be a flexible mode of learning since it gives freedom to the learner to decide when they want to gain knowledge and also makes them feel like their own valuable time is in their control (Yengin, Karahoca, Karahoca, & Ozcinar, 2010). The new time calls for a very different approach to e-learning like many other changes that were also kicked in due to the pandemic (Shukla, 2020). COVID-19 pandemic brought in a huge impact on the current state of India, as an example, which has affected India's nascent stage of the e-learning platform and has, therefore, impacted India highly than other humanitarian man-made tragedies (Shukla, 2020). The impact of COVID-19 was not only seen in the education sector, but it impacted all sectors around the world. Because of this, the educational sector was the most affected one all around the world (Shukla, 2020). India felt the blow of the pandemic in a very difficult way, and it is still fighting to survive since as a developing nation, resources have always been low and still under a crunch (Shukla, 2020).

The awakening and adoption of online learning that increased during the pandemic, all happened for good reasons. E-learning and online methodologies made knowledge and information accessible to all (Bureau, 2022). Alongside accessibility and approachability, better learning opportunities came through at these low times and created a space for learning that has never existed before (Bureau, 2022).

11.1.1 E-Learning

In today's 21st century, one of the most crucial roles' is played by technology itself where all kinds of professionals, educators, and learners are reflecting upon knowledge-based out of technology and e-learning (Kumar Basak, Wotto & Bélanger, 2018). E-learning is a pioneering training approach in an environment that is interactive and flexible resulting in maximum learning. Teachers use it in the form of one-on-one learning, and group learning, content-based learning, video-based learning, etc. In developing countries like India, the imminent scope of e-learning technology seems optimistic and affluent because technology users in every sector of society have started using the combination of advanced resolutions, techno-based skills, and traditional methods of working which offers numerous prospects and profits. As we move more toward coursework that has an online format and web-based learning, we are agreeing to settle for a framework of web-based interaction (Northrup, 2022). The e-learning approach to teaching–learning is economic in terms of being time and cost-effective. Learners get an opportunity to have the best learning experiences in the universities of their choice and dream and get better acumen and info. The futuristic space is indicating a positive change with entering into an era when traditional education will have to play a very diminutive role in learning. It has been stirred further in the form of M-learning (mobile learning) and D-learning (digital learning). The basic perspective of M-learning (mobile learning) which is an extension of e-learning, is in the form of mobility of the three aspects, namely, technology, learning, and the learner, whereas the perspectives of D-learning (digital learning) are technology, digital content, and instruction. E-learning consists of more text and graphics-based instructions. D-learning uses technology to connect, collaborate, curate, and create locally and globally by using digital content. Where e-learning is understood to be an alternative to conventional traditional learning through technology and the internet, M-learning at the same time allows all learners to inculcate some traditional aspects as well like downloading e-books and PDFs for future reference or printing (Kumar Basak, Wotto, & Bélanger, 2018).

11.1.2 Evolution of E-Learning

If we know where is the change happening and how to adapt ourselves to grab the change, it will make us ready to use e-learning to the fullest. There are many factors responsible for the paradigm shift in technology, e.g., demographic changes, changing industry expectations, and the new era of generation Z, who is looking forward to transform their lives completely by using the power of knowledge. For this purpose, although classroom learning is impactful it is not the only option. Now, internet and technology-based learning programs are comparatively more effective, which help teachers and knowledge leaders to train their students in several creative and innovative ways, i.e., innovative use of technology to satisfy the needs of all. Now the question arises why we have started thinking in the line of technology now. The first computerized training program began in the 1960s which led to the birth of the user graphic interface (UGI) and the mouse in the 1970s when the use of blackboard and chalk was prevalent in the classrooms. Digitalization in the 19th century has digitalized all resource management that has also modernized all HR functions too and provided a competitive advantage to all (Mazurchenko & Maršíková, 2019). The era of personal computers started in the 1980s when whiteboards, markers, slides, and overhead projectors were in the limelight. Owing to the vast changes in the education sector, the first digital natives were born in the 1990s, and were comfortable with the PowerPoint presentation computer-assisted instruction (CAI), web-based teaching–learning tools, etc. With the exposing growth of the IT economy, a greater number of IT-based solutions were provided to the students by the company in the form of e-learning courses in the 2000s. Society adopted these advanced and innovative new environments in the form of e-classrooms and video-conferencing quickly and consciously by the 2010s. Productivity of technological knowledge when got challenged by unprecedented natural, societal, and economic forces, gave a new turn to the paradigm shift for e-learning in the 2020s when people started using innovative, on-demand, and practical approaches and pedagogies in the form of collaborative learning, blended learning, etc., intending to reach to the unreached and it has now become an indispensable part of our life.

Reflection on e-learning during the COVID-19 times has created a different response toward all learners and teachers. The pandemic created a rushed gateway for education from primary, secondary, and higher classes' digitalization which increased bullying and misinformation that requires global efforts to curb the digital age and gap (Jackman, Gentile, Cho & Park, 2021). The World Bank Group (WBG) is known as the largest financer of all educational courses in any developing world has been working on educational programs in more than 80 countries to enlarge the learning opportunities for all (Edtech Team, 2022). The World Bank also estimated levels of poverty in the field of learning and education by measuring 10-year-old children's level of understanding of reading and learning skills (World Bank Group, 2022). After COVID-19 became widespread and was declared a pandemic, more than 180+ countries mandated the closure of schools, affecting billions of learners and millions of educators which eventually brought forward a hybrid system of learning and teaching through e-learning (2022). From statistical data gathered, as of April 20200, around 5 billion internet users have been recorded worldwide and this number is approximately 63% of the total global population (Johnson, 2022). It was also highlighted that approximately, 4.65 billion users from this pool were social media users (Johnson, 2022). From a pool of data, it was suggested that among the top users of the internet were China, India, and the United States. From a February 2020 report, China was the biggest user with more than a billion users (Johnson, 2022). The debate on distance learning and e-learning has come a long way from traditional beliefs, but both together keep students in contact with their educators and both create a balance between their advantages and disadvantages, creating a huge space for learners with learning disabilities (Petretto et al., 2021). Studies based on opinions during pandemic emergencies in remote regions have been studied as well, where students have reported that due to a lack of online infrastructure and the number of educators not enough for the number of learners to contribute to the failure of class conversations and studies being affected hugely (Wu, 2021).

11.1.3 Objectives of E-Learning

It has come to light that classroom training is the most effective and traditional way of teaching and learning across the world, but it is not the only option now during the onset of the pandemic and therefore, it is much to classify the objectives of an unconventional method of e-learning (Bilal, 2022). Unless the objectives are not defined, we cannot find the track to finish our tasks at hand. E-learning helps us to achieve the following objectives:

a. To enhance the quality of teaching–learning,

b. Meet the learning style or needs of students,

c. Improve the efficiency and effectiveness of e-learning,

d. Improve user accessibility and time flexibility to engage learners in the learning process,

e. To determine the program outcome,

f. To organize the content in a logical sequence,

g. To evaluate learners' performances in context with achieving program objectives,

h. To analyze the characteristics of the learners,

i. And, to provide continuous feedback for the improvement of learners.

11.1.4 Categories of E-Learning

E-learning is being used nowadays for designing and developing educational course materials that are accessible to learners through various digital platforms. The integration of modern learning with conventional practices provides teachers with new strategies important for 21st-century learners (Figg & Jaipal Jamani, 2009). To make it more attractive and easily understandable, these courses are being developed innovatively in the form of simulation, story-telling, etc.

Categories of e-learning can be summarized as follows:

a. **Informal Learning**: It is a lifelong process that begins with our need for the information and its use at any stage. This dynamic and adaptable kind of learning leads us to search and explore the information with the help of different search engines like Google, Bing, etc. along with personal information management tools like wikis and blogs.

b. **Blended Learning**: The most popular integrated form of learning, namely, blended learning is a combination of traditional classroom learning and e-learning, beyond the classroom settings, which helps the learner by providing an opportunity to learn at their own pace anytime anywhere. Hence providing the tools to meet the challenges of the new world of learning, it combines not only the learning environment but also various delivery methods associated with the different web-based apps with face-to-face interactions. This multi-delivery approach optimizes all learning outcomes and the efficient cost of all content delivery makes it a well-blended approach (Kumar et al., 2021).

c. **Social Learning**: The information that is circulating among the community members at the local and global level contributes to the transfer of tacit knowledge which helps in encountering the challenges produced by the unstable and unprecedented environment. E-Learning is social in nature and hence relies upon the community.

d. **Shared Learning**: The emerging trends of e-learning have brought people across the globe on one platform through mutually shared knowledge and enhanced learning opportunities. It productively strengthens the learning society and forms a much-advanced knowledge economy. Knowledge management is a sequential process that helps in creating the environment for the community members to share, distribute, and adapt the information exchange procedure.

e. **Learning Networks**: It is a system of networks utilized to develop and sustain relationships with people and support each other in the acquaintance of knowledge. It increases the opportunity for the members to connect online from any place and to share their knowledge and expertise. Pen and paper-based education systems had a limited scope having a greater number of inadequacies whereas e-learning-based personal learning networks will create easily accessible connectivity and gain contemporary knowledge to keep themselves updated.

11.1.5 Technological Advancements of E-Learning

The new technologies which are in huge demand, and giving a new dimension to the e-learning concept are as follows:

a. **Video-Streaming and Producing their Content**: It is a simple and comfortable way for a teacher to make their videos and put them online. It is cost-effective as well as easily accessible to the students.

b. **Web Development with Open Sources**: In modern times, many types of open sources are being popularized among creators and designers due to their less expensive nature in terms of cost, time, and energy. Grabbing the attention of the people is an alternative way for every stakeholder in the education sector. E.g., Linux (OS), Firefox (browser), APACHE (server solutions), MySQL (database solutions), JAVA and PHP (programming languages), Open Office (open-source applications), Moodle (open-source products), etc. The world wide web was developed from a system of interlinked documents that would be accessed only via the internet (Salisah, 2022).

c. **Audio Files**: Now, technological advancements have made e-learning more convenient on mobile audio broadcast thus giving freedom to students to use it through MP3 files which are assisted by file-sharing services like bit-torrents, etc., podcasting for students using the services like iTunes can be carried easily in smartphones (2022).

d. **Blogging**: Blog services are used by teachers and students to publish their innovative ideas and lessons without the need for any technological knowledge. Multimedia tools like video/audio/image inserting tools etc. have given wings to the creativity and innovation of the learners with minimum effort.

e. **Internet Speed**: Nowadays, a broadband connection is available at a lesser cost with unlimited data to strengthen the internet infrastructure so that multimedia-based online applications can be used frequently which makes the e-learning approach reach the unreached. The easily accessible and affordable internet services have created competition; therefore, markets are pushing manufacturers to look for ways to exist in this global digital world (Amladi, 2017).

f. **Mobile Learning**: Now students have 24×7 access to e-learning services via the internet over mobile phones where they get an opportunity to use free applications at negligible cost, e.g., Google and Android have shrunken the world to the smallest.

g. **Global Society 5.0**: WWW, the world wide web, a system of interlinked hypertext documents which can be accessible by us via the internet, was first designed and tested in 1989 by a single universal document identifier (UDI). This web 1.0 was static and monodirectional mainly a read-only web, which means it was not interactive. It aimed at publishing the information for anyone at any time, just to mark the online presence of the user. The main protocols of web 1.0 were HTTP, HTML, and URI. Then slowly came into existence web 2.0 (wisdom web) which was a people-centric, participative, and read–write web, which means it was bidirectional having more interaction with less control. The major feature of web 2.0 was that it was highly collaborative and helped achieve collective intelligence. After this, the new version of web 2.0 evolved in 2006, known as web 3.0 or semantic web. It is similar to the Global file system or global database in which the web is readable not only by humans but by machines also. It means machines now overpowered humans. The major difference between web 2.0 and web 3.0 was that web 2.0 was concerned with the contextual creativity of users and producers whereas web 3.0 referred to the linked data sets. The present time is known as the era of web 4.0 also known as the symbiotic web where the reciprocity between the machine and humans was symbiotic in nature. It resulted in the construction of more powerful interfaces, e.g., mind-controlled interfaces. It has cleverer, faster, analyzer, decisive and more commanding interface. The journey thus started from read-only has now reached to read-write-execution-concurrency web involving every stakeholder in a collaborative manner which is parallel to the human brain having a web of highly dynamic and smart interactions. Advancements in technology are leading us toward using artificial intelligence (AI), virtual reality (VR), augmented reality (AR), etc., giving us real-world experiences in the form of multi-dynamism. It is a concept of the futuristic society aiming at achieving a high level of confluence between cyberspace and physical space.

In the information society, instructions or information were gathered by using the network and were analyzed by humans, however, in Society 5.0, cyberspace in the form of multi-faceted virtual space has connected people, things, and systems from the sensors in the physical space and then gathered in cyberspace. This big data is analyzed and optimized by using AI and the result is feedback to humans in physical space (Net at Work, 2020). It made human beings free from tedious tasks flawless with excellence and thus optimized the overall social and organizational systems. It is presumed that it will contribute to economic development by successfully solving major social challenges besides achieving sustainable development goals. Japan has become the first country to have a human-centered society by incorporating advanced technology innovatively and practically. At present, most organizations have started using AI, multiple intelligence (MI), cloud computing, personalized internet of things (IoT), robotics, neural sciences, etc. which has brought a revolution in the technological industry. It is well in use by farmers, teachers, doctors, industrialists, scientists, commercial e-platforms, and many more. It has highly influenced the nature of e-learning and its effect can be visualized clearly in the rapidly changing society.

11.1.6 Methodology of Active E-Learning

E-learning has proven and shown potential in driving a better environment for learners across the world (Oye, Salleh, & Lahad, 2012). The output from e-learning through electronic information and communication technologies (ICTs) is very huge and hold a big impact. E-learning is actually looking at bridging the gap between work and learning that is faced many times in workplace technologies (Oye, Salleh, & Lahad, 2012). The future of educational activities will mainly be based on active digital methodologies which, when incorporated appropriately with the help of instructions and ICT, result in stimulating learners' hypothalamus by which they will be motivated and engaged creatively, the outcome of which gets reflected in long term. E-Learning deals with the transfer of knowledge in an interactive mode where the learners absorb, assimilate, and consider the information and translate it into new knowledge whereas active e-learning supports brainstorming, making students to think, discuss, challenge, and collaboratively analyze the information through various activities. As per the e-learning design model developed by (Zane L. Berge, 2022) to achieve the learning outcomes of e-learning, the three parameters, i.e., 'Learning goals', 'Learning activities', and 'Evaluation' must be in alignment with each other. It is possible by creating a learner-centered learning environment by using the e-content developed by e-educators. To create an active e-learning environment, the learning curriculum has to be framed keeping in view how to develop the cognitive, psychomotor, and affective domains of the child which can be made possible by focusing on the 'active', 'interactive', and 'reflective' learning of the learners including the interdependence of all these three parameters upon each other. It will help the learner in analyzing the information and assisting in learning by generating new ideas. It also develops the interest of the learner in learning and provides a real environment by using the pre-designed content in context with meaningful learning. This kind of constructivist environment encourages the learners to explore their ideas and meanings, thus helping them to get engaged through self-motivation and construct new knowledge (2022). To make the e-learning environment more effective, it is essential to plan various pre-learning activities to make the content interesting and learner-centered to achieve the required result. Before moving ahead with the concept of the e-learning environment, it is necessary to understand the difference between the two types of environments, i.e., the learning-centered environment and the situated-learning environment. Active methodologies on digital technologies have been found to foster more participation of students in the learning process (Mattar & Ramos, 2019).

Environment Type – I: Pre-learning environment:

To achieve the desired outcome of the e-learning approach, the pre-learning environment plays a crucial role means of pre-learning activities. Through these activities, learners are provided with a detailed explanation of the content and the organization, its priorities, timeline, and the responsibilities of the learner, the content developer, and the course. The effectiveness of pre-learning tools can be measured too and used as indicators to show how interactive pre-learning environments enhance learning capability (Fortugno & Zimmerman, 2022). The expected behavioral outcome of the course material is made clear through the concise instructions so that the learner becomes confident enough to start the e-course.

AIR-Learning: Once the pre-learning is created and adopted by the learner, the process of learning becomes easy and joyful because the learner is now involved in 'AIR-Learning', i.e., active learning, interactive learning, and reflective learning.

i. **Objectives of Active Learning**:

  a. It compels the learners to read, speak, think deeply, organize, and reproduce the content on their own in a constructive way.

  b. The learner can generate new knowledge equipped with enhanced higher-order thinking and analytical skills, leading to synthesizing and self-evaluating himself based on the feedback received from the teachers.

  c. It inculcates the values of trust, respect, flexibility, and responsibility, making them challenged and risk-taking for their significant learning and less dependent on direct instructions.

ii. **Objectives of Interactive Learning**:

  a. Adding information to the existing knowledge by performing activities and then cognitively processing the information.

  b. Self-interaction of the learner begins to integrate with the whole learning and helps in retaining the overall information.

  c. To overcome the monotony of isolation, pre-learning activities of e-learning provide the opportunity to the learners for teamwork and collaboration. It helps them in developing interpersonal relationships and to achieve their learning goals. As mentioned by Vygotsky (1986), learning takes place in a social context by making meaningful interactions with people.

d. The instructor, in the form of the scaffolder, helps the learners to fill the gaps faced by the learner with the content, within the learner's social/peer interaction which adds value to the learning through appropriate feedback and evaluation communicated to the learner.

iii. **Objectives of Reflective Learning**:

a. Reflective learning allows the learner to think contextually, analyze systematically, apply meticulously, and produce effectively.

b. Active reflecting learning involves the experience gained by the learner such as logical and dialectical thinking which, in turn, enhances intellectual competence.

Environment Type – II: Learning-centered environment:

a. Academic ownership of learning is transferred from the teacher to the learners.

b. More emphasis is given to students' timings.

c. Individualized learning cannot be catered to.

d. An ample number of learning resources are provided through the internet, instructors, institutions, organizations, etc., supporting the e-curriculum. Learners can opt for the required learning paths according to their own choice.

Environment Type – III: Situated-learning environment:

a. The purpose of this is to create a learner-centered environment that helps the learners to actualize the meaning of the concept and the purpose of learning it. They should know why to know 'why'?

b. It can be achieved only when the learning is situated under varied circumstances.

c. We are going to discuss the interaction in context with e-learning, therefore educationists support framing an active e-learning environment by making them practice the e-learning techniques of SQ3R.

'S' stands for 'survey' or research of the reading material.

'Q' means 'queries' dealing with the inquisitive minds of learners which help them to clarify their doubts or confusions.

3 'R' represents 'reading', 'recalling', and 'reviewing' the content, thus making the learners self-interactive and research-oriented. It means that the learner is interacting with the e-course content in the form of one-way communication from the author of the content to the learner by getting thoroughly engaged in the process of self-construction of skill and competency, i.e., the learning outcomes.

Objectives of the Chapter

The following are the objectives of the chapter:

1. To meet the learning needs of students and to enhance the quality of teaching–learning.
2. To improve user accessibility and time flexibility to engage learners in the process of effective learning.
3. To promote an environment of continuous learning and to help in building a safe and secure user-friendly environment cum to assist the teachers and administrators in planning effective strategies leading to better learning outcomes.
4. To help the learners in achieving digital literacy, along with safeguarding them from cyberbullying, online protection, digital accountability, and digital well-being.

Organization of the Chapter:

The rest of the chapter is organized as follows: Section 11.2 elaborates on emerging trends of e-learning from the past to the pandemic period. Section 11.3 highlights the DLMS. Section 11.4 stresses digital citizenship. Section 11.5 concludes the chapter.

11.2 EMERGING TRENDS OF E-LEARNING FROM PAST TO PANDEMIC

Revolutionary changes caused by changes like e-learning with changes in perceptions, trends, attitudes, values, and thinking processes, especially in the last 2 years of the pandemic period have greatly affected the mode of the teaching–learning process (Berge, 2022). The changes were also experienced in terms of the teaching–learning process, pedagogies, curriculum, the cognitive, affective, and psychomotor skill sets which involve the stakeholders like parents also. Recently, in a study by Debattista, it was demonstrated that there must be a rubric for e-learning that would help in driving the expectations and feedback of learners (Daniels, Sarte, & Cruz, 2019). In the process where the change was the only constant, unseen, unexpected, frequent but inevitable, the monologue changed to multilogue, AV rooms are now digitalized, and Avs with more advanced AI, MI, and AR. It is the best way of exploring machines without being a machine. These changes can be observed concerning the below-mentioned aspects/areas:

a. **Teachers**: Now instructors have emerged as an extra innovative mentors. Their apprehension, professional, ethical, and knowledge sensitivity have immensely improved. Teachers no more remain the only source of information but are the torch-bearer to reach the destination, i.e., educational excellence using the advancement in the field of technology. Flipping classrooms has made teacher–student environments very interactive and personalized to achieve positive goals easier (Bergmann & Sams, 2022).

b. **Learners**: The possibility for lifelong learning accomplished with the help of skillful and innovative utility of technological know-how is assisting them to transform into chance takers, flamboyant, artistic, progressive and advanced, creative thinkers, communicative, innovative, collaborative, knowledge-seeker, inquisitive who are proactive and futuristic, who realize how to analyze effectively, having the deep understanding of the information. Its systematic, personal, and social implications are viewed from both, the short-term as well as the long-term perspective. A new innovation in the learner-based system was how the emotion detection system creates a system of deep learning in the virtual environment where facial expression in real time provides a framework for intelligent recommendation (Thakur & Mandal, 2022).

c. **Changing Trends in Creating a Problem-Based Curriculum**: It involves differentiated instructions. These instructions are personalized, integrated, and authentic in a real-world context providing quality assurance for the learners, of the learners, and the learners guided by their scaffolders which were one of the distant dreams in the past. Out of many studies conducted, on technology, e-learning can be considered to achieve quality education at the maximum and studies have also proved it to be highly consistent (Doculan, 2016).

d. **Changes in Pedagogical Approaches**: The new face of pedagogy is transformed from the traditional chalk-and-talk method to the synchronous and asynchronous way of dealing with the transaction of the content, whether it is hybrid or blended. It provides an opportunity for the learners for aesthetic, artistic, scientific, social, and cultural growth and to become independent learners by learning at their own pace. They are preparing themselves for the questions yet to be asked and answered, issues yet to be thought and resolved, technology yet to be invented and dreamed of, values yet to be learned and taught, and netiquettes yet to be adaptive and adopted. It requires a great deal of self-assessment through introspection. Rommetveit (1987), in his book, described adapted and shaped indigenous methods for educators and students for illuminating insights into learning. Versatile methodological analysis like experimentation, experiential way by using play way, gamification, sudoku, etc. via creative, challenging, fruitful, and competency-based education using AI, MI, VR, AR, etc.

e. **Assessment**: The 360-degree assessment recommended by NEP (2020), can be achieved through e-learning. The focus should be on assessing the learning of learners, not on examining their mistakes. It can be done by answering the question; of what is being internalized by them, how it is transforming the way they are communicating, and how successfully they can formulate and correlate the known with the unknown via machine learning, surfing, finding, sequencing, and deciding and finalizing.

f. **Implementation**: Implementation of all the above-mentioned points should neither be an old engine in a new car nor be a new engine in a dented and painted old car, looking new. Rather it should be a new engine in a brand-new car with all the scope of modifications and improvisations based on the need, and the path to be taken with the futuristic goal. Any language is not ready-made too but its eternal being and inherent use make it a living occurrence (Vološinov, 2000).

11.3 DIGITAL LEARNING MANAGEMENT SYSTEM (DLMS)

It is an organized and systematized e-learning software to facilitate learning for sustainable support to the entire online educational ecosystem. It is used for tracking scores and to ensure that the students' standards for each course are maintained. They are platforms that enable users to create good online courses; and also, the students to be able to access these courses. These platforms are of different types and they have different capabilities. Some are free, while others are offered at a fee.

11.3.1 Digital Learning Management System: Definition

Digital: This means the electronic technology that helps in generating, storing, and processing the data.

Learning: To fulfill the learning objectives, it is used as a tool to handle the delivery, administration, automation, and analytics of learning material.

Management: A well-executed management of the instructions as a part of documentation and data-based management plays an important role in a learning management system (LMS).

System: The system which handles the overall incoming and outgoing teaching–learning instructions within a single Edu-ecosystem to facilitate the learners. www, i.e., the World Wide Web came into existence in 1989 and has drastically changed everything in education.

The Utility of a LMS cannot be restricted only to the educational field, but it is equally important for and has overpowered the life of the common man also by becoming an essential and integral part of their life.

11.3.2 Components and Functions of Digital Learning Management System (DLMS)

The area of e-learning has become enormously vast, embedding in itself, productive e-teachers, quality e-content, responsive learners, outsourcing, and partnering.

a. **E-Teachers**: E-learning has provided an opportunity for teachers to enhance their potential by focusing on the productivity of both, the lessons and the learners. It could be made possible only because e-learning is a very flexible and dynamic approach that helps in providing academic ownership to the learners which is very valuable to them in the form of efficient learning through online platforms. E-teachers are well awarded for using development tools following the rapidly changing technological demands, having left with no other option than to adapt and adapt to the new environment of the learning platform. The role of professional educators is very important since it creates a personalized classroom experience for each student and this is a nice time to observe their learning ("The Four Pillars of F-L-I-P", 2022). The changes are initiated in the form of digital experiences based on electronic data transfer, online analysis, decision support, document management system, etc. As such, the e-learning system transformed the teachers as well as the learners to become more powerful and impactful. Now 21st-century educators can better be known as the adapter, the communicator, the learner, the visionary, the leader, the role model the collaborator, and the calculated risk-taker who can transform the traditional teaching–learning environment into a collaborative and sustainable environment. E-learning can be understood as assisted learning with the computer but student-centered where knowledge is digitally received (Jethro, Grace & Thomas, 2012).

A survey on educational bodies around the world was conducted where 59 countries that participated demonstrated complete flexibility and resilience during tough times of the pandemic and also came up with great established strategies for providing education too. (Reimers & Schleicher, 2020). The pandemic affected many in the world and most students and youth and their learning, but the UNESCO worked around different ministries of the world to create lesser sustainable recovery from this catastrophe ("Education: From disruption to recovery", 2022). Technology-based online learning has encompassed the basic use of the internet and how materials can be used by teachers and learners to regulate results (Malik & Rana, 2020).

b. **Quality Learners**: Once the academic ownership is transferred to the students, it makes them more responsible and accountable to achieve their learning objectives toward quality education. The parameters to be taken into consideration to create an e-learning environment are curriculum development, course structure, pedagogies, evaluation and assessments, and implementation of all these, by following an individualistic approach through the realistic models. E-learning has created a digitized environment that has impacted the traditional distance learning education forum, and this is a challenge more than a problem faced by many teachers and learners (Peters, 2000).

Responsiveness to the change: The stakeholders need to have a more futuristic perception of the technology by being the follower of the change because eventually, they have to adapt the way anyway. The integration teaching aspect of e-learning adopted classroom combines both theoretical study and all experimental studies for the whole practical experience (Ding, Ni, Li & Xue, 2017). Creating awareness, counseling, and organizing training sessions can ease the process of internalizing the change.

a. **Outsourcing Services**: For technological education to be imparted by any institution can be the best solution for those who are lacking the appropriate resources, e.g., lesson designs, IT-based infrastructure, online support services, marketing services, software development, etc. Like, in many studies performed by the University of Finland, it was observed that thousands of students did not pass their introductory programming courses around the world until adequate outsourced courses were added (R, 2021).

b. **Partnering**: As mentioned by Kuo et al. (2009), partnering in the area of financial collaboration, research, lesson designing, and teacher resource sharing through webinars, conferences, etc. for joint ventures has become an integral part of the

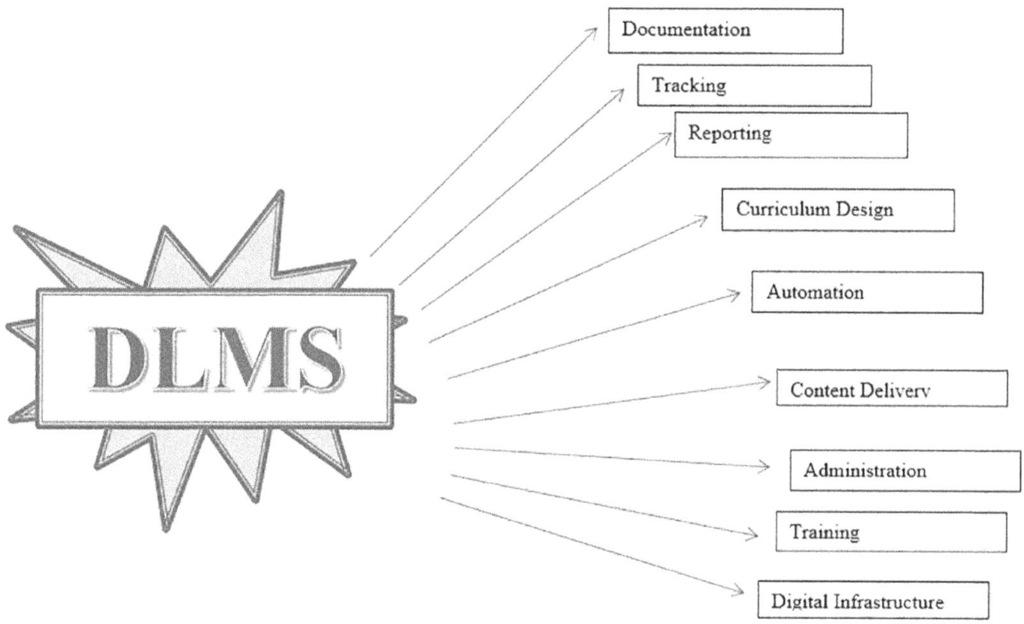

FIGURE 11.1 Framework of DLMS.

modern education system. All effectively designed education materials will facilitate the achievement of learning outcomes (Brown & Voltz, 2005). For making the partnering plan more effective and achievable, the leadership and project management skills of the administrators play a pivotal role (Figure 11.1).

Functions of DLMS

The following are the functions of DLMS:

1. Delivery and tracking of tasks, quizzes, padlet, Mentimeter, online exams, etc.
2. Managing the social learning platforms e.g., discussion boards, discourse forums, etc.
3. Tracking students' progress, their learning analysis, and other administrative works.
4. Sending automatized information to students, teachers, and parents.
5. Preparing students' reports and portfolios in the form of detailed reporting.
6. Study material delivery of synchronous and asynchronous modes of learning.

The importance of LMS is to support National Education Policy (NEP 2020) to a great extent by involving technology in education and keeping in view the interest, potential, profile, and skill level of the learners thus emphasizing the individualistic approach in education.

It is a system that manages all electronic learning within a single entity.

To achieve their learning goals, it is extensively used by schools and universities, organizations and cooperation, MOOCS platforms, etc. Discussions on hot topics like having virtual laboratories, technological solutions, student learning, and tracking assessments would go hand in hand with online education and learning (Kats, 2010).

11.3.3 Integration of Instructional Tools into DLMS

E-learning is a kind of active digital learning in which the knowledge is imparted by using web technology as its fundamental technical infrastructure. A LMS facility acts as an essential tool in the whole teaching and learning process for higher education mostly to cultivate and improve student skills (Isaias & Issa, 2013). The innovative methodology consists of critical tasks of instructional design to enhance the outcome by providing automated support by using instructions with the help of web technology via its analysis, design, documentation, evaluation, and implementation.

As mentioned by Oye, Salleh, and Lahad (2012), the methodology of e-learning can broadly be divided into three sections:

A. Curriculum tools and Interaction in learning

B. Digital library tool

C. Knowledge representation tool

11.3.3.1 Curriculum Tools and Interaction in Learning

It builds up a meticulous and standard environment that helps in supporting the classroom teaching–learning process, especially at the initial stage of selection. The function of digital library tools is to collect and explore information by facilitating effective and efficient access to resources. Learner-tutor interaction and learner-learner interaction are the basis of computer-supported collaborative learning (CSCL), which makes the learners more motivated and engaged so that they actively participate because of the relatively user-friendly system and online active group interaction, which is one of the objectives of e-learning to perceive higher self-reported learning. It also promotes problem-solving, analyzing, and decision-making, thus providing increased opportunities for development. Science is mostly, a laboratory-based experience and a huge challenge is for each student to participate which should be an essential part of the curriculum and an interaction tool that must be in place (Yadav, 2019).

The curriculum tools are mostly used at the high school and the university level to facilitate class activities by selecting and organizing the material. To make it more significant, it is integrated with additional supporting tools such as discourse forums, online quizzes, etc., which aids in collaboration and evaluation.

Tripolar Relationship of a Commercial Curriculum Tool:

It is an integration of tools concerned with the students, teachers, and administrators (Figure 11.2).

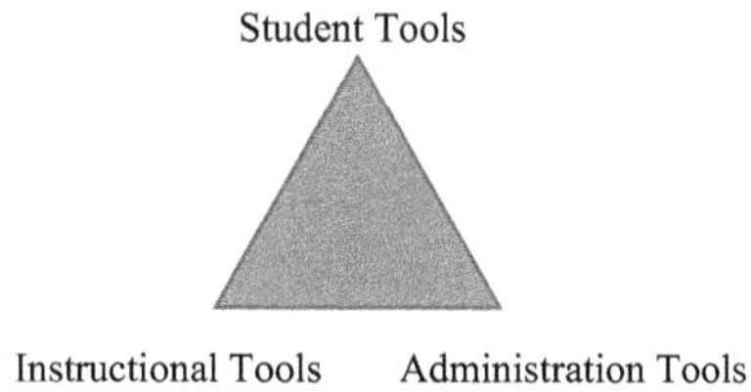

FIGURE 11.2 Tripolar relationship of a commercial curriculum tool.

Student Tools:

i. To browse the learning material for reading, assignments, projects, etc.
ii. Teamwork and sharing in the form of asynchronous and synchronous bulletin boards and discourse forums.
iii. Self-assessment and evaluation with the help of assignments designed by the teachers.
iv. Web communication technology and blackboard are more suitable for self-dependent and collaborative learning.

Instructional Tools:
These are the tools that help in course content and lecture delivery. They are also collaborative and communicative tools that help in a homework assignment and its assessment.

Administration Tools:
These tools comprise several advanced tools which are mainly used by the system administrators. There are 21 Windows administrative tools that are used by the administrator for monitoring and control.

The instructional tool, administrational tool, and students' tools are all mutually dependent on each other and support each other in achieving the learning objectives.

11.3.3.2 Digital Library Tool
Besides supporting the curriculum, Digital Library Tools (DLTs) focus on the exploration of information, and collection of data of that particular information searched by locating the resources. Pandemic time has encouraged the use of DLT to a great extent facilitating the learners to locate the right kind of information amidst a large amount of digital material through searching, browsing, and exploring the concerned material.

11.3.3.3 Knowledge Representation Tool (KRT)
It develops the visualizing power of the learners by reviewing, capturing, or developing new knowledge. KRT helps in constructing a multi-dimensional logical meaningful conceptualization and skill development by actively involving both, the learner and the facilitator.

11.4 DIGITAL CITIZENSHIP

During the past couple of years, the social participation governing human rights and dignity through the responsive and responsible use of ICT has rapidly turned people technology-oriented, because remote and hybrid learning has been shifted to e-learning. The concept of digital citizenship was existing earlier also but the focus was more on harnessing the same for our day-to-day activity and the approach was superficial having a limited scope. It was focusing on safety, security, and legality by protecting the password and hiding one's identity when using intellectual property while citing the sources. Now the emphasis has been democratized ensuring that the learners are more confident and empowered to use digital tools and platforms by taking responsibility and accountability toward society. As the digital communication channel went through a revolution and modern technologies are progressing speedily, the positive effects are being transitioned into society so all technologies can be optimally utilized ("*What is Digital Citizenship and Why is Digital Citizenship Important in Education*?", 2022). The concept of digital citizenship refers to owning the use of technology responsibly which provides the ability to provide a fruitful and positive result from its use and learning. The ethical use of technology would further be included in what digital citizenship means. An example to support this concept is the online exchange of goods online where electronic participation happens in a society where people communicate wisely and ethically through the internet with each other ("*What is Digital Citizenship and Why is Digital Citizenship Important in Education*?", 2022).

As mentioned by Karen Mossverger, digital citizens are the people who use the internet regularly and effectively by the following netiquette. Mike Ribble put forth the following three principles for the technology users, making them responsible digital citizens; respect, educate, and protect. Digital citizenship has positively engaged users with digital technology with more ease and confidence. Its importance has emerged in the form of digital citizenship education (DCE) which enhances their digital literacy thus ensuring their online safety, digital health and well-being, digital responsibility, and cyber security by creating and consuming the digital content governing human affairs. The need for digital citizenship in the form of safeguards against technology dangers was desperately felt to protect the children from being the victim of its negative aspects. Ribble (2017) stated that digital citizenship is an appropriate method, which prepares students for the optimal and responsible use of technology by following the code of conduct and norms so that they can create a positive and constructive digital image. There are many existing dangers of technology as well that provide a threat to our cyber world (LillyWhite & LillyWhite, 2022). Data suggested in 2022 that 80% of the fraud in the UK is cyber-enabled; therefore, it could be said that digital citizenship was compromised (LillyWhite & LillyWhite, 2022). The process of digital transformation is a whole revolutionary process where transitions happen around technology that is at the center (Rachida, 2020). The AI, blockchain is an automated part that bolsters digitalization (Rachida, 2020).

11.4.1 Management of Digitally Trained Human Resources

The people who are not digitally enabled to adapt to the traditional LMS prove ineffective in terms of finance and time constraints also. With the recent advancement of technology, it is universally accepted that learning is difficult if people will not keep themselves updated and connect the learning into a lively online ecosystem that is mobile responsive and has digitally advanced technological infrastructure. Nowadays, organizations are integrating their HR system with the enterprise resource planning (ERP), which is a business management software dealing with the financial, operational, reporting, manufacturing, and supply chain of the organizations. It helps the stakeholders to automate the time-consuming daily assignments.

During the pandemic, professionals were left with no other choice than to get acquainted with the new normal digital phenomena hence willingly or unwillingly, technological professional learning has become an ongoing process of overall learning with and from others, learning to teach by using digital technologies, it has changed the classroom system drastically. This transition to digital learning could be made possible by sharing ideas using modern tools and resources and cocreating resources. One of the modern tools created and used was an experiment that used a facial detection tool for the detection of a learner's state of mind during an online learning session (Mukhopadhyay et al., 2020). In 2020, this paper paved a path to assessing a far more complex challenge for an educator and learner in this online digital system. The convolution neural network (CNN) model, a supervised deep learning algorithm, chosen by the authors for facial expression identification and emotion detection has recognized four types of emotions, i.e., confusion, dissatisfaction, satisfaction, and frustration under a six-image frame, and it was considered to provide a 65% accurate data. If more studies and experiments are conducted to sufficiently assess a learner's state of mind in the form of other facial features like eyebrow movement, blinking of eyes, and other micro-expressions. It was a great success for the online digital world to read the learners' state of mind virtually which was quite difficult during the offline mode of teaching–learning. It can bring a revolution in the field of education by helping counselors and teachers by means of assessing the learners' state of mind through their expressions to resolve the social, emotional, and psychological problems of children and save them from juvenile delinquency by diagnosing, detecting, and solving issues related to their behavior. Besides its contribution to the field of education, it can be utilized in other sectors of society also, especially in cases of corruption, lawbreaking, wrongdoing, misconduct, etc.

All these advancements including IoT have reshaped the functioning of human resources. Human resources can be trained in professional practices by facilitating easy communication, efficient handling of a huge amount of data for operational and strategic decisions, assessment of employee-performance data and employee recruitment process, and enhancing the security of employee data and records. All these technological explosions have proven the fact that electronic learning (E-learning) will soon be replaced by digital learning (D-learning). There are some limitations and implications always to an alternative method; therefore, proposing studies with rubrics is a good method for higher education institutions (Debattista, 2018).

11.4.2 Importance of Digi-Edu-Citizenship

The following points lay the strong foundation to understand the Importance of Digi-Edu-Citizenship:

1. Creating a positive image for the learners so that their potential can be tracked and judged by the employers on their social media profiles.

2. It makes the students successfully practice respect and responsibility for themselves and also for others in an online environment.
3. It helps children to utilize digital apps and social networking sites safely, ethically, effectively, and responsibly to protect them from the adverse effect of cyberbullying using smartphones, laptops, tablets, computers, etc.
4. It incorporates the competencies, potential and positive behaviorism to live with sustainability in the digital society.
5. Digital citizenship also promotes equity, equality, and inclusiveness.
6. It helps teachers in their professional development.
7. Counseling of parents about digital citizenship gets reflected automatically in their children also.

It is possible by creating a LMS by training the students in basic digital skills, e.g., creating and saving strong passwords, avoiding plagiarism, addressing cyberbullying, determining to visit a safe website only, protecting from identity theft, and exploring the search engine more effectively. In 2017, a Deloitte study showed that there is a lack of global understanding of workers in different organizations which creates a lack of foundations in global science (Little, 2017). When this lack of foundational science is bridged, more organizations will come in support of the digitization of the world (Little, 2017).

The framework of Global Citizenship Education (GCE), as mentioned by Global Citizen Education Solutions, can be understood in a better way in terms of the four below-mentioned parameters which are interconnected and associated with each other:

a. GCE for diversity and inclusion covering the aspects of (i) gender equality education, (ii) anti-racist education, and (iii) multicultural education.
b. GCE for community participation which consists of (i) public education, (ii) education for development, and social justice education
c. GCE for commercial participation including (i) education for entrepreneurship, (ii) education for developing 21st-century skills, and (iii) intercultural education.

GCE for managing resources which supports (i) education for sustainable development, (ii) environmental science education, and (iii) nature studies or education for the conservation of resources.

11.4.3 Factors Affecting Digital Citizenship

The factors that impact Digital Citizenship are as follows:

a. **Easy Access:** The equal opportunity of easy access to technology must be a right for every learner.

b. **E-Commerce**: To protect everyone from fake propaganda, the buying and selling of online materials, or opportunities for courses and employment, knowledge of e-commerce is essential.

c. **Community and Collaboration**: To promote digital citizenship, students must be taught how to maintain a digital relationship which is possible only through the knowledge of society and a collaborative approach through the e-learning platform.

d. **Etiquette Cum Netiquette**: Knowing the code of conduct and norms before stepping into the technological world make the students aware of leaving their positive footprints.

e. **Digital Literacy**: It encourages the students to differentiate between fact and fiction while reading or writing using online sources.

f. **Health and Welfare**: This area deals with the physical and psychological health of the users especially considering screen time, cyberbullying, etc. Technology can be a friend or a foe, depending on its usage. It connects people psychologically in the virtual world and at the same time, keeps them apart in the physical world. If this lacuna or gap is not bridged properly, it will create a ditch that will affect the objective of e-learning adversely.

g. **Cyber Laws and Litigation**: The people who use gadgets and devices must be aware of the law and litigation to protect themselves against cybercrime as these legal rules and practices include the ethical responsibility of using technology.

h. **Rights and Responsibilities for Sustainability**: These are concerned with the freedom availed by people in the digital as well as the real world because everyone is entitled to and responsible for one's ideas and opinions to be channelized appropriately and it depends on the understanding of one's rights and responsibilities concerned with the ethical values to sustain in the digital ecosystem as well.

i. **Security and Privacy**: Despite protecting the system and gadgets from the threats like viruses and bugs, one of the most challenging tasks is to keep ourselves safe and protected from fraudulent sites and apps which is an important aspect of training people for digital citizenship.

All of the above-mentioned factors are contributing to creating awareness to inculcate the concept of digital citizenship which helps every member of the digital society interact with each other in the digital world through positive and secure interaction.

Since we are becoming a global community and the involvement of youth using digital technology is accelerating manyfold, we need to bridge the digital skill gap because there are risk factors arising with the use of digital technology and to cope with this lacuna, literacy of the digital citizenship in the form of imparting knowledge about the roles, responsibilities, and skills for navigating through digital life is an integral part of the digital world.

In support of the findings regarding the increasing number of techno-savvy people across the globe, recent studies made by Joseph Johnson in May 2022 also emphasize internet demographics and use. It was mentioned that according to the global digital population as of April 2021, Asia was found to be the region having the largest number of online users numerically 2–8 billion, and Europe, with 744 million techno-users was in the second position. Among the countries like China, India, and the USA where the number of internet users is the maximum, China stands at the top with a billion people using the internet followed by India with its internet users count of around 658 million. The organization for economic cooperation and development (OECD) Learning framework 2030 presumes digital literacy to be the core fundamental competency to achieve futuristic educational goals. In the current scenario, this kind of literacy demands to development of and dissemination of the e-curriculum through the e-learning approach.

11.5 CONCLUSION

The comprehensive and large-scale transformation of the K-12 education system in the form of transition from the analog to the digital world is now in progress and it is the demand of the time for the same to occur inclusively, irrespective of the inequality. Then only our target of achieving the sustainable development of the digital economy would be possible. This development will enhance the ability to transform the life of every individual by providing digital skill education. 21st-century educators can play a pivotal role in creating a democratic digital global society by mentoring learners and empowering them with the required digital skills. Digital competency-building training programs for educators as well as for the work forces much required to link with digital skills.

The online and offline options are no longer existing as separate entities in the form of the digital world but they have been incorporated and amalgamated into our daily lives in such a way that it has become an indispensable part of our continued existence in the digital ecosystem. Global efforts for digital skill education and training must be coordinated on an urgent basis to make the learners succeed on the digital edge while reducing the risks and inequality through implementing an appropriate DLMS.

To meet the above-mentioned expectations beyond the periphery of the classrooms, the adoption of the e-learning approach can be considered as an added advantage because of the flexibility of learning at any place and anytime with the optimum accessibility having better learning outcomes toward the creation of a future-fit digital world. The chapter concludes with a succession of theoretical and real-world facts of action that might augment these comebacks to the advantage of the individual as well as the society.

REFERENCES

Amladi, P. (2017). HR's guide to the digital transformation: Ten digital economy use cases for transforming human resources in manufacturing. *Strategic HR Review*, *16*(2), 66–70. doi: 10.1108/shr-12-2016-0110.

Bandyopadhyay, S., Thakur, S., & Mandal, J. (2022). Emotion detection for online recommender system using deep learning: A proposed method. *Innovations in Systems and Software Engineering*. doi: https://doi.org/10.1007/s11334-022-00437-7

Berge, L. (2022). Active, interactive, and reflective eLearning. Retrieved 26 June 2022, from https://www.researchgate.net/publication/234737912_Active_Interactive_and_Reflective_eLearning.

Bergmann, J., & Sams, A. (2022). Flip your classroom reach every student in every class every day (pp. 120–190). Washington, DC: International Society for Technology in Education. - References - Scientific Research Publishing. Retrieved 26 June 2022, from https://www.scirp.org/(S(oyulxb452alnt1aej1nfow45))/reference/ReferencesPapers.aspx?ReferenceID=1791200.

Bilal, S. (2022). E-learning revolutionize education: An exploratory study. *E-Learning: A Boom or Curse*. https://www.researchgate.net/publication/280862765_E-Learning_Revolutionise_Education_AN_Exploratory_study.

Brown, A., & Voltz, B. (2005). Elements of effective e-learning design. *The International Review of Research in Open and Distributed Learning, 6*(1). doi: 10.19173/irrodl.v6i1.217.

Bureau, B. (2022). Why e-learning is the future of education globally. Retrieved 26 June 2022, from https://www.businessworld.in/article/Why-E-Learning-Is-the-Future-of-Education-Globally/22-04-2021-387414/.

Daniels, M., Sarte, E., & Cruz, J. (2019). Students' perception on e-learning: A basis for the development of e-learning framework in higher education institutions. *IOP Conference Series: Materials Science and Engineering, 482*, 012008. doi: 10.1088/1757-899x/482/1/012008.

Debattista, M. (2018). A comprehensive rubric for instructional design in e-learning. *International journal of Information and Learning Technology, 35*(2), 93–104. doi: 10.1108/IJILT-09-2017-0092.

Ding, X., Ni, X., Li, L., & Xue, Z. (2017). The exploration and practice of CDIO-based integration postgraduate teaching mode. *2017 2nd International Conference on Education, E-Learning and Management Technology (EEMT 2017), Xi'an University of Posts & Telecommunications, Xi'an, China.*

Doculan, J. A. D. (2016). E-learning readiness assessment tool for philippine higher education institutions. *International Journal on Integrating Technology in Education, 5*(2), 33–43. doi: 10.5121/ijite.2016.5203.

Edtech Team, World Bank Group (2022). Retrieved 26 June 2022, from https://www.worldbank.org/en/topic/edutech/brief/edtech-covid-19.

Education: From Disruption to Recovery. (2022). Retrieved 26 June 2022, from https://en.unesco.org/covid19/educationresponse.

Figg, C., & Jaipal Jamani, K. (2009). Engaging 21st century learners and differentiating instruction with technology. *Teaching and Learning, 5*(1). doi: 10.26522/tl.v5i1.297.

Fortugno, N., & Zimmerman, E. (2022). Learning to play to learn: Lessons in educational game designed. Retrieved 26 June 2022, from http://www.ericzimmerman.com/texts/learningtoplay.html.

Gupta, S., & Panigrahi, S. (2021). National Education Policy-2020: Rethinking assessment in higher education. *Scholarly Research Journal for Humanity, Science & English Language, 9*(46), 11256–11263.

Isaias, P., & Issa, T. (2013). E-learning and sustainability in higher education: An international case study. *The International Journal of Learning in Higher Education, 19*(4), 77–90. doi: 10.18848/2327-7955/cgp/v19i04/48673.

Jackman, J., Gentile, D., Cho, N., & Park, Y. (2021). Addressing the digital skills gap for future education. *Nature Human Behaviour, 5*(5), 542–545. doi: 10.1038/s41562-021-01074-z.

Jethro, O., Grace, M., & Thomas, A. (2012). E-learning and its effects on teaching and learning in a global age. *International Journal of Academic Research in Business and Social Sciences, 2*((1), 203–210.

Johnson, J. (2022). Internet users in the world 2022 | Statista. Retrieved 26 June 2022, from https://www.statista.com/statistics/617136/digital-population-worldwide/.

Kats, Y. (2010). *Learning Management System Technologies and Software Solutions for Online Teaching.* Hershey, PA: Information Science Reference.

Kumar, A., Krishnamurthi, R., Bhatia, S., Kaushik, K., Ahuja, N., Nayyar, A., & Masud, M. (2021). Blended learning tools and practices: A comprehensive analysis. *IEEE Access, 9*, 85151–85197.

Kumar Basak, S., Wotto, M., & Bélanger, P. (2018). E-learning, M-learning and D-learning: Conceptual definition and comparative analysis. *E-Learning and Digital Media, 15*(4), 191–216. doi: 10.1177/2042753018785180.

LillyWhite, S., & LillyWhite, S. (2022). What is digital citizenship? | The basics for teachers - FutureLearn. Retrieved 26 June 2022, from https://www.futurelearn.com/info/blog/what-is-digital-citizenship-teacher-guide.

Little, G. (2017). Deloitte human capital trends in perspective: The science of organization design and operation. *SSRN Electronic Journal*. doi: 10.2139/ssrn.2937125.

Malik, S., & Rana, A. (2020). E-learning: Role, advantages, and disadvantages of its implementation in higher education. *JIMS8I - International Journal of Information Communication and Computing Technology, 8*(1), 403. doi: 10.5958/2347-7202.2020.00003.1.

Mattar, J., & Ramos, D. (2019). Active methodologies and digital technologies. *International Journal for Innovation Education and Research, 7*(3), 1–12. doi: 10.31686/ijier.vol7.iss3.1156.

Mazurchenko, A., & Maršíková, K. (2019). Digitally-powered human resource management: Skills and roles in the digital era. *Acta Informatica Pragensia, 8*(2), 72–87. doi: 10.18267/j.aip.125.

Mukhopadhyay, M., Pal, S., Nayyar, A., Pramanik, P., Dasgupta, N., & Choudhury, P. (2020). Facial emotion detection to assess learner's state of mind in an online learning system. *Proceedings of the 2020 5th International Conference on Intelligent Information Technology.* doi: 10.1145/3385209.3385231.

National Education Policy. (2020). Available at: https://www.education.gov.in/sites/upload_files/mhrd/files/NEP_Final_English_0.pdf

Net at Work: Management. (2022). Retrieved 28 June 2022, from https://www.netatwork.com/team-category/management/.

Northrup, P. (2022). *A Framework for Designing Interactivity into Web-Based Instructions* (2nd ed., pp. 31–39). Englewood Cliffs, NJ: Educational Technology Publications, Inc.

Oye, N.O., Salleh, M., & Lahad, N. A. (2012). E-learning methodologies and tools. *International Journal of Advanced Computer Science and Applications, 3*(2). doi: 10.14569/ijacsa.2012.030208.

Peters, O. (2000). Digital learning environments: New possibilities and opportunities. *International Review of Research in Open and Distance Learning, I*(1). [Online].

Petretto, D., Carta, S., Cataudella, S., Masala, I., Mascia, M., & Penna, M. et al. (2021). The use of distance learning and e-learning in students with learning disabilities: A review on the effects and some hint of analysis on the use during covid-19 outbreak. *Clinical Practice & Epidemiology in Mental Health, 17*(1), 92–102. doi: 10.2174/1745017902117010092.

Rachida, A. (2020). Digitalization for human resource management. *Marketing and Management of Innovations*, (1), 245–255. doi: 10.21272/mmi.2020.1-20.

Reimers, F., & Schleicher, A. (2020). Schooling disrupted, schooling rethought: How the Covid-19 pandemic is changing education. *OECD Policy Responses to Coronavirus (COVID-19)*. doi: 10.1787/68b11faf-en.

Rommetveit, R., Vygotsky, L., & Kozulin, A. (Eds.). (1987) Thought and language. Cambridge, MA: MIT Press, pp. lvi+287.

R. (2021). Exploring the use of automated learning tools in teaching and learning process. *Journal of Education, 23*(6), 177.

Salisah, T. (2022). World wide web: From web 1.0 to web 4.0 and society 5.0. Retrieved 26 June 2022, from https://medium.com/@tuhfatussalisah/world-wide-web-from-web-1-0-to-web-4-0-and-society-5-0-48690a43b776.

Shukla, A. (2020). *Impact of COVID-19 & Pandemic Lockdown in India: Rejuvenation of Education System*. India: Eureka Publications.

The Four Pillars of F-L-I-P. (2022). Retrieved 26 June 2022, from https://flippedlearning.org/wpcontent/uploads/2016/07/FLIP_handout_FNL_Web.pdf.

Vološinov, V. (2000). *Marxism and the Philosophy of Language*. Cambridge, MA: Harvard University Press.

What is Digital Citizenship and Why is Digital Citizenship Important in Education?. (2022). Retrieved 26 June 2022, from https://www.scientificworldinfo.com/2020/03/what-is-digital-citizenship-in-education.html.

Wu, S. (2021). How teachers conduct online teaching during the covid-19 pandemic: A case study of Taiwan. *Frontiers in Education*, *6*. doi: 10.3389/feduc.2021.675434.

Yadav, A. (2019). Differentiating instruction 21st century classrooms. *SSRN Electronic Journal.* doi: 10.2139/ssrn.3512793.

Yengin, İ., Karahoca, D., Karahoca, A., & Ozcinar, Z. (2010). Being ready for the paradigm shifts in e-learning: Where is the change happening and how to catch the change? *Procedia - Social and Behavioral Sciences*, *2*(2), 5762–5768. doi: 10.1016/j.sbspro.2010.03.940.

Index

Note: **Bold** page numbers refer to tables and *italic* page numbers refer to figures.

For Product Safety Concerns and Information please contact our EU representative GPSR@taylorandfrancis.com
Taylor & Francis Verlag GmbH, Kaufingerstraße 24, 80331 München, Germany

www.ingramcontent.com/pod-product-compliance
Ingram Content Group UK Ltd.
Pitfield, Milton Keynes, MK11 3LW, UK
UKHW050057200926
475702UK00019B/2895
* 9 7 8 1 0 3 2 3 4 4 5 9 1 *